3RD EDITION

DOOLIN'S TROUBLE SHOOTERS BIBLE

Air Conditioning
Refrigeration
Heat Pumps
Heating

DALLAS

DOOLCO, INC.

TECHNICAL PUBLICATIONS

OTHER SERVICE MANUALS BY JIM DOOLIN

Auto Air Conditioning

Residential Cooling (Part One)

Residential Cooling (Part Two)

Residential Gas Heating

Window Units

Commercial Refrigeration

Frost Free & Conventional Refrigerators

La Biblia Doolin Para El Tecnico Reparador

DOOLCO, INC.
11258 Goodnight Lane, Suite 105
Dallas, Texas 75229
1-800-886-2653

doolco@swbell.net   www.doolco.com

© DOOLCO, INC., 1996, 3rd Edition

ISBN 0-914626-11-6

Printed and bound in the United States of America by Taylor Publishing Co., Dallas, Texas

INTRODUCTION

The refrigeration industry (man made ice) is just now in its second century. The need for man made ice was initially developed for the preservation of food. The perceived need of air conditioning for human comfort began to develop in the 1920s and 1930s primarily in public buildings. As the technology developed in compressors and refrigerant gases, the need for air conditioning in homes became possible. One of the first housing developments in the U.S. to have central cooling built-in was in St. Louis, Missouri in the early 1950s. Since the 1970s almost all homes and automobiles built in the U.S. have had air conditioning installed by the builder or manufacturer.

As we approach the 21st century there are few people in the U.S. whose lives are not directly affected by air conditioning and refrigeration on a daily basis. From food preservation and the food service industry, to commercial and residential cooling, the hospital and medical industry to space flight, we live with it daily. Today, in third world countries, refrigeration for food preservation purposes has penetrated much further than air conditioning for human comfort. The technology is available, however many third world countries cannot afford it. People who study U.S. population movements are well aware that were it not for air conditioning today, many areas of the South and Southwest such as Dallas, Fort Worth, Houston, Phoenix, Los Angeles, Miami and Atlanta would never have developed as dramatically as they did. Without residential cooling, these large cities would instead be large towns with populations of about 25% of present day populations. At the present time there are approximately 1,200 community colleges and trade schools offering courses in air conditioning and refrigeration. It is estimated that probably 50,000 men get into the trade each year by this method of training.

Almost all fifty of our states now require some form of licensing for ACR self-employment. Licensing requirements will vary from one state to another. The best way to find out what the requirements are in your state is to contact your nearest refrigeration supply house. To obtain license information in Texas you should contact the Texas Department of Labor & Standards, P.O. Box 12157, Austin, Texas 78711 or phone (512) 463-2904.

THE PUBLISHERS

FOREWORD

The first and second editions of *Doolin's Trouble Shooters Bible*, first published in 1963 have now sold nearly 200,000 copies. Though J. H. Doolin is deceased now, his training philosphy continues through the Trouble Shooters Bible. As current President of Doolco, Inc., I was involved with the founding of the company in 1963. The task of revising the Doolin's Trouble Shooters Bible is now done by a long time associate of Mr. Doolin's by the name of Bob Dixon. Bob Dixon has worked as a serviceman in the ACR trade since the mid 1950s. When Mr. Doolin operated a trade school in Dallas, Bob Dixon was his chief instructor. Bob's vast knowledge of the service business and a rare ability to teach the skills necessary to become a good serviceman are now carried over to the Trouble Shooters Bible. His communication skills have enabled us to keep the reputation that the Doolin's Trouble Shooters Bible is "the best" ACR service manual in existence.

JAMES P. DOOLIN
PRESIDENT
DOOLCO, INC.

CONTENTS

CHAPTER 1

BASIC PRINCIPLES OF REFRIGERATION

This is your training section, and it contains all of the theory you will receive while you are learning to service air conditioning and refrigeration equipment. We are not going into the theory of refrigeration in great detail. It is enough that you have a reasonable understanding of the physical laws that are the basis for the refrigeration cycle. This information will not repair any refrigerator or make you one dollar. However, it will lay a foundation for you to build a wonderful trade on. You will be a more efficient trouble shooter if you have this information in the back of your head. Keep it in the back of your head and your hands will travel faster and surer.

What you are about to read here is contained in books that would fill the shelves of a large library. These books are highly technical and full of scientific words and phrases. With all due respect to the authors of these books and the scientists who furnished the material, we are going to condense the whole lot down to a few pages of readable language. The scientists who contributed the most to this library were Charles, Dalton, and Boyles. These three scientists discovered certain physical laws which cannot be disputed. These laws are much like Newton's Law of Gravity. For the purpose of this course and with due respect for these scientists, we are going to call these laws Doolin's Laws. I will lay down these laws, and we will examine them and tie them into the refrigeration cycle; then we will leave them, for they will not repair refrigeration equipment.

LAW NO. I

There is no such thing as "cold" until the thermometer registers 460 degrees below zero. Everything above this is heat.

We use the term "cold" every day. We say, "The box is getting colder." This is not exactly true. In fact, the box simply has less heat inside.

Here is the temperature scale as we know it plus the lower scale. **(Fig. 1-1)**

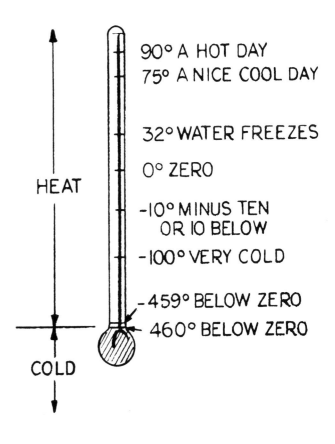

90° A HOT DAY
75° A NICE COOL DAY

32° WATER FREEZES

0° ZERO

-10° MINUS TEN
OR 10 BELOW

-100° VERY COLD

-459° BELOW ZERO
-460° BELOW ZERO

HEAT

COLD

(Fig. 1-1)

In plain language refrigeration is a problem dealing with heat. Since I have already said everything is heat down to 460 degrees below zero, that is where cold begins, there is still some heat at 459 degrees below zero. This is so cold—you notice I use the word "cold;" yet, we know better. This is so cold that if you were exposed to it for a few minutes and I touched you with a screwdriver, your body would shatter like glass.

You may never have given it much thought; but a refrigerator in your home, the one you keep your food in, is a box where there is a mechanical unit taking the heat out of the inside and everything that is stored inside. The unit is not putting something into the box which was not there before. The refrigerating unit is not imparting some quality to the inside. All it is doing is pulling the heat out and getting rid of it on the outside of the box. The insulation in the box keeps heat from getting back in too fast. Then there is a constant struggle for the machinery to get the heat out of the box and to try to keep enough of it out so that there is less heat on the inside than on the outside. The reason water in your ice cube trays turns to ice is that the heat was pulled out of the water until it reached 32 degrees, and then it turned solid. As a matter of fact, an ice cube is very hot if we remember that cold begins at 460 degrees below zero.

I know this does not repair refrigerators. But again, we are just getting acquainted with some facts that will help you to be a master serviceman.

If you are straight on this first law that to refrigerate means to remove heat, then let's go to *Law No. II*.

LAW NO. II
Heat is ever ready to flow to anything which contains less heat.

This statement is just as true as your saying that water will run downhill or seek its own level. It's the old law of gravity. For instance, if I pour water into the high end of this tube, you will agree that it will run over on the short end of the "U" until it levels off at the gravity line. You know this will happen, and it is a true law of gravity.

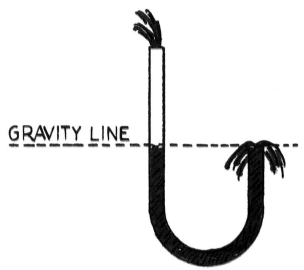

GRAVITY LINE

It is just as true as the fact that heat will always start to flow to anything with less heat. If I had two anvils and one had less heat than the other, the heat in the warmer one would flow over to the cooler one until both were the same temperature.

Let's examine the two anvils in **Figure 1-2**.

If one were 60 degrees and the other 80 degrees in temperature, the 80 degree anvil would immediately give up some if its heat to the other until they were both equalized at 70 degrees. You can no more stop the flow

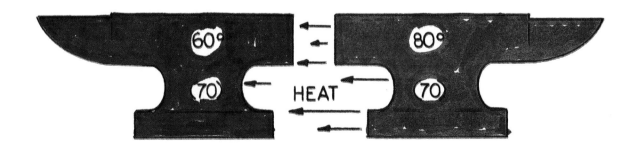

(Fig. 1-2)

of this heat than you can stop water from running downhill. But you can dam water up. Likewise, you can dam up heat with an insulation barrier; but this will not stop the heat flow. It will go on through the insulation if it takes ten years. All the insulation in your home refrigerator is doing is slowing down this flow of heat enough so that the refrigeration unit can get ahead of it. Nothing can stop the flow of heat. All that can be done is to slow it down; that is, the rate at which it moves. Everything is a conductor of heat—air, steel, wood, copper, even flesh. Some materials are better conductors than others.

The Egyptians had an air conditioning system which utilized the law of heat flow in the Pharaoh's palace two thousand years before Christ. Here is the way it worked. Every night 3,000 slaves came down to the palace at sundown and removed the main wall of the palace. The wall was 15 feet high, 10 feet wide and 60 feet long. It weighed over a thousand tons. One side of this huge marble block was polished and fitted into the palace ballroom. The rest was just rough stone. Using rollers, the slaves hauled this huge block out into the Sahara Desert where it stayed all night. The desert was cold at night, and the heat in the stone flowed out into the cold desert air. At the crack of dawn, they dragged the stone back to the palace and cemented it back into place. All day long while it was 130 degrees on the roof of the palace, this stone soaked up the heat in the palace. It is guessed that the temperature in the palace stayed about 80 degrees all day. Imagine, 3,000 men doing what roughly one 100 hp motor would do today.

Actually, we use the same principle today. But instead of 3,000 men to do the job, we use a mechanical system; and the refrigerant gas carries the heat outside where we get rid of it. Simpler yet, modern refrigeration today soaks up heat where it is not wanted and carries it out where it does not matter and gets rid of it. It's the same principle as the stone and the slaves, only we use refrigerant and a pump. The heat is ever ready to flow to anything that has less heat. Your body gets cold when you are improperly dressed not because as grandma always said, "The cold will penetrate!" but because your body heat flows out through your clothes.

Now we have laid down two laws. *Law No. I,* that refrigeration is a problem dealing with heat all the way down to absolute zero at 460 degrees below zero. *Law No. II,* that heat is fluid or in a constant state of motion always ready to move to anything with less heat, or "cooler," if you please. With these two laws in mind, we are going to invent a refrigerator after we get one more

law stated.

This third law may give you some trouble, but I believe we can get it across. At first it may not seem to have as good a connection with refrigeration as the first two laws. But after we understand it thoroughly, you will see the possibility of our inventing a simple refrigerator together utilizing these simple three laws.

LAW NO. III

Anytime a liquid changes to a gas or vapor, it must give up its heat; and the heat is carried off in the vapor.

To say it another way, whenever a liquid changes its state from a liquid to a vapor, heat is pulled out of the liquid that is still left; and this heat is carried off in the vapor.

This is the law of changing states of liquid. You may never have given it much thought, but when your skin is wet and you dry off very rapidly, the liquid is changing to a vapor. The heat is being pulled out of the flesh of your body; and you are chilled. The heat has actually left in the water vapor. This law is simple if you will stop and think a minute. Have you ever had gasoline, drip gas, or carbon-tet on your hands? Did you notice the cold (correction, heat loss) from your hands?

At the risk of belaboring a point, I want to be sure you nail this law down. It is the basis of all the refrigeration in the world, natural or mechanical (except the sun and the earth relationship—solar heat).

Take two buckets of water for instance. We put a tight lid on one bucket and leave the other open. Set them both out in the yard. The bucket that is evaporating, that is, drying up, will be 20 or 30 degrees cooler than the closed bucket. Stick a thermometer down in the water in the open bucket. It should read approximately 70 degrees. Take the lid off the other bucket and stick a thermometer in the water. It should be almost as hot as the sun's heat on the lid. The water that was evaporating was cooler because the heat in the residual water left in the bucket was leaving in the water vapor carried off by evaporation .

For example, suppose you set two buckets of gasoline out in the yard. You have the same amount in each bucket. You know that gasoline evaporates even faster than water. The gasoline left in the open bucket just before it all evaporated away would be 40 or 50 degrees cooler than the closed bucket where the vapor could not get out and carry away the heat. (Fig. 1-3)

This is an iron bound rule. Any liquid capable of evaporating gives off a vapor, and the heat necessary to make the liquid evaporate is carried off in the vapor.

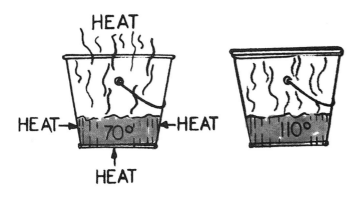

(Fig. 1-3)

The heat necessary to make water boil or evaporate into steam is always contained in the steam. In a steam boiler the heat from the furnace passes on through the water and is carried off in the steam.

Summing this up, any liquid capable of changing into vapor and from vapor back into liquid is a *refrigerant*; that is, it moves or displaces heat. It takes on heat changing from a liquid to a vapor, and it gives up this same heat out of the vapor when it changes back into a liquid.

Men using these three laws as the basis of their experiments invented modern mechanical refrigeration. First, they knew they had a problem dealing with heat—*Law No. I*. They had to get rid of heat, not put cold inside of a box. Second, they knew that heat was fluid and would always move immediately over into something less hot, like a big hunk of ice. Third, they knew the law of changing states of liquid. With these laws in mind, mechanical refrigeration came into being.

Heat was the problem, not cooling. If the heat could be displaced, it would be cool because there was less heat. The pioneers in refrigeration knew that heat would flow immediately to anything with less heat. They knew also that if there was anything inside of a box colder than the foods stored there, heat would go to this colder thing. They reasoned that if they insulated the box, they could hold back heat trying to get in from outside. They knew that ice would work with these first two laws; but they were after something better. So they brought the third law into action. They reasoned that if all this were true, then what was needed was some kind of liquid in a container which would evaporate carrying the vapor outside of the box through a chimney. This vapor would carry the heat inside of the box to the outside. This was pretty good reasoning since it is almost exactly the basic principle of modern refrigeration. Here is a picture of a refrigerator much like the first one built. (**Fig. 1-4**)

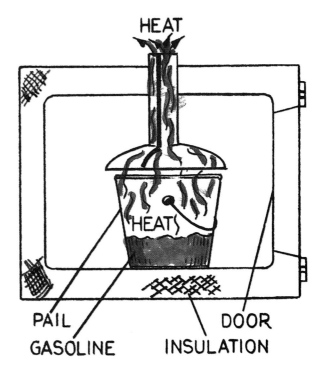

(Fig. 1-4)

The pail is full of gasoline. As the gasoline evaporates, the fumes go up into the hood and out through the chimney. These gasoline fumes are loaded with heat which was pulled out of the liquid left in the pail. More heat is coming through the side of the pail into the liquid and on out in the vapor. This crude box would actually be approximately 20 degrees cooler on the inside than the outside because of the heat being pulled out of the interior.

But what's wrong with this unit? First, it wouldn't keep meat. Second, some escaping fumes might stink up the box. Third, you would always be setting in a fresh bucket of gasoline, so there would be too much waste. You wouldn't want to strike a match to see what was in your refrigerator either. But suppose we solder the hood on top of the bucket. Now we are rid of the stink, but we will have to fill the bucket through the chimney. We have improved our refrigerator; but it still requires that we keep replenishing the gasoline, and it is not cold enough. We can improve that by getting an old plumber's pump and sucking the fumes off at a faster rate. But why waste the vapor fumes which contain the heat? Why not pump the fumes out and then pump them into a coil like a whiskey still and get the heat out. This will condense the vapor back into gasoline.

Now we have a refrigeration cycle complete. Of course, we will have to pipe the fumes or vapor to the pump, then to the condenser coil, then back to the pail inside of the box.

Here is our improved version of our original refrigerator. **(Fig. 1-5)**

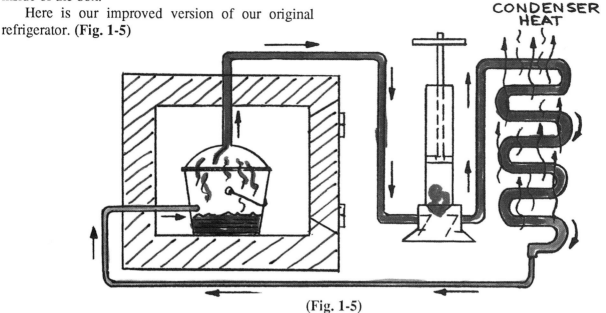

CONDENSER HEAT

(Fig. 1-5)

We have gone this far. Why not improve our refrigerator once more. We will add an electric pump and condenser fan to blow the heat out of the condenser coil. **(Fig. 1-6)**

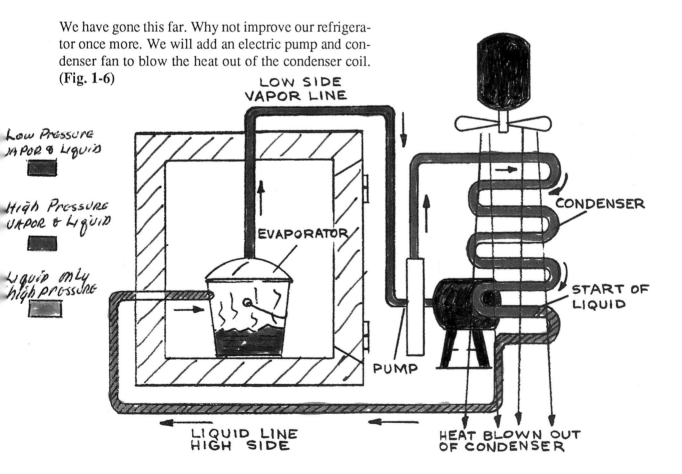

Low Pressure
VAPOR & LIQUID

High Pressure
VAPOR & LIQUID

Liquid only
high pressure

LOW SIDE
VAPOR LINE

EVAPORATOR

PUMP

CONDENSER

START OF
LIQUID

LIQUID LINE
HIGH SIDE

HEAT BLOWN OUT
OF CONDENSER

(Fig. 1-6)

The crude refrigerator would actually work and preserve food inside of the box. But we can go one step further and improve on our evaporator (pail) and the type of refrigerant (gasoline) we are using. Gasoline is dangerous.

Starting with the refrigerant (gasoline) we should know something about the characteristics or peculiarities of liquids which are suitable as refrigerants. First we must keep in mind that all liquids that are capable of evaporating and condensing are in a sense of the word capable of taking on heat or giving up heat and are therefore refrigerants. Water is a refrigerant, but its range of boiling or evaporating is too high. To better understand refrigerants, we should understand that when any liquid is evaporating, it is simply doing a slow boil. When it is boiling, it is simply evaporating at high speed. Here is an important point: the rate or speed at which a liquid evaporates determines how much heat will be absorbed and how fast the heat will be taken on and carried off in the vapor.

Some liquids like water evaporate very slowly unless heated up to 212 degrees. At 212 degrees water boils or evaporates very rapidly, but it is useless as a refrigerant. Gasoline evaporates much easier. High test, drip, or casing head gas boils very easily and will actually freeze anything in contact with it when it is evaporating rapidly. Carbon-tet is a pretty good refrig-

erant. Propane and butane are excellent refrigerants. But propane and butane are highly inflammable and explosive; so is gasoline. So the ideal refrigerant would be a liquid that would evaporate very fast or would boil at a very low temperature and still not be dangerous. The refrigerants you have to work with today meet these requirements. They boil easily (change to a vapor) and they are non-toxic and non-inflammable. To improve the evaporator, we will use a length of tubing and let the liquid boil inside of the tubing instead of the pail. We will attach fins to the tubing to increase the heat transfer surface.

Now to get back to our refrigerator, let's bring it up to date by improving the evaporator, the condenser, and the refrigerant. Suppose we find a certain liquid that will boil so easily that when it boils it is below freezing. Freon will do this. Now here is our modern air conditioner using Freon as a refrigerant. A window unit for example: (Fig. 1-7)

In this crude drawing you see the basic refrigerating machine which utilizes the principles we have been reading about. First we have the evaporator where the liquid changes to a vapor and when changing to a vapor, pulls the heat out of the copper tubing and fins, which get very cold. These fins and tubes in turn take the heat out of the air being circulated over them.

Going step by step through the cooling or evaporat-

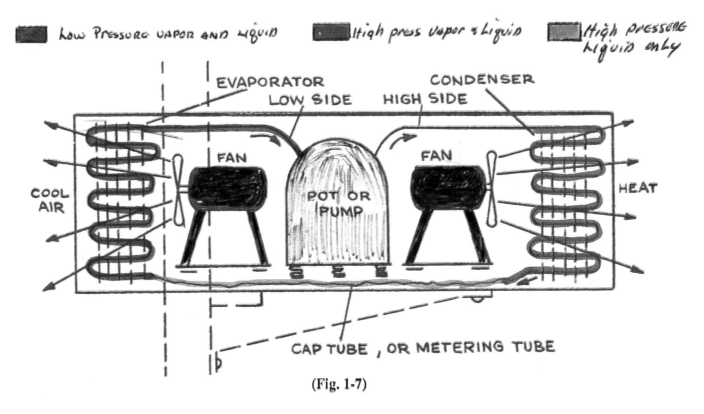

(Fig. 1-7)

ing end of refrigeration, we find in a window unit, for example, that first, the tube gives up its heat to the vapor *(Law III)*. The heat in the fins flows into the tubing and on into the vapor. The heat in the air going over the fins gives up its heat to fins, then the tube, then the vapor. The heat in your body flows out into the air with less heat, and you are air conditioned. Remember heat flows through everything—copper, steel, wood, and even insulation.

Now going one step further, we have vapor leaving the evaporator loaded with heat. This vapor is sucked back to the pump, and from there it is compressed in the high side or condensing side where the same heat that was soaked up inside of the house is blown out of the hot condenser on the outside of the house. As soon as the heat is blown out of the fins and tubes of the condenser, the vapor gives up its heat through these fins. This is the reverse of the evaporator. The vapor cools and turns back into a liquid, which falls to the bottom of the condenser and starts feeding back over to the evaporator to expand, or boil, and again change from a liquid to a vapor and pick up another load of heat. Around and around we go from liquid to vapor, vapor to liquid. This is a constant feeding and pumping process. There you have the refrigerating cycle. This refrigeration cycle is the same the world over for all mechanical refrigeration. The method of cooling may vary, and the method of condensing the gas may vary; but the refrigeration cycle is the same in the largest air conditioning plant in the world or in your home refrigerator. By utilizing the laws we have studied and some copper tubing, a refrigerant, and pump, with fans to help, we have constructed a refrigerating device that is the basis of all refrigeration. If this can be called theory, then this is the theory of mechanical refrigeration. We are not going to manufacture Freon, copper, or electric pumps. It is enough that we understand this simple merry-go-round. A liquid changing its state from liquid to vapor and vapor to liquid soaks up heat which just naturally likes to move into something less hot. This same heat is blown away outside of the house by a fan blowing through the condenser.

We have learned several new terms and from now on we will use them correctly. A serviceman always calls the cooling coil the evaporator. The ice cube maker in your refrigerator is an evaporator. The main cooling coil in the Empire State Building is an evaporator. All of the evaporator and the suction line including the intake side of the pump are the low side. The high side or condensing side is always called the high side or condenser.

In this section we will start with a window unit to get straight on the refrigerating cycle and the terms or terminology. Fortunately, a window unit air conditioner is a perfect teaching aid. If a good teacher set out to design a unit which would best illustrate the mechanics and principles of refrigeration; he could not design a better teaching aid than a window unit. For this reason, we will be referring constantly to window units for illustration.

There are no mysteries connected with the refrigeration cycle, nor are there any strange quirks that are not understandable. Refrigeration is an exact science and very simple to understand once you have these basic laws.

Had a good blacksmith had this written section in his possession many years ago, he could have made a fortune, especially if he had a yen to experiment. This blacksmith would have reasoned that if all this were true, he had best be getting to work and trap some liquid and start it around and around until he pulled all of the heat out of areas where people did not want heat like refrigerators, meat cases or in the case of air conditioning, the heat in a room. He would probably have used iron pipe, a very simple hand-driven pump, and a refrigerant like casing head gas or gasoline.

Up to this point we have been studying the refrigeration cycle in its simplest form, and by now we should have a pretty good understanding of what is going on in refrigeration and why it is going on. Now let's get to the business at hand and refine or improve on this refrigeration cycle and its components or parts. The modern systems are not complicated but simply improved. Not one of you would hesitate to tell me that basically the Model T Ford operated on the same principle and for the same purpose as the new cars of today. You might tell me that there are certain refinements in a new Ford; but a Ford still has four wheels and a motor; it still uses fuel; and it is still designed to transport you from one place to another. This would apply to refrigeration as well, whether it be air conditioning, domestic, commercial, or industrial. Some systems are old fashioned, not as fast or as well refined control-wise; but, basically, the same old job is to remove heat.

In the beginning the suction pumps were simple one-cylinder pumps with piston, crankshaft, and flapper valves. These pumps were belt driven by pulley wheels and were not much different from a simple air compressor used in a garage today. They had suction and discharge valve reeds much like the following

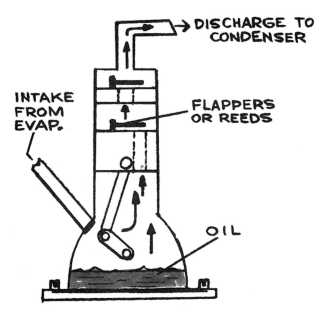

INTAKE FROM EVAP.

DISCHARGE TO CONDENSER

FLAPPERS OR REEDS

OIL

(Fig. 1-8)

drawing. (Fig. 1-8)

These pumps were far from perfect even though they did a respectable job of pumping. The seal which keeps the gas from leaking out around the crankshaft end was a constant source of trouble. When this seal leaked, the gas escaped. Today, with the exception of very large pumps and certain open type units, this pump has in most cases been replaced with the sealed hermetic pot in order to eliminate this seal leak. This sealed pot will be analyzed in detail later.

The original evaporators were little more than closed kettles or boilers much like the milk pail on the inventors unit. The gas simply boiled inside the boiler. The liquid level in the boiler was controlled by a float. The original condensers were just air or water-cooled coils of pipe. They had no fins.

Today we have high speed pumps, copper tubing with attached fins for better transfer of heat, wonderful refrigerants like Freon 12, Freon 22, and blower fans for moving air faster and quieter. *But the refrigeration principle remains the same.*

The modern evaporator today is a continuous copper tube running back and forth with hundreds of fins bonded to the tubing. These fins look much like the radiator of your car and serve the same purpose. They speed up heat transfer. In almost every case these modern evaporators have a fan or blower which forces air over the cold coil in order to transfer the heat more

rapidly. The liquid is still doing the same old thing inside the copper tube; that is, boiling and changing to a vapor, even though the boiling process is confined in a small copper tube. In the evaporator of your home refrigerator, this tubing is bonded to the whole ice cube maker. There are no fins. The metal can holds the ice cube trays. Take a look and you can see where the liquid comes in and the vapor goes out. You can even hear it boiling, gurgling and exploding into vapor as it enters the evaporator to pick up a load of heat.

The modern condenser is constructed much like the evaporator. It, too, has a fan or blower, if it is air cooled like a window unit. The air blower over the condenser tubes carries away the undesirable heat and thereby causes the vapor to turn back into a liquid which returns to the evaporator to soak up more heat. This is a continuous process. The condenser on the unit of your home box may not have a fan. It may depend upon good air circulation to get rid of the heat. Look and see.

The pot, the condenser, the evaporator, and the fans on a window unit are the essential parts of the window unit. These parts are necessary to carry out the function of heat transfer from within to without.

The next important part of this simple refrigeration unit that has not been mentioned yet is the cap tube or metering device. All refrigeration units require at least a cap tube or thermostatic expansion valve to measure the right quantity of liquid into the evaporator in order to keep the evaporator at a desired temperature.

Now let's discuss something that we must understand in order to understand the use of gauges and the control of the desired temperature in the evaporator. This temperature is controlled in the evaporator by metering in the liquid with a cap tube or expansion valve. We mentioned earlier that many liquids are suitable for refrigerants. That is, many liquids are capable of evaporating and condensing back again into a liquid. We said that some liquids were considered too dangerous to be used as refrigerants. Also, we said that some liquids boiled more readily than others. Now let's examine the boiling point of some of these liquids. We can start with water. Water boils at 212 degrees. Do you agree? Ordinary gasoline boils at about 150 degrees. High test gasoline will boil at about 130 degrees. **Caution!** Don't ever boil gasoline. I am just using gasoline- as an example. Carbon-tet will boil at approximately 120 degrees. In fact, some liquids will boil at such low temperatures you could take a bath in them while they were boiling; and the temperature would be about right for bathing. You must remember that when liquids are boiling, they are taking on heat as the speeded-up evaporation takes place and the heat is carried off in the vapor. If this is so, then does it not stand to reason that any liquid which boils very readily or at a very low temperature would make a good refrigerant, regardless of its properties or what it is made of chemically, so long as it is safe and will not freeze itself solid?

Chemical companies have manufactured a certain liquid called refrigerant which boils so easily that it boils even below freezing. Remember that water boils at 212 degrees. Freon 12 boils at 21.7 degrees below zero. Just imagine a liquid that is chemically manufactured so that if you warm it up to 20 degrees below zero, it will start to boil. If you had an open bucket of Freon 12 that you didn't want to boil away, you would have to store it at 21 degrees below zero. If it warmed up to 20 degrees below zero, the bucket of Freon 12 would boil away.

All liquids that are capable of evaporating (boiling) have a particular rule which applies to them. They boil at a temperature governed by the atmospheric pressure on their surface. Water boils at 212 degrees at an atmospheric pressure of 14.7 psi, pounds per square inch. The absence of atmospheric pressure is a vacuum. So you see, there is a direct relationship between the boiling points of liquids and the atmospheric pressure. This is true of all liquids that are capable of being boiled. I know at first reading this may be hard to understand, but don't worry about it. We will go over it until you understand it. Again, if Freon has a certain boiling point in relationship to the pressure on it or the vacuum it is in, then you may already suspect the reason a serviceman uses gauges to read the pressure in the system. This temperature pressure relationship is always constant. If Freon has a certain pressure caused by its own boiling, then it has a certain temperature at this exact pressure. This temperature pressure relationship of refrigerant gas is a fixed relationship just like the multiplication tables. If the steady pressure on Freon 12 is 30 psi, then it has an exact certain temperature. If the steady temperature of Freon is 34 degrees, then its pressure is exactly so much. One cannot be without the other. In the old multiplication tables, we say 2 x 2 equals 4; and 4 divided by 2 equals 2. Likewise, a refrigeration serviceman should have a temperature-pressure chart so he can read temperature to pressure or pressure to temperature. Just remember, any time you have a certain pressure on your gauges, you can look at your chart under the column of the kind of refrigerant used and it will tell you the temperature. After a little practice you will know

most of your maintained evaporator pressures. Likewise the temperature.

Now let's go back to the metering device we mentioned which meters the liquid refrigerant into the evaporator. Let's take one type, the cap tube. This little tube is designed to let just so much liquid Freon into the evaporator. The principle it works on is to choke back the liquid through a small hole until only the right amount can get through. Now we come to our first lesson in refrigeration where a cap tube is used between the condenser and the evaporator in order to maintain a certain pressure in the evaporator. You are not interested in the engineering design of a window unit. However, to service one and charge it with refrigerant, you must understand what the designer was striving for and at what pressures he expects the unit to operate. By knowing this, you can charge the system and know when it is operating properly. First, the designer wanted the evaporator to be as cold as possible without freezing up the coil at 32 degrees F; therefore, he decided to maintain a temperature in his coil of 40 degrees. So he looked on a temperature-pressure chart and ran his finger down the left hand temperature column until he came to 40 degrees. Then he looked straight across to pressure on this line in the column under F-22, the kind of refrigerant he decided to use. **(Fig. 1-9)**

TEMPER-ATURE °F.	REFRIGERANT			
	12	22	717	*500
12	15.8	34.9	25.6	21.2
13	16.5	35.9	26.5	21.9
14	17.1	36.9	27.5	22.6
15	17.7	37.9	28.4	23.4
16	18.4	38.9	29.4	24.2
17	19.0	40.0	30.4	24.9
18	19.7	41.1	31.4	25.7
19	20.4	42.2	32.5	26.5
20	21.0	43.3	33.5	27.3
21	21.7	44.4	34.6	28.2
22	22.4	45.5	35.7	29.0
23	23.2	46.7	36.8	29.8
24	23.9	47.8	37.9	30.7
25	24.6	49.0	39.0	31.6
26	25.4	50.2	40.2	32.4
27	26.1	51.5	41.4	33.3
28	26.9	52.7	42.6	34.3
29	27.7	54.0	43.8	35.2
30	28.5	55.2	45 0	36.1
31	29.3	56.5	46.3	37.0
32	30.1	57.8	47.6	38.0
33	30.9	59.2	48.9	39.0
34	31.7	60.5	50.2	40.0
35	32.6	61.9	51.6	41.0
36	33.4	63.3	52.9	42.0
37	34.3	64.6	54.3	43.1
38	35.2	66.1	55.7	44.1
39	36.1	67.5	57.2	45.2
40	37.0	69.0	58.6	46.2
41	37.9	70.5	60.1	47.2
42	38.8	72.0	61.6	48.4

TEMPERATURE PRESSURE CHART

Gage Pressure—Bold Fig.

Vacuum—Italic Figures

(Fig. 1-9)

There he found that he would have to maintain a pressure of 69 pounds in the evaporator. So he cut the cap tube to a certain length (Law of Hydraulics) and since he had a pot or pump running at a steady pumping rate, he knew the tube would let just the right amount of gas through to boil at a steady pressure of 69 psi.

This is as simple as a water pump with a discharge pressure which you could maintain by pinching down on the end of the garden hose. The designer of a window unit has pinched down on the supply of Freon to the evaporator until he can maintain a steady pressure, since the pump is pulling this Freon vapor out of the evaporator at a steady rate. As a serviceman, you could put your gauges on a window unit and confirm this. Assuming the unit was fully charged, you would find a steady pressure on your gauge at about 69 psi, give or take a few pounds. **(Fig. 1-10)**

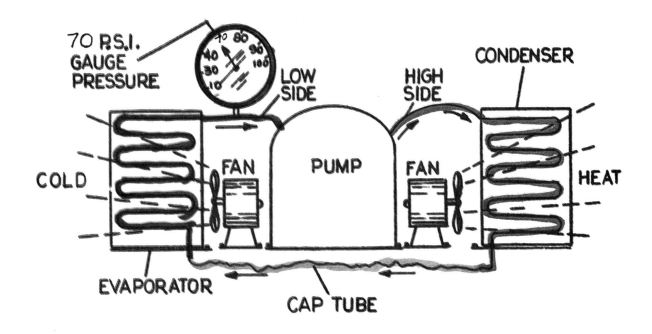

(Fig. 1-10)

Now we have the same old pot, condenser, evaporator, and cap tube. That is all there is to the refrigeration cycle in a window unit. I could have told you that when you put your gauges on a window unit, you would find the gauge would read 69 psi if the unit is charged. If it reads 30 psi, you are short on gas. But this would not be teaching you the whole truth, and I want you to understand what is going on inside of a unit and why it works. For understanding these fundamental facts, you will be a far better serviceman than one who does work automatically never knowing *why*.

While you are driving your car or just sitting thinking, think about the three laws and the peculiar rules which apply to all liquids. Many everyday happenings will fit into the same rules. You can feel the heat in steam coming out of a teakettle. It will burn you. This is artificially applied heat from a gas burner that went on through the kettle into the water and out into the steam.

Remember that in the strange chemical makeup of refrigerants, they are doing the same thing. Yet they need no gas burner. They boil at low temperatures and take on heat the same way, and the heat goes off in the vapor. Yet everything may seem extremely cold.

Pumps, evaporators, condensers and cap tubes are ordinary everyday terms to the air conditioning serviceman.

Now if we do some thinking on what has been said here so far, it would be possible to see that the main working tool of a serviceman is the gauges. The gauges tell the serviceman just what is going on inside. You will admit that if you had a pressure gauge on a water faucet, you could say that if it showed pressure that more than likely that was the water main pressure for that area. Suppose it were a very low pressure, then you would have to say that you would try to find out if there were something wrong or if that were all the pressure available to that area. Suppose we found a very weak pressure, then we might say that we were short on gas. Suppose we found extremely high pressure in the evaporator, then we could say that the pump was not sucking out the expanded or evaporated gas. That is, the cap tube was feeding the gas over, but it was not being taken out as fast as it should be leaving the evaporator. Bad valves might cause this.

Not only does the gauge manifold you will own serve as a trouble shooter's main tool, it also is designed to permit the use of an extra hose for charging the unit. Here is a simple drawing of a serviceman's gauges and manifold, with hoses. (Fig. 1-11)

If the water pressure available to your home was too low and you called the city water department, more than likely a man would come out and turn on the water and estimate its force by using his eyes. A better water man would bring along a gauge and read the actual pressure in your water system. Then suppose he knew the design pressure down at the street corner was so much. Then couldn't he estimate that your water was ample or being starved between the house and the street corner where the main was. You could compare this to your putting the gauges on any type unit. If you know the pressure the designer had in mind when he built the unit and the kind of refrigerant and you have a temp-press chart, you could certainly say whether the pressure was up to snuff or not. You should compare the gauges to a flashlight in the darkness. You can walk in the darkness without a light, but you can travel so much faster and surer with a good flashlight.

■

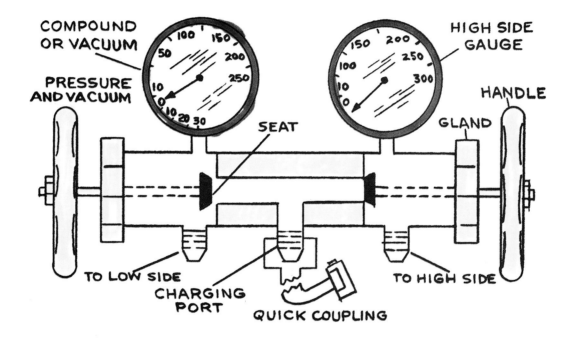

(Fig. 1-11)

TEMPERATURE PRESSURE RELATIONSHIP

TEMPER-ATURE °F.	REFRIGERANT 12	22	717	*500
-60	19.0	11.9	18.6	—
-55	17.3	9.1	16.6	—
-50	15.4	6.0	14.3	—
-45	13.3	2.6	11.7	—
-40	10.9	0.6	8.7	7.9
-35	8.3	2.7	5.4	4.8
-30	5.4	5.0	1.6	1.4
-25	2.3	7.5	1.3	1.1
-20	0.5	10.3	3.6	3.1
-18	1.3	11.5	4.6	4.0
-16	2.0	12.7	5.6	4.9
-14	2.8	13.9	6.7	5.8
-12	3.6	15.2	7.9	6.8
-10	4.4	16.6	9.0	7.8
-8	5.3	18.0	10.3	8.8
-6	6.2	19.4	11.6	9.9
-4	7.1	20.9	12.9	11.0
-2	8.1	22.5	14.3	12.1
0	9.2	24.1	15.7	13.3
1	9.7	24.9	16.5	13.9
2	10.2	25.7	17.2	14.5
3	10.7	26.6	18.0	15.1
4	11.2	27.4	18.8	15.7
5	11.8	28.3	19.6	16.4
6	12.3	29.2	20.4	17.0
7	12.9	30.1	21.2	17.7
8	13.5	31.0	22.1	18.4
9	14.0	32.0	22.9	19.0
10	14.6	32.9	23.8	19.7
11	15.2	33.9	24.7	20.5

TEMPER-ATURE °F.	REFRIGERANT 12	22	717	*500
12	15.8	34.9	25.6	21.2
13	16.5	35.9	26.5	21.9
14	17.1	36.9	27.5	22.6
15	17.7	37.9	28.4	23.4
16	18.4	38.9	29.4	24.2
17	19.0	40.0	30.4	24.9
18	19.7	41.1	31.4	25.7
19	20.4	42.2	32.5	26.5
20	21.0	43.3	33.5	27.3
21	21.7	44.4	34.6	28.2
22	22.4	45.5	35.7	29.0
23	23.2	46.7	36.8	29.8
24	23.9	47.8	37.9	30.7
25	24.6	49.0	39.0	31.6
26	25.4	50.2	40.2	32.4
27	26.1	51.5	41.4	33.3
28	26.9	52.7	42.6	34.3
29	27.7	54.0	43.8	35.2
30	28.5	55.2	45.0	36.1
31	29.3	56.5	46.3	37.0
32	30.1	57.8	47.6	38.0
33	30.9	59.2	48.9	39.0
34	31.7	60.5	50.2	40.0
35	32.6	61.9	51.6	41.0
36	33.4	63.3	52.9	42.0
37	34.3	64.6	54.3	43.1
38	35.2	66.1	55.7	44.1
39	36.1	67.5	57.2	45.2
40	37.0	69.0	58.6	46.2
41	37.9	70.5	60.1	47.2
42	38.8	72.0	61.6	48.4

TEMPER-ATURE °F.	REFRIGERANT 12	22	717	*500
43	39.7	73.5	63.1	49.6
44	40.7	75.0	64.7	50.7
45	41.7	76.6	66.3	51.8
46	42.6	78.2	67.9	53.0
47	43.6	79.8	69.5	54.2
48	44.6	81.4	71.1	55.4
49	45.7	83.0	72.8	56.6
50	46.7	84.7	74.5	57.8
55	52.0	93.3	83.4	64.1
60	57.7	102.5	92.9	71.0
65	63.7	112.2	103.1	78.1
70	70.1	122.5	114.1	85.8
75	76.9	133.4	125.8	93.9
80	84.1	145.0	138.3	102.5
85	91.7	157.2	151.7	111.5
90	99.7	170.1	165.9	121.2
95	108.2	183.7	181.1	131.3
100	117.1	197.9	197.2	141.9
105	126.5	212.9	214.2	153.1
110	136.4	228.7	232.3	164.9
115	146.7	245.3	251.5	177.4
120	157.6	262.6	271.7	190.3
125	169.0	280.7	293.1	204.0
130	181.0	299.3	—	218.2
135	193.5	319.6	—	233.2
140	206.6	341.3	—	248.8
145	220.3	364.0	—	265.2
150	234.6	387.2	—	282.3
155	249.5	410.8	—	300.0
160	265.1	434.6	—	318.7

* Reproduced by permission of Carrier Corp.

Gage Pressure—Bold Fig. — Vacuum—Italic Figures

(Fig. 1-12)

Look at -20F on the left column. Now look across to the Refrigerant -12 column on the same line and you will see 0.5 psi, that is 5/10 of a pound pressure. Here Freon-12 starts to boil and make its own pressure. Suppose we want an air conditioning unit to have a non-freezing evaporator pressure 32 degrees +. We look at 34 degrees in the left column and we see that on the same line under the window unit refrigerant F-22, we have 60.5, the working pressure in the evaporator.

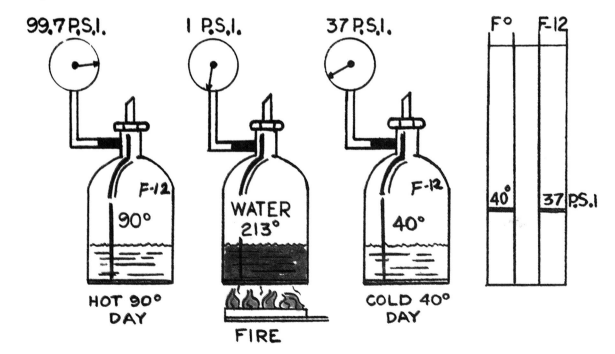

CHAPTER 2

FREON AND TEMPERATURE-PRESSURE RELATIONSHIPS

We have stated that if a liquid refrigerant boils of its own accord, then it must boil up a pressure, unless the vapor escapes. This is true. It does just that. Stop and reflect for a moment and you will understand this important fact. Water boils at 212 degrees and starts giving off steam. If this same water were in a closed boiler or cylinder and you added a little more heat to the fire and brought the temperature of the water up to 213 degrees, you know that it would bring up a little steam pressure, say about one pound per square inch. The same thing happens to Freon, only it does not have to be but one degree warmer than 20 degrees below zero; and it starts to raise a little Freon vapor pressure of about one pound psi, just like the 213 degree water raising a steam pressure of one pound psi Once we get used to this crazy boiling point of Freon, we have the key to a great deal of understanding. I know it is difficult to picture a liquid so crazy that it starts to boil all over the place if it ever gets warmed up to 20 degrees below. You might ask how cold is this stuff generally. The answer would be that it is never cold until you let the stuff boil, or let some vapor escape. It is always quiet. You could never have it in your possession unless it were in a strong metal pressure container like a service cylinder. While it is in this cylinder, it has boiled up a pressure that has the stuff stabilized at the temperature of the place where you have the cylinder stored or the temperature of the drum wherever it is. In other words, you can never have Freon in a cylinder or drum without its being a certain pressure; and that pressure will be just what your chart tells you it should be if the drum is stored or located in a place where there is a certain temperature. If you had a drum of F-12 in your garage and I came by and asked you what the pressure was in that drum, your answer would be that you did not know; but if I would tell you the temperature of the weather that day, you could look on your chart and tell me the exact pressure if we assume that the drum was about the same temperature as the weather that day. If you walked over to this drum and let a little vapor pressure escape, the liquid in the drum would begin to boil immediately and the drum would start getting cool (heat going off in the vapor). Just as soon as the drum warmed back up, the pressure would be right back where it was before you cooled the drum down. Refrigerants are wonderful liquids. Anytime there is a change in the pressure of a refrigerant, there will be an automatic change in the temperature of the liquid and whatever it is contained in. The butane and propane tanks you see being used with a stove for cooking and heating around rural homes work for the same reason. As a matter of fact, propane is a far better refrigerant than some of the older refrigerants, but it is extremely dangerous and must be handled with great caution and must never be used as a refrigerant. Actually, the butane and propane tanks are boiling off vapor to be used as fuel; and the pressure in these tanks out in the yard will be high in the daytime and lower at night. That is, their pressure will vary with the temperature of their surroundings.

If all this is true about these liquids that boil so easily, then does it not stand to reason that if you can maintain any certain pressure on a liquid refrigerant, it will have to be a certain temperature? And if it is a certain temperature, it will have to be a certain pressure. Your reasoning is right, but how does this concern you? You are concerned with pressure because we are going to service a Freon circulation system, and we expect it to maintain certain temperatures for us. Said again, we are going to service certain equipment that has been manufactured to maintain certain pressures, thereby creating certain temperatures.

We, as servicemen, are concerned with the pressure of Freon in a system because it tells us whether or not the unit is operating properly, whether it has a charge of gas or is short on gas. Nail this down now and never forget it. "We do not make pressure in the system by changing the design." The engineers that design refrigeration equipment have already determined what pressure should be in the evaporator and condenser. All we do is see if that pressure is as it should be or has gone wrong some way. Those engineers have already determined the size or style of the metering device and coils. They have worked out all the problems where a pump pumps

(Fig. 2-1)

at a certain speed and capacity knowing that this pump will handle just so much vapor on the high side or low side. This is called balancing the system. The work of engineering the metering device in relation to the pump speed and capacity is no more complicated than rigging a small motor driven pump with a garden hose connected to the discharge and the outlet end of the garden hose brought around and hooked back up again to the suction of the pump. **Fig. 2-1**.

With this system filled with water, we start the pump circulating. I would then have you stand out about the middle of the garden hose where you could see both gauges, and then I would have you kink the hose and bleed through just enough water so that the pressure gauge on the discharge side between you and the pump came up higher than the section of hose between you and the pump on the suction side. Suppose I told you to kink the hose until the pressure came up to 80 pounds and the suction side went down to a vacuum. You could do it, and that is all there is to causing a pressure drop on Freon circulating in a system. But use a cap tube or valve for metering the gas instead of a kink in the copper line. Remember if there is a certain pressure maintained on the low side (evaporator) then it will have to be a certain temperature. We said that the ideal temperature of the evaporator in a window unit would be a pressure

just high enough to get the job done and not freeze moisture on the coil. We look at the chart in **Figure 1-12** and we see that it is about 69 psi for units using F-22. This pressure will be good and cold but not freezing. The pressure in the evaporator of a household refrigerator was designed to be about 9 or 10 psi using F-12, and this will be about zero temperature.

After this discussion you might say to me, "if all of this is true, then if a man knew just about the right pressure for every evaporator, he could tell whether it was operating properly if he had a gauge to put on the unit." This would be correct. However, there is more than just the evaporator pressure to consider. To know why this pressure is to be so much is important; but it is more important to know at any given instant what it should be. Remember the low side or pressure drop side starts the instant the gas leaves the end of the restrictor (cap tube). This low pressure side continues until the vapor enters the suction valves of the compressor and is compressed in the cylinder by the piston to start its journey through the high side (condenser) back to the end of the cap tube to start the expansion or boiling again.

The kink in the garden hose can be compared to all refrigeration metering devices; except that they are, of course, not so crude. The cap tube can always be

20

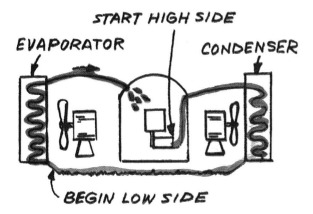

START HIGH SIDE

EVAPORATOR   CONDENSER

BEGIN LOW SIDE

identified by its small size; actually, the small size is the equivalent of the kink. The thermostatic expansion valve is a better metering device than a cap tube, but far more expensive. It is more complicated than the simple cap tube. You could compare the thermostatic expansion valve to the carburetor on an expensive automobile. You know the carburetor on the auto is designed to meter in the right amount of gasoline as it is needed; that is, when you press on the accelerator, it responds. The

thermostatic expansion valve responds when its power element goes to work; and basically, it is a metering device to meter refrigerant into the evaporator as it is needed. The cap tube is a very crude metering device compared to the expansion valve. You might say that a cap tube on your automobile would be the equivalent of taking off the carburetor and running the gas line from the tank directly to the manifold and letting it drip into the hole where the carburetor used to be mounted. Your car would probably run after a fashion, but it would run at a speed equal to the amount of the drip, regardless of the load or speed you wanted to travel. Once you set this drip, that would be the operating speed of your engine. The same applies to the cap tube used on any refrigerating device; it just meters in so much of a certain kind of refrigerant regardless of the particular load on the evaporator at any given time. Now let's see what a thermostatic expansion valve does when it is installed on refrigerating equipment. The liquid line from the condenser is hooked to the valve, and the outlet side of the valve is connected to the evaporator. (**Fig. 2-2**)

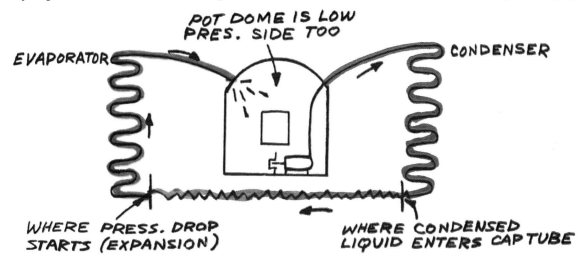

POT DOME IS LOW
PRES. SIDE TOO

EVAPORATOR   CONDENSER

WHERE PRESS. DROP
STARTS (EXPANSION)

WHERE CONDENSED
LIQUID ENTERS CAP TUBE

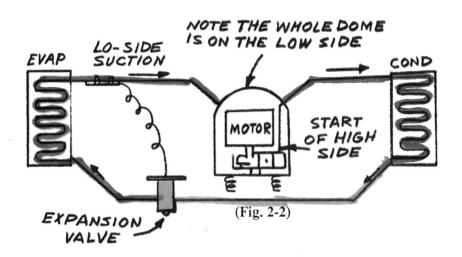

NOTE THE WHOLE DOME
IS ON THE LOW SIDE

EVAP   LO-SIDE
SUCTION

COND

MOTOR   START
OF HIGH
SIDE

EXPANSION
VALVE

(Fig. 2-2)

The power element, as it is called, is the little line running out of the top of the valve up to the other end of the low side line where it comes out of the evaporator. On the end of this little tube, there is a small round bulb or cylinder which is attached tightly to the suction line going back to the compressor. This little bulb is sensitive to the temperature of anything it is attached to, and it causes the valve to react accordingly. The original expansion valves were automatic, and some of them are still being used for certain types of equipment. You could compare the old automatic expansion valve to the regulator on an oxygen acetylene cylinder. Once set, it would just meter so much gas into the evaporator as it was used or sucked back to the compressor. Its operation was fixed, once set, and it functioned much like a cap tube—constant feeding device. The modern thermostatic expansion valve has a power element added to it, and it will meter refrigerant to the evaporator in just the quantity that the little power bulb makes it deliver. Examine this simple drawing of thermostatic expansion valve, and then we will go into it a little further.

You will note that the valve has an inlet and outlet. The liquid from the condenser enters the valve on the inlet side and leaves through the outlet side of the valve to the evaporator. The instant the liquid feeds through the small orifice controlled by the needle, the liquid begins to expand or boil. This place is always the coldest point in a system. Examine the valve, you will find that when the pressure is increased on the diaphragm, the needle will open and permit a greater flow of refrigerant; or a decrease in diaphragm pressure will let the spring lift the needle to decrease the flow. (**Fig. 2-3**)

The order of flow and attachment is shown in **Fig. 2-4**.

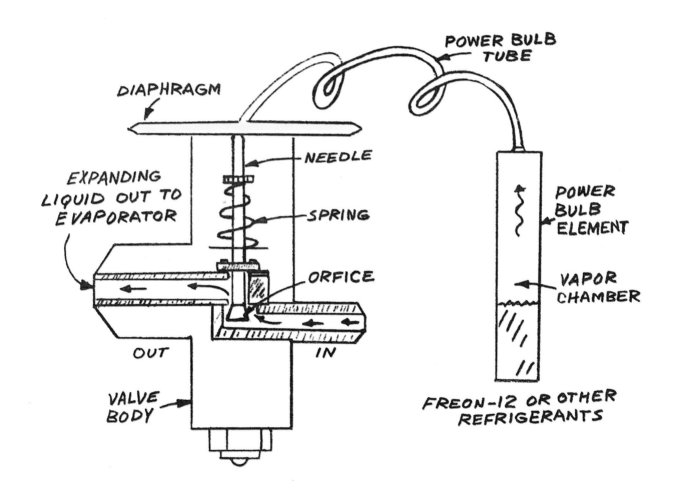

(Fig. 2-3)

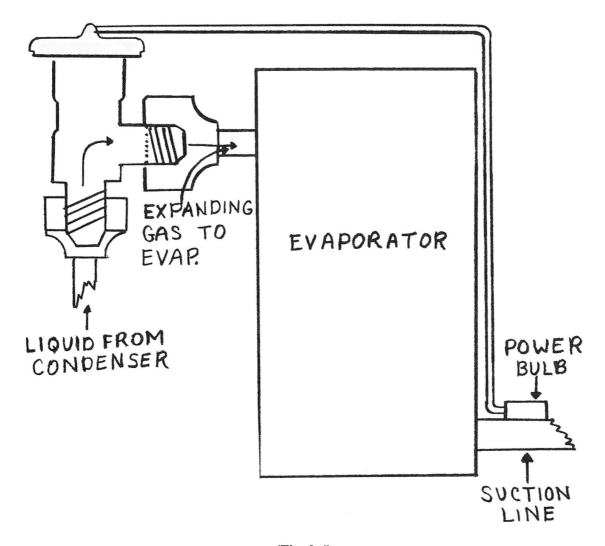

(Fig. 2-4)

On many expansion valves you can unscrew the power element from the main body of the valve. Here is the way the power bulb gets its power to push the needle down; remember a spring generally pushes the needle back up when the power bulb cools down and loses pressure. The power bulb is actually a small cylinder of refrigerant; that is, it is about the size of half of a pencil and hollow. It has a small charge of refrigerant inside, approximately a thimble full. The small tube is hollow and runs down to a flat diaphragm made of two thin round discs of metal sealed together around the edges. (Fig. 2-5)

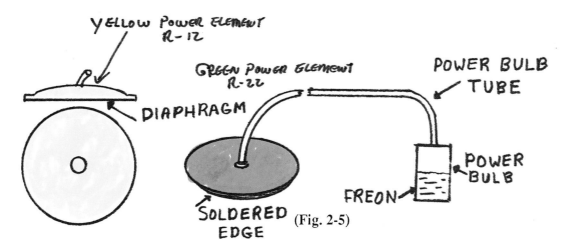

(Fig. 2-5)

The principle that this valve works on is just what we have been studying about when we have said that Freon would always have a certain pressure on it if it were a certain temperature. We have the liquid refrigerant trapped in the power bulb; it cannot get away, and all it can do is expand and contract as the temperature changes on the power bulb. The diaphragm being very thin, though, flexible metal, it will move downward when the pressure builds up in the bulb. This pressure in the bulb will travel down the small tube and cause the diaphragm to expand like filling an inner tube. When this expansion takes place, the needle is pushed down into the body of the valve.

You could say that this pressure works very much like the way the tires on your automobile build up pressure when they are hot and shrink when the weather is cold. Ordinary thermostats work on the same principle, whether they be used for heating plants or refrigerators. The difference between the thermostat and the thermostatic expansion valve is that the thermostat diaphragm operates to move a flip-flop switch turning off the electrical power, and the thermostatic diaphragm on a valve operates to depress a needle.

To go one step further, you should understand what caused the expansion valve to meter in more or less refrigerant as it is needed. The power bulb is attached to the suction line where the line comes out of the evaporator. Usually the bulb is clamped on or strapped on to this line with bolts so that there is a very tight union between the bulb and the suction line. When the vapor changes temperature (if it should) in the suction line, the change will be transmitted to the bulb. The liquid charge in the bulb will immediately respond to this change in temperature. If the suction line warms up, the power bulb will warm up; and the liquid refrigerant in the bulb will begin to expand and raise the pressure; and cause the valve to open. In the case of the bulb suddenly cooling down, the pressure in the bulb will drop and cause the valve to close.

You know that this change in pressure will cause the diaphragm to expand or contract. Now the question is; what causes the suction line coming out of the evaporator to be different temperatures at different times? Here is the answer. Suppose someone opened the door on a large refrigerator which utilized an expansion valve on the evaporator. The warm air let in by opening the door would make contact with the evaporator and would immediately cause the boiling liquid in the evaporator to take on more heat. This heat, however slight, would cause a change in the temperature of the

copper pipe carrying the vapor back to the pump. The power bulb of the valve being attached to this pipe would change temperature also. This temperature change in the bulb would be transmitted in pressure to the diaphragm, which would in turn change the needle position; and the change in the needle would let in more liquid to take care of the heat which was let in by opening the door. The same thing would apply if the person put a large package of warm meat into the refrigerator. In the case where the box does not have any more load put on it, it would continue to pull down; and as the vapor got colder and colder, the amount of liquid refrigerant feeding into the evaporator would be decreased.

One thing must be made perfectly clear right now. The cap tube and the expansion valve are two different things doing the same job. One is an expansive valve, very accurate; and the other is a cheap restrictor tube doing the same thing.

Should you open the door on a refrigerator that operated with a cap tube feeding the evaporator, there would be no change in the amount of liquid being fed to the evaporator, even though the load had increased. You could compare this to running a straight fuel line to the manifold of your auto engine with no carburetor. The cap tube in refrigeration is a straight metering device which meters at a constant rate. Simple as it is, it does a fair job where loads remain fairly constant; and a long pull down occurs only when you first start up, like putting a new refrigerator into operation. Most household freezers and home refrigerators operate with cap tubes. The cap tube does a fair job of feeding the liquid when it is used on the right type of job. But why do manufacturers use a cap tube when an expansion valve will do a much more accurate job of metering gas? Simple, a cap tube costs about twenty cents; but a good thermostatic expansion valve may cost more than twenty dollars. Expansion valves are priced according to the size and horsepower of the equipment. This is quite a difference in cost; and if you will reflect for a moment, you will realize that there is not that much importance connected to the metering. Take, for instance, a home freezer designed to operate at zero using F-22 with a cap tube. When the box is first installed, it will be warm inside and the evaporator bonded to the inside of the case will be starving for liquid; but when the box is down to zero where it will stay for months, the tube is feeding just the right amount of refrigerant. This amount would be equal (see your chart) to about enough liquid to boil at a pressure of 1 or 2 psi, which was the designed working pressure. Again, you could compare this to a straight tube feeding the manifold of your auto engine. The engine would be starved for fuel when you were trying to get up speed. You might even have to slip the clutch in order to get up to a running speed with this drip feed. But once you get rolling, the fuel feeding rate would be satisfactory for a certain speed as long as you maintained this speed. Had an expensive expansion valve been installed on the freezer, it would have been feeding gas into the evaporator at over 75 psi when the unit first started up, but would be leveling off at 5 psi or a little lower when the box was ready to begin its off-and-on cycle. A valve, much like a carburetor, would adjust to the load. In short, engineers design cap tubes to meter correctly during the time the box is to run the most and the best; but expansion valves will respond to almost any condition.

Figure 2-6 shows two illustrations of expansion valves. The one to the left shows the direction of flow and the working mechanism of the valve. The valve to the right shows the temperatures and pressures as the valve is put into operation. Notice the F1 pressure of

48.8. This is so because the suction line where the remote bulb is located is at 52°F and the corresponding pressure would be 48.8. The power element opposes the spring pressure F3 which is 11.8 pounds of spring pressure. When the suction line gets colder, the remote bulb gets colder. This drops the pressure in the power element, letting the spring pressure push the valve stem closer to a closed position. When the remote bulb gets warm, pressure is increased on the power element diaphragm, therefore opposing spring pressure and pushing the seat open. The expansion valve is a slow modulating valve, so slow that you will hardly realize a change in pressure on your low side gauge.

At this point it is time to review what has been said earlier: that is, that all refrigeration equipment is doing the same job, getting rid of undesirable heat. This is true and will hold good throughout this training section. The job of removing heat may be done in many different styles, but the same basic machinery will be used. The units will have a pump, condenser, evaporator and some kind of metering device. The particular heat to be removed may vary from air conditioning people for

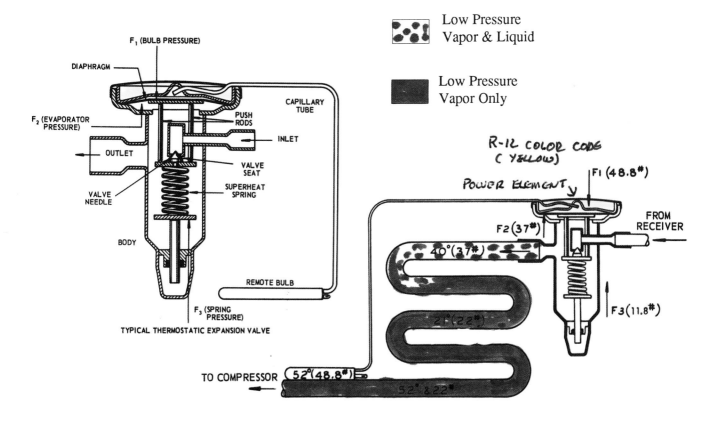

(Fig. 2-6)

25

human comfort to air conditioning meat, milk, and cheese and other things that give up their heat through the air that surrounds them in a refrigerated space. The air in turn carries the heat to the evaporator to be taken outside of the refrigerated area and there dissipated. Where air is not used to carry heat to the evaporator, we use water; for instance, in a drinking fountain or brine. Brine is sometimes used to transfer heat from the product to the evaporator.

Now that we have the transfer of heat method down pat, we should examine some of the different styles of equipment that do the work. In the air conditioning field, we have the big units installed in office buildings, hotels and hospitals. These units are generally kept up by operating engineers or maintenance men. The units are simple and differ from a window unit only in size and method of condensing and evaporating. In some cases, they are steam driven rather than electrical. The

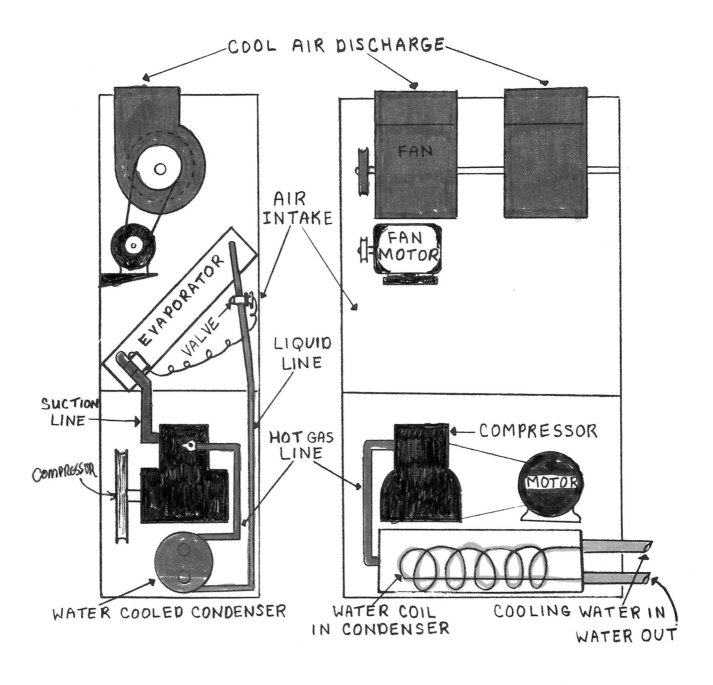

(Fig. 2-7)

condensing equipment is usually water cooled, and the evaporators may be direct expansion into a large finned coil with air blowing over it; or the evaporator may chill water or brine to be pumped to large coils, which are steam coils in the winter and cold water coils in the summer. The big installations are almost always known as remote systems; that is, the condenser and evaporator are located some distance from the main pump. However, they could be packaged or closely coupled. Next in order of size, we have the packaged units. These units are always completely contained in a package or shell. You see them standing against the wall in groceries, restaurants and clubs. The packaged unit may discharge the cool air directly out of the top or front of the unit, or the unit may be concealed in a back room, and the refrigerated air be carried by ducts to the area to be conditioned. These packaged units are simple to service and offer plenty of opportunity for the serviceman to show his stuff. Taking a look inside of the package, we find the evaporator in the center or midway up on the package. The evaporator blocks off the upper half of the air conditioner from the lower half where the pump is located. The air is circulated over the evaporator by a squirrel cage blower located in the upper half of the unit. The condenser is generally a shell and tube, or coil and tank, located beside or underneath the pump. **(See Figure 2-7.)**

This condenser depends upon a supply of cool water to condense the hot vapor instead of the air-cooled finned coil that we have studied so far. The cool water may come from the city water mains and be wasted to the sewer after it picks up the heat in the condenser, or it may be recirculated through a cooling tower and used over again. We will examine these systems in greater detail when we start trouble shooting them. The pump in a packaged unit will be an open type, belt driven; or it may be a sealed pot, or semi-hermetic. No matter, they are just pumps; and in most cases, they will have service valves for the serviceman to attach his gauges. We will attach the gauges to one of these units shortly.

Next in size is the remote installation; or in the case where it is installed in the home, we call it the central system or split system. Here the evaporator is located some distance from the pump and condenser. **(See Figure 2-8.)**

The condenser and pump, called the condensing unit, is generally located out beside the house; and the evaporator is located on the outlet side of the central heating furnace. Thousands of these combination heat-

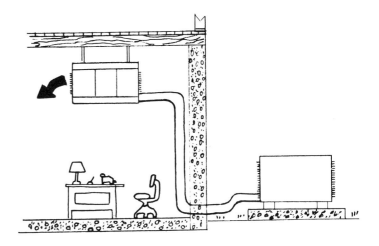

(Fig. 2-8)

ing and cooling systems are being installed all over the U.S.A. Next in order, we have the window unit. You have already taken a look at a simple window unit drawing, and you understand the principle it works on. Later, we will go into great detail on these units, which are a breed all their own.

Leaving the air conditioning field, we go into the commercial refrigeration field. Remember that when you refer to air conditioning, you are referring to conditioning air for the specific purpose of comfort cooling human beings. But the job of cooling down a large grocery store walk-in box may also be a job of conditioning air; but here you are conditioning air so that it may in turn soak heat out of the contents of the box like meat, cheese, and milk. Commercial refrigerators may or may not be self contained systems like the packaged air conditioner; that is, they may have their units directly under the case itself. Take reach-in dairy boxes for instance; many of these cases located in grocery stores have the pump directly under the box. You can feel the hot air blowing out of the air-cooled condenser, which is located right beside the pump. The evaporator will be up inside of the reach-in box and will generally be a blower type, or as it is called by servicemen, a forced evaporator. The days of the huge evaporator coil hanging inside of a box (depending upon its size rather than the forced air over the coil) are gone. Nowadays, we use small evaporators with a fan to force the circulation of the air in the box through the coil. We use the same thing in a large walk-in storage box in the back of your grocery store. The evaporator coil will be suspended from the ceiling of the box over in one corner, and it will have a blower fan behind the coil circulating the air within the box over the evaporator at all times. In fact, these blower fans never stop; they even run when the pump is not running to make sure that the evaporator is always exposed to circulating air in the box. The open type help-yourself cases have forced evaporators, and the evaporator may be down under the trays of meat. These cases are designed for air to circulate gently over the merchandise by low speed blowers. The operation of these self-service open type cases depend upon the principle of cool air being dense or heavy. Cool air will lie like water in a horse trough inside of the guard rails and thereby keep the product cool. This type of case is very tricky and depends upon a very careful placement of the merchandise in order to keep it properly refrigerated. The condensing units on practically all commercial refrigeration cases, except those that are self contained, are located outside of the store or in a back storeroom. Back there, you have the motor-driven pump and a water or fan-cooled condenser.

Take a look at this water cooled condenser on a commercial unit. (Fig. 2-9)

You have progressed far enough now that it would be timely to get the gauges on a unit for a careful examination of the conditions within the system. Remember the simplest of all units to attach gauges to are those open-type belt-driven pumps where the manufacturer had you in mind when he designed the pump. The manufacturer has placed on the pump two service valves in order for you to make a quick connection with your gauges without losing any gas or doing any damage. Taking a look at a simple belt driven pump, you will find the serviceman's service valves on the suction inlet and the discharge outlet. These are more commonly called the high side and the low side. (**See Figure 2-10**)

There is nothing complicated about these service valves. They are simple and serve a very useful purpose for the serviceman. The main feature of these valves is that they shut off like any other ordinary valve, and still they have another port or entry that can be shut off with a backseat operation. You always connect your gauge hose to this small opening which is very plain in **Figure 2-11.**

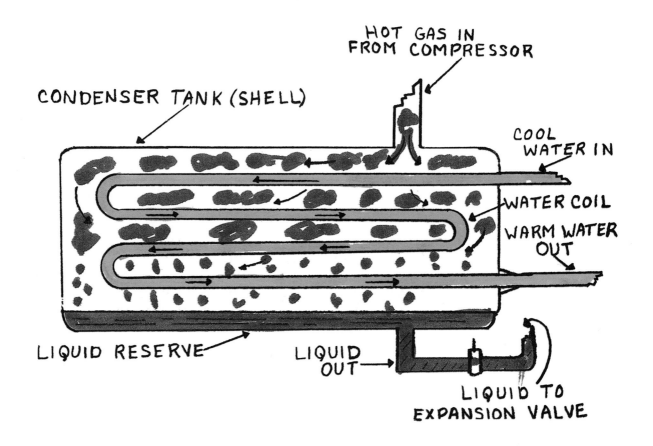

(Fig. 2-9)

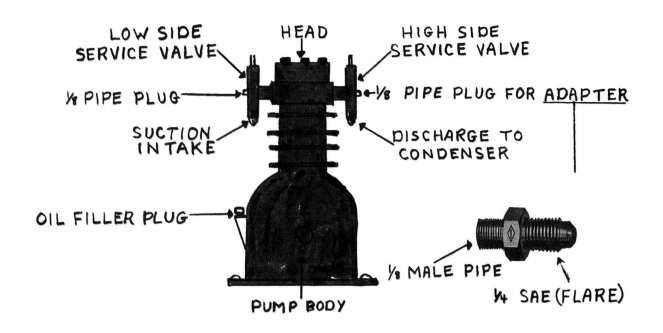

(Fig. 2-10)

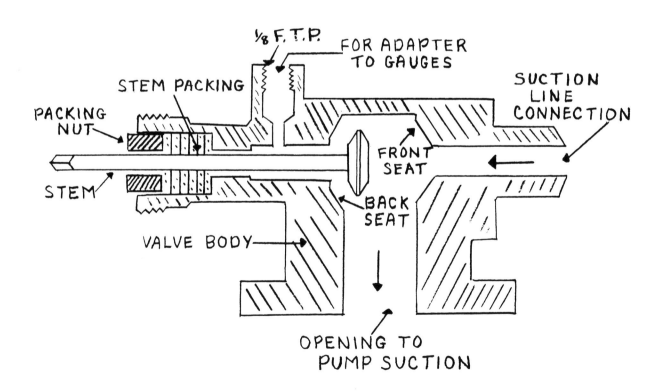

(Fig. 2-11)

Many units do not have service valves for the simple reason that they are expensive.

Here is the order for attaching the gauges to the high and low side of any conventional open-type pump with service valves.

Take the valve stem cover off the low side service valve. Backseat the valve tightly, take out the 1/8" pipe plug in the gauge port. Screw in tightly a 1/8" pipe to 1/4" flare adapter. Attach the quick coupler on your low-side gauge to this flare adapter. Un-backseat the valve stem one or two turns. Pressure will immediately show on your compound gauge. Crack one of the hose fittings and bleed off the air that was in the hose. (Never let air into a system). You are now ready to see and analyze the trouble by looking at your gauge. You may attach the high-side gauge to the high-side the same way, and you can read the head pressure at the time you are reading the evaporator pressure on the low side. You may want to attach your drum to the charging hose and be ready to charge in gas if it is needed. Remember, it *does not matter* whether you are attaching your gauges to a household refrigerator which has service valves or to a one-thousand ton unit, you will have the same pressure-temperature relationship that we have been studying. Of course, the press-temp will vary with the job to be done. For instance, a home freezer using F-22 will have, or should have, an evaporator pressure on the low-side of 5 psi, give or take a few pounds. Likewise, your gauges on a one-hundred ton locker plant using F-22 should show a low-side pressure of 5 psi, give or take several pounds one way or another.

Remember this, when reading your pressure temperature chart, do not convert the box temperature to pressure. It is the refrigerant temperature you convert to pressure. In home refrigerators and freezers the refrigerant temperature will be approximately 20° to 30°F colder than the box temperature. I think you can understand that the evaporator will have to be colder than box temperature to remove heat. For example, let's say the freezer is operating normally at 0°F box temperature. If the refrigerant temperature is 30°F colder than the box temperature then the refrigerant temperature would be -30°F below zero. Now convert -30°F to pressure and you would have an operating pressure of 5 psi.

For our first example, we will take a home freezer charged with F-12 using an open-type belt-driven unit with service valves on the pump; note F-12 instead of F-22. This freezer has a thermostatic expansion valve for the metering device.

Here is our freezer with gauges attached as seen in

Figure 2-12)

Suppose the complaint on this freezer was that it was not cold enough. After you placed your thermometer inside the case, you find that the box is running steady at 20 degrees above zero. A home freezer should be zero degrees. The customer was correct. The thermometer confirmed that the box was not cold enough. Now you attach your gauges. Immediately, you note that the pressure on the low side is in a vacuum. You hook up your service cylinder of F-12 to the charging hose and purge the hose of air.

We know we need gas since the pressure was not up to the temp-press reading for 20 degrees above zero. See chart. Now we can confirm this shortage of refrigerant by watching the reaction on the low side pressure gauge after we give it a little gas. We crack the charging valve on the manifold and let some of the Freon we have in our service drum boil into the low side of the freezer through our charging manifold. Immediately there is a change in the system being in a vacuum. Now it is slightly higher than before. It will continue to come up every time you give the unit a shot of gas until the pressure is equal to the chart reading for zero using F-12; approximately 0-2 psi. Of course, it may go above this pressure if the box has not yet pulled down to zero. Suppose the pressure went to 21.1 psi after you had added gas. This would not be unusual, if you remember the box was 20 degrees above zero when you started. See your chart. If the valve is working perfectly when you have a charge, the pressure will be whatever the temperature of the refrigerant is in the evaporator. This would not be so with a cap tube. A cap tube would feed until your gauge reads 0-2 psi and no more; remember the design pressure.

Many of you should suspect by this time that if this pressure holds true on your gauge, you would not have any pressure drop across the line from the evaporator to the pump where your service valve and gauges are located. True, there is a pressure drop, but we will discuss this later when you will understand it more fully. Right now I want you to assume that there is always a well balanced system with no pressure drops. After we learn what is happening in a perfect unit, we will study the simple calculations for making allowances for small line losses and load pressures.

Before we go on, you should have straight in your mind just what has taken place in the above freezer. Most trouble shooting does not involve even this much gauge reading; however, we want to get used to reading pressure in terms of temperature.

Here are some questions that need answering.

How do you purge a hose? By looking, you will see that the middle hose on a manifold can go out through either the high side port or the low side port by opening either manifold stem. If you have pressure in the unit, purge back out through the manifold. If you have no pressure on the unit (empty) purge down to the service valve on the unit from your service cylinder of Freon.

Use good judgment. That is, whichever way you have to blow a hose to get the air out of it, blow it. If you happen to have a vacuum on the unit then that will require some consideration before you open any thing that might permit the vacuum to suck in air.

You might by this time ask me, "How do you pull the temperature down in a box if the evaporator pressure is exactly what the temperature is? That is, how does the

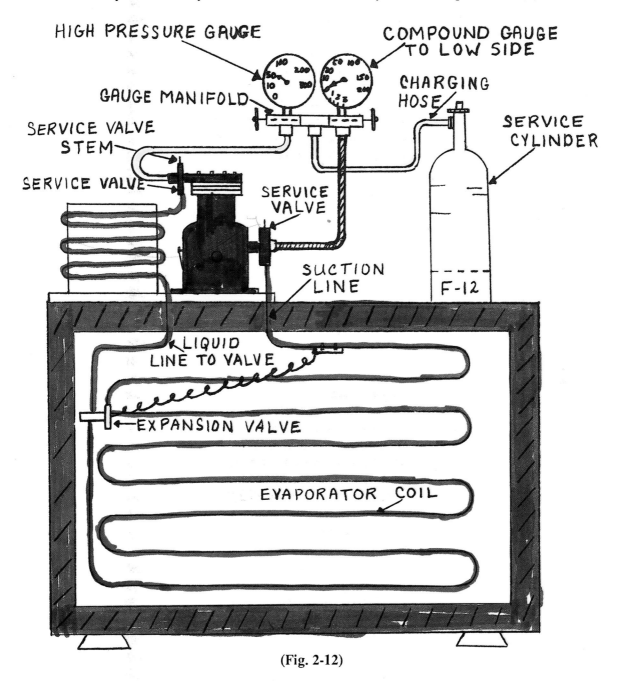

(Fig. 2-12)

Note: The serviceman's gauges are reversed as to how you would normally see them. Don't let it confuse you. They still read pressures as they should.

box get any colder?" That's a good question, but if you will remember the refrigerant temperature in the evaporator is always colder than the box temperature (20° to 30°F colder). So, if you have a box temperature of O°F and a pressure of 0 to 1 psi the refrigerant temperature is still cold enough to absorb heat and will continue to do so until the thermostat setting cuts the compressor off. (0 to 1 psi converts to approximately a -20°F refrigerant temperature in the evaporator). The cap tube is designed to feed a constant amount of liquid into the evaporator and will not vary. It will not throttle, open up, or do anything but just feed as it is designed. Does the cap tube feed the same amount of refrigerant into the evaporator if the box is say at 30°F as it does when the box is at O°F? Yes it does. However the evaporator pressure or low side pressure is higher at 30°F than it is at 0°F. It is not because the cap tube is feeding more. It is because the higher temperature in the box causes the pressure to be higher. Remember, we have already said that a rise in temperature of a refrigerant has a relationship to its rise in pressure. One or the other cannot make a change without directly affecting the other. So we know now that a box at 30°F would have an evaporator pressure of approximately 10 psi, but upon pulling down to 0°F box temperature the operating pressure would be around zero pressure, using R-12.

In the case of the expansion valve, the valve is always leading the load at any given time by ten or fifteen degrees or equivalent pressure. This is called superheat setting and is set or fixed at the factory. Superheat is a fancy term for describing the simple lead the valve holds over the load at any given time. We will go into this more fully later on. However, it would be well to note at this time that if the expansion valve always has a superheat lead on the work being done, then it has no bottom; that is, it could continue to pull the load on down indefinitely or to any temperature within the limit of its design. True, it will do just this because it is a superior metering device.

Do all these gauge pressures have to be so accurate for any given situation? Absolutely not, give or take a few pounds more or less, this will always be the rule. We are attaching considerable importance to these gauge pressures for the reason that we want to understand them. If your automobile tires call for 24 psi of air by the manufacturer, would you dare to put 30 psi in the tires? You bet you would, and very often the case might be that the tires would have more than 24 psi simply because the weather was extremely hot.

■

Note: This receiver built to stand upright.

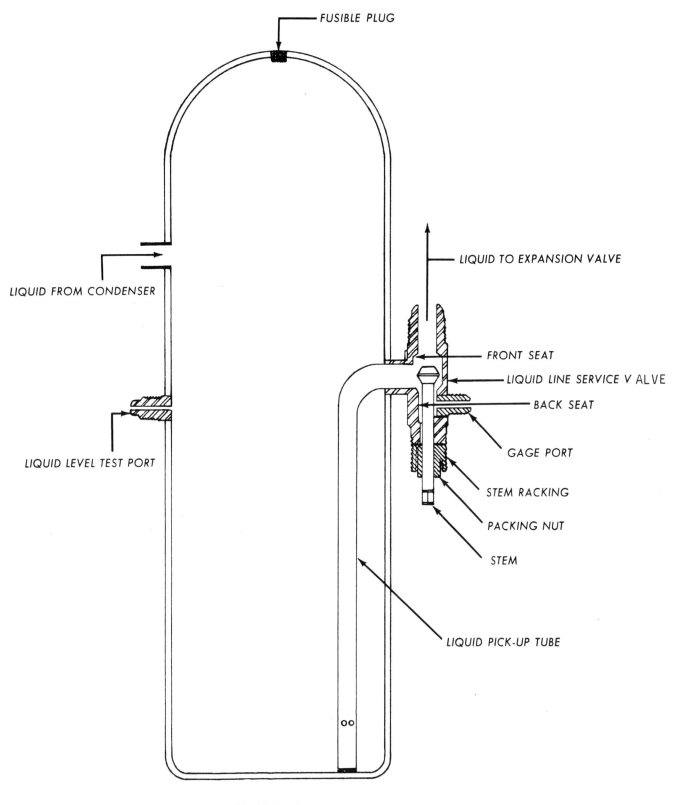

FUSIBLE PLUG

LIQUID TO EXPANSION VALVE

LIQUID FROM CONDENSER

FRONT SEAT

LIQUID LINE SERVICE V ALVE

BACK SEAT

GAGE PORT

LIQUID LEVEL TEST PORT

STEM RACKING

PACKING NUT

STEM

LIQUID PICK-UP TUBE

Liquid Receiver
And Liquid Line Service Valve

APPROXIMATE OPERATING PRESSURES

TYPE OF SYSTEM	LOW SIDE R-12	HIGH SIDE R-12	LOW SIDE R-22	HIGH SIDE R-22	SPACE TEMP
Window Units	37 psi	175 psi	69 psi	275 psi	78F
Central System	37 psi	175 psi	69 psi	275 psi	78F
Air Conditioners (Using R-500)	46 psi	225 psi	-	-	78F
Larger Air Conditioners	37 psi	175 psi	69 psi	275 psi	78F
Water Cooled Air Conditioners	34 psi	130 psi	65 psi	225 psi	78F
Home Refrigerator (Conventional)	9 psi	130 psi	-	-	40F
Home Refrigerator (Dual Temp & Frost Free)	4 psi	130 psi	-	-	Freezer 0 F Food Comp. 38F
Home Freezer	0 psi	125 psi	5 psi	230 psi	0 F
Walk-in Meat Cooler	12 psi	160 psi	-	-	34F
Walk-in Produce Cooler	16 psi	170 psi	-	-	42F
Walk-in Freezer	0 psi	150 psi	5 psi	240 psi	0 F
Walk-in Freezer (Using R-502)	18 psi	210 psi	-	-	-10F
Meat Display Open-Forced Air	9 psi	160 psi	-	-	34F
Meat Display Closed-Forced Air	12 psi	160 psi	-	-	34F
Dairy Case Open-Forced Air	16 psi	170 psi	-	-	38F
Beverage Box	12 psi	170 psi	-	-	34F
Water Cooler	18 psi	150 psi	-	-	Water Temp 48F

CHAPTER 3

CHARGING WINDOW UNITS
AND REFRIGERATORS

You should by now have a pretty good idea as to what is going on inside of a refrigerating unit. You understand the physical laws relating to the refrigeration principle. You understand that a pressure drop must occur in the evaporator in order for these laws to begin to work. You know that the device for making this pressure drop is the cap tube or expansion valve. You know that the cap tube is a cheap version of an expansion valve. You know that this cap tube permits the pump to clear the low side and keep it ahead of the high side so that the liquid Freon will expand or boil into this lower pressure area. You know that the expansion valve is an expensive control device which derives its power to control from a power bulb charged with the same kind of refrigerant that it is controlling. With this basic information in your mind, we can proceed much faster to master the art of working on refrigeration equipment.

Before we explain pressure drop and superheat more fully, we should again consider the condenser. The condenser generally is a trouble-free part of air-cooled units as long as it is kept clean. In addition to being clean, an air-cooled condenser should be located so that it has at all times a good supply of fresh reasonably cool air for the fan to circulate over the finned surface of the coil. See **Figure 3-1**. This is not so with a water-cooled condenser. This type of condenser can be a constant source of trouble. A water-cooled condenser has the same relationship to a refrigerating unit that a radiator has to an automobile engine. We will treat with water-cooled condensers in our trouble shooting section later on.

In the case of an air-cooled condenser, the service-man's principal concern is that it be clean so that the pump will work against a proper head pressure. In order to understand what would be a proper pressure, you should understand that there is a temp-press relationship on the high side as well as in the low side evaporator. You may have guessed this fact already. The condenser head pressure will always be directly related to the temperature of the air blowing over it if it is air cooled. If water cooled, the head pressure will have a relationship to the temperature of the water which is cooling the condenser coil. Look on your temp-press chart and see if you can read head pressure in terms of temperature by using these examples. Suppose air on the outside of the house were being circulated over a condenser and there was perfect heat transfer. If that air outside of the house were 95 degrees, then the head pressure should be approximately 169 psi for F-12.

Suppose the unit were water cooled and the city water temperature on this particular day were 70 degrees, then the head pressure would be approximately 117 psi. If you were using R-22 in a system, then under the same conditions the head pressure would be approximately 200 psi air cooled and approximately 200 psi water cooled. Assuming perfect heat transfer, then you would have these head pressures for the temperatures above. In figuring your head pressure, always add approximately 20° or 30°F to the cooling medium, whether it be air or water to derive at the approximate pressures. See your pressure-temperature chart. Example, outside air is 95°F + 20°F = 115°F. Converted to pressure the chart reads 146.7 psi. Remember the actual pressure reading on your gauges can vary several pounds.

If the weather in the first instance changed a few degrees during the day, likewise the head pressure would change. The one catch to this is that manufacturers do not always build condensers that are perfect heat

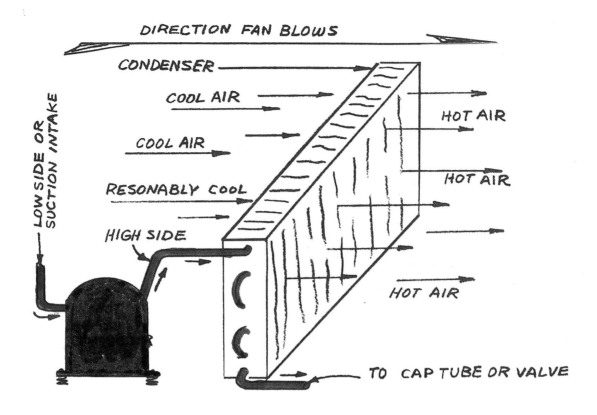

DIRECTION FAN BLOWS

CONDENSER

COOL AIR

COOL AIR

RESONABLY COOL

LOW SIDE OR SUCTION INTAKE

HIGH SIDE

HOT AIR

HOT AIR

HOT AIR

TO CAP TUBE OR VALVE

(Fig. 3-1)

transfer coils. Like the radiator on your automobile, the water does not always return to the motor block as perfectly cool as the air blowing through the radiator. In the case of a refrigerating unit condenser, let's say that the actual metal that the condenser is made of will more accurately reflect the exact temperature of the hot vapor inside; and this temperature will be equal to the pressure. For a general rule, you may always say that any reasonably good condenser will get rid of the heat within about forty pounds pressure or equivalent degrees. So in the first instance, where you had air at 95 degrees going over the condenser, if you assume about 70 percent condenser efficiency, the coil would be about 120 degrees; and the true pressure on your gauge would be about 157 psi. See your chart. The point about condensers to always remember is this. The hot gas is going to condense at one pressure or another, and it will condense when the pressure is equal to the temperature of the cooling air. See your chart. Said another way, if the condenser cooling air is going to stay at a certain temperature, then the gas will condense all day long at the particular pressure for that temperature. Summing this up, we must remember that the condensing and

evaporating of liquids is a two-way street. The same laws applying to the evaporator coil apply to the condenser coil. The only difference being that in the evaporator, the temperature is made by pressure drop; and in the condenser the pressure is governed by the temperature of the outside cooling agent. Finally, your temp-press chart is good for both the high side and low side.

Now to clear up superheat and pressure drop, "superheat" is one of the most overworked terms in the refrigeration industry. I will not attempt to explain just what the engineers mean by superheat when they describe the factory setting on a thermostatic valve. Here is my version, and it will take care of any problem you will ever encounter involving superheat. And this would be a problem you would not encounter once in five years. My rule is, leave valves alone. Do not attempt to adjust the superheat unless you are manufacturing expansion valves. In the first place, it is not practical; and in the second place, if the superheat needs changing on a valve, the whole valve needs replacing. Superheat is an unbalanced adjustment to the valve made by the factory so that it will always lead the work load going over the evaporator. Suppose a valve had a fifteen degree superheat setting adjusted into the valve at the factory. This means that the vapor coming out of the evaporator will have picked up fifteen degrees of heat on its way through the evaporator. In other words, this means the valve is set or unbalanced so that it will always be leading the cooling job by fifteen degrees or more. The chemistry related to the expanding quality of Freon and its ability to absorb heat is very complicated and has to do with its saturation point relative to its state at any one given moment. You just remember that the valve is always ahead of the load. Load here means the job of cooling.

Taking it step by step, let's suppose that we start out with a warm walk-in box in a grocery store. This box has a blower coil over in one corner, and the metering device is an expansion valve. We have just installed the box, and it is bound to be warm. The minute we start the pump and the evaporator fan, the gas starts expanding in the coil; and the coil will cool down to fifteen degrees below the temperature of the air being circulated over it. Suppose the air in the box were exactly 80 degrees the instant we started, now the temperature of the coil is being held by the valve to 65 degrees at this instant; that is, fifteen degrees cooler than the air flowing over the coil. Now suppose the box gives up one degree of heat, and that one degree of heat has already left the box in the vapor line. Immediately the coil will be one more degree cooler; that is, it will be 79 degrees in the box; and the coil will be 64 degrees. This lead will always be on the valve and will continue as long as the unit is running and the thermostatic valve is operating. Suppose you did not believe the valve were holding the coil cold enough. More often the case would be that the pump was worn out and was not getting the vapor out of the coil. Don't blame expansion valves for your troubles. Many servicemen who are mystified by the expansion valve make it a spanking boy for other failures in the equipment. Put this down as a hard and fast rule for servicemen. *Don't tamper with the valve.* If you do not trust it, put in a new one; and you will not have any doubts. More often than you can believe, the valve will outwear all of the equipment attached to it, including the unit.

Pressure drop is a loss in a line between the beginning of the flow to the pump and the actual pressure when it reaches the pump. Take for instance a pump that is removing refrigerant vapor and the vapor pressure starting toward the pump is 40 psi.

By the time it reaches the pump, the intake pressure on the pump will have diminished to about 35 psi. Then the pressure on the low side gauge would read about 35 psi. This is caused by friction loss in the suction line from the evaporator to the pump. Elbows and sharp bends will cause even greater losses where the vapor has to turn several corners before it gets to the pump. In reading your low-side gauge, your good judgment will prevail. You will know, if the pump is very close to the evaporator, there will be practically no loss in the line. In the case of your pump being a hundred feet from the evaporator, the loss could equal more than half of the pressure in the evaporator in the beginning. Nevertheless, make estimates for line loss. One-half psi will do for household refrigerators; one psi for window units—not important; five psi for large package units; five to fifteen pounds for commercial remote installations. Just deduct this from your reading. For instance, you read thirty pounds pressure on your low side gauge connected to a pump in back of a grocery store with the evaporator located way up in the front of the store approximately fifty feet from the condensing unit. Your evaporator pressure could be about thirty-five pounds.

When the equipment is tested at the factory and if the engineers see that there will be a problem with pressure drop, this is what they will do. They will install an expansion valve with an external equalizer line, that will sense pressure drop. See **Figure 3-2.** The valve can now sense pressure drop and signal for the valve to respond accordingly and tend to overcome the pressure drop. If the external equalizer line were not there the valve would feed erratically and decrease the capacity of the system. The equalizer line is an open line from the suction line back to the expansion valve.

Before you begin to work on any refrigeration equipment, you should locate the pump, condenser, evaporator and metering device. It would be a good idea to locate the electric disconnect switch also, just in case you should need to turn the unit off very quickly. No matter what size the equipment happens to be, look it over carefully before you go to work. In the case of bigger units in back rooms or packages, take your flashlight and examine the unit. If you do not find the main parts of the system, hunt until you do find them. You may have to trace copper lines running out of the unit where only the pump is located. The lines will take you to the condenser and the evaporator, too. The condenser and the evaporator will be somewhere. You cannot fail to locate them if you hunt for them. It may take a moment or two of thought to figure out how a particular condenser does its job getting rid of heat, but you will understand it if you understand the purpose of

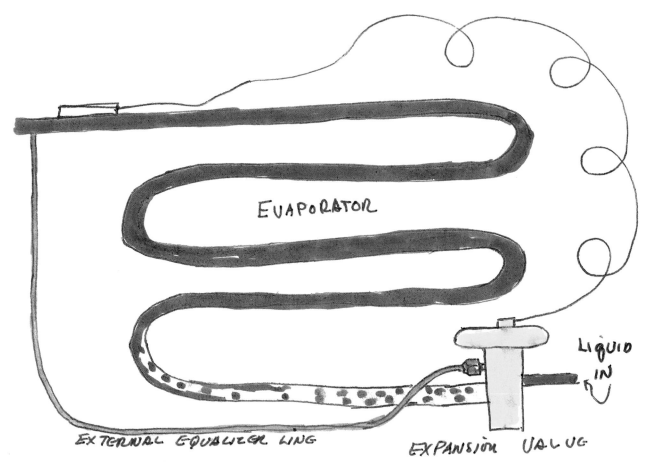

EVAPORATOR

LIQUID IN

EXTERNAL EQUALIZER LINE

EXPANSION VALVE

Fig. 3-2

the condenser. You may have to study the type or style of the evaporator for a few minutes to find out just what the designer had in mind when he planned the evaporator. It may have a fan sucking air over the coil; or it may be a loose coil of pipe down in a container cooling pie dough. No matter, it will be an evaporator; and you know what goes on inside an evaporator. This you know, no matter what the style or shape of the evaporator happens to be.

Locating the pump will be an easy problem. Every system has to have a pump. It may be located right in front of you as is the case in a window unit; or it may be outside of the house in the case of the central system; or it may be in a back storeroom with lines running out one side of the storeroom to the evaporator in another room or up in a duct with another line running out to the condenser on the roof. The pump may be belt driven or sealed up with the motor and pump on the inside of a welded-up dome; or it may be semi-hermetic, that is, sealed up inside of a dome or cast iron body which can be unflanged or unbolted to take the whole pump apart.

The metering device will, of course, be near the evaporator no matter where the evaporator is located. It will be one of two kinds, either a cap tube or expansion valve. You know the job the metering device has to perform and why. You know there will be a pressure drop from the liquid line through the evaporator so there will be a change of temperature in the evaporator. You know that this pressure drop is related to the pump clearing out the vapor or expanding Freon in the evaporator at a steady pumping rate. The pump pumps at a steady pace; therefore, the valve or cap tube may be designed to maintain any given temperature wanted, whether it be only reasonably cold as in the case of the air conditioner, or extremely cold in the case of a low-temp freezer.

Once these components or parts of the system are located, you should be able to say to yourself, "Now what is the job to be done?" Does it cool meat, air condition people, freeze ice cream or cool pie dough? Answer the question yourself. See the chart in **Figure 3-3** for the range of temperatures required to do certain

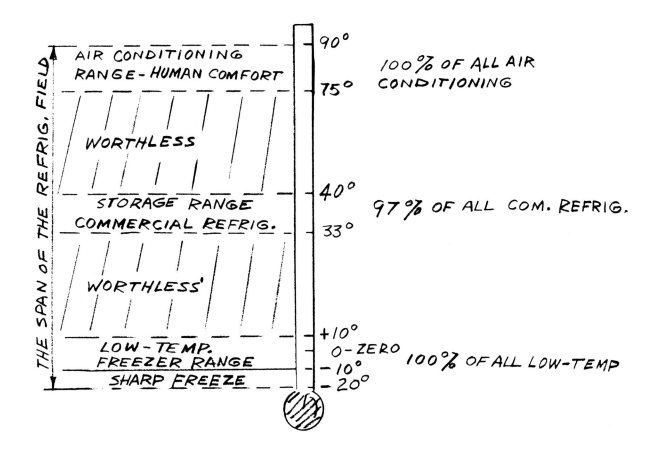

THE SPAN OF THE REFRIG. FIELD

AIR CONDITIONING RANGE - HUMAN COMFORT — 90°

75°

WORTHLESS

STORAGE RANGE — 40°
COMMERCIAL REFRIG. — 33°

WORTHLESS

LOW - TEMP. — +10°
FREEZER RANGE — 0 - ZERO
— 10°
SHARP FREEZE — 20°

100% OF ALL AIR CONDITIONING

97% OF ALL COM. REFRIG.

100% OF ALL LOW-TEMP

Fig. 3-3

jobs. You will note that there are only three areas of re-frigeration which cover 99 percent of all cooling jobs.

You now have a fair idea of the temperature you want in various evaporators. Your next move will be to lay hands on the equipment. Do not ever put your hands into a dangerous trap like a vee belt drive or a high speed fan which you can hardly see. Feel cautiously, but feel. The first thing you should feel of is the suction line coming out of the evaporator, and next you should feel of the high side line coming out of the pump. Be careful feeling of the high side. Do it like a woman feels of a hot iron until you can tell whether it is too hot to hold or not. These two operations of feeling of the low side suction and the high side discharge tell you more than all the other indicators in or about the unit. In fact, the gauge pressure, when you read it later on, will merely confirm that which you have already sensed by your feeling of the lines. The high side will tell you immediately if it is doing any work or not. There can be no heat without some work being done. You can never have heat on the high side of a pump unless it is picking it up from the

vapor coming from the evaporator, unless the building were burning down. The intensity of this heat on the high side will tell you approximately how much work is being done. This feeling of the high side is a very important procedure in service work. If you found it extremely hot, you could say one of two things to yourself: either it is doing a terrific job of getting the heat out of the load area; or maybe the condenser is not getting rid of the heat. You could confirm this by seeing whether the condenser is dirty, or the fan or water pump are not working properly. If you felt of the motor in your automobile and it was extremely hot could you not say right then that either it had been working very hard; or the water and radiator were not doing their job?

When you feel of the suction line on the unit near the evaporator or the pump, you are making the most important move that can be made by a serviceman when servicing commercial equipment or air conditioning. The temperature of the suction line will tell you more than the gauges can tell you. Other factors may affect the pressure readings, but the temperature of the suction

39

line is a sure fire tattletale as to what is going on inside of the evaporator or low side.

Now here is an obstacle we must get over, even though it may give you considerable trouble. The way this is going to be said will lead you to believe that all you have been studying has lost its meaning, but this is not the case. The temperature of the suction line coming out of the evaporator should be cool, and in many cases, very cold; and it should be cool or cold all the way back to the pump unless the pump is located a considerable distance from the evaporator. Up until now, and we will still say it, the job of the evaporator is to pick up heat and send it to the pump and condenser; and the more heat it can pick up in the vapor inside, the better. All right, you might have thought that if the evaporator had soaked up considerable heat and it was on its way to the pump and condenser, then the return line or suction line to the pump should be warm with this heat. This is not the case. The vapor is loaded with heat, but the line and the vapor are still very cool. You remember our first law that there is heat down to absolute zero. Then couldn't a suction line that was very cool still have lots of heat in it? While it is true that the suction line is cool or cold to the human touch, actually it is the same temperature as the vapor inside and is loaded with heat. Even though the vapor has a load of heat, there are still a few fine particles or specks of boiling refrigerant in the vapor trying to pick up more heat; although it is not getting the heat in the evaporator, but out of the area where the suction line is running back to the pump. In fact, the vapor is still capable and willing to take on more heat right up to the point where it enters the pump unless the pump is so far from the evaporator that the suction line has picked up all the heat it can hold in the expanding vapor inside. When this is the case, the suction line will be the same temperature as the room where it is located.

Here is another approach to this same problem. If you remember, it was mentioned in chapter two that the coldest point in an evaporator is right where the liquid begins to expand out of the valve or the end of the cap tube. Right at this place the Freon is in such quantity and state of boiling that you could hold a blow torch on the line coming right out of the expansion valve or cap tube, and the line would probably never warm up. The line would stay cold in spite of the intense heat. Said another way, the blow torch would not heat the first inch of tube coming out of the valve to the evaporator. This particular area of the evaporator would just stay cold with the flame right on the pipe. If you tried this experiment, the suction line back by the pump would be considerably

warmer than the normal running condition for no other reason than the expanding Freon would be getting a head start on soaking up heat from the blow torch even before it got through the evaporator. Take the blow torch away; and the Freon would take on most of its heat from the evaporator; but it would still be capable of taking on more heat, unless the evaporator were so large that every last particle of liquid Freon exploded and boiled within the evaporator; then you would have a warm suction line.

Summing up this cool suction line business, we can say that there are still a few particles of Freon willing and able to take on a little more heat through the walls of the suction line, even though the heat comes from the outside of the box. Now here is the significance of this fact. If this be true, then a serviceman can always say to himself after feeling of a cold suction line, that there is a colder place up the line somewhere; and it must be the evaporator. This would be correct. There is a colder place up the line, and it is the evaporator because we have already stated that the coldest place in a system was right where the gas left the metering device and entered the evaporator, so all it could do from there on is get warmer all the way back to the pump. Of course, you realize that if the pump were several hundred feet from the evaporator, the suction line at the pump would be about the same temperature as the surrounding air.

If we are sure we understand that a cool suction line near the pump can be loaded with heat picked up in the evaporator, we are ready to go back again and discuss the importance of a serviceman's feeling of the suction line.

The designer of the equipment knew that there would be some spill over of expanding Freon from the evaporator, and he wanted this spill over in a modest quantity since he knew that at least the whole evaporator would be getting expanding liquid. Now keeping this in mind, does it not stand to reason that even if you can't see the evaporator by feeling of the suction line, you will have a hunch what is going on in the evaporator by the condition or quantity of spill over. Suppose the suction line were just room temperature and the pump was very close to the evaporator as in the case of the package unit. Then you could say right then, that if the evaporator were only three feet from the pump, the last row or two of tubes feeding out of the evaporator would not be cold enough; that is, the coil would be starved because a three foot section of suction line would be too short a distance to pick up enough heat to make a suction line warm. Therefore, the evaporator cannot do all of the work it is capable of doing.

In **Figure 3-4** you see that the whole coil has an ample supply of boiling Freon in order to make the total surface of the coil work. You will note that the coldest spot is that section of tube right where the liquid begins to expand out of the valve. It is extra cold here simply because there is more rich liquid changing from a liquid to a vapor. The changing of this liquid Freon to a vapor takes place all down through the coil. Many small drops or particles of Freon do not explode when they enter the evaporator, and these drops are carried all the way down to the last foot of the evaporator coil and further before expanding. You could rightly reason at this point that the perfect coil would be one that was so designed that the Freon always boiled in every section of the evaporator in the same quality and quantity. Expensive evaporators with special multiple distributor lines coming out of the valve are more nearly perfect coils. That is, a perfect coil is one that would expand all liquid at an ideal pressure; and when the liquid left the evaporator or load area, it would not be capable of taking on any more heat because it was already fully loaded for the quantity of Freon used to get the job done; and the temperature of the suction line would be warm.

Take a look at **Figure 3-5**. You will notice that there is some refrigeration, but the coil is not being fully utilized, and it is in a starved condition. This will, of course, make the suction line warmer; and this is the key to knowing when a unit is short on refrigerant. If there is not enough liquid to feed the whole coil when it is ready to take on more liquid, the coil will simply expand what liquid is available; and the suction line will not have any spill over and will consequently be warm—that is, finished as far as any more expansion work is concerned. Now you can see the value of the expansion valve. The valve will try to control the quantity of Freon boiling or expanding in the evaporator to the point where the whole coil is getting its share of boiling Freon, and not too much will spill over out of the evaporator and out of the box to pick up heat where no heat is needed to be picked up. It's just this simple. There is no use in having a lot of Freon expanding in the suction line running outside of a box because the condenser will have to handle this heat just like the heat soaked up in the box. However, it is always best to have a nice cool suction line in order to be sure that the whole evaporator is cold.

Now you can contrast or compare the expansion valve with the cap tube in relation to this control of the quantity of Freon in the evaporator. You know that the expansion valve will start closing off the minute too much rich liquid contacts the power bulb—that is, the power bulb gets too cold right where the suction line leaves the box. This closing off of the valve will cut down on the particles of Freon which would have been carried on outside of the box by the suction line. We call this a flood back. If the valve closes too much, of course, the power bulb will warm up with the change and immediately start feeding again. Actually, if you had a thermometer strapped to the suction line right where the power bulb is located, you would note that the temperature would rise and fall all the time. This is called hunting because the valve is hunting. This means that the valve is always trying to settle down like the governor on an engine, and it will settle down when everything is operating right, and no one is opening the door to the box or changing the load.

Now, with the information you have on what is going on inside the evaporator, stop and reflect for just a moment on what would take place in a suction line where a cap tube was used for a metering device. Let's go through this stage by stage. First, we know that a cap tube is a small restrictor tube feeding at a steady fixed rate to the evaporator. We said that when it started out to feed the evaporator, the evaporator would be starved because the designer had sized the cap tube to feed the right amount of liquid to take care of the lowest temperature which was wanted in the box. For instance, a home freezer would be expected to go to zero and hold there or a few degrees below. So using F-22, the cap tube would be designed to feed enough liquid in the evaporator to maintain a steady boiling pressure of 2 to 5 psi. See your chart. (Remember the refrigerant temperature runs 20° to 30°F colder than box temperature.)

Now, you should ask yourself just what happens if the thermostat does not shut the box off at zero and the unit continues to run with the cap tube feeding at a steady pressure and the box at zero. Here is what would happen. The cap tube would continue to feed liquid, and that liquid would spill out of the box in the suction line, and the suction line all the way back to the pump would frost up since there would be boiling Freon inside. Therefore, we can say right here and now, and you can put this down for future reference, whenever a cap tube is used to meter gas into any kind of refrigeration device, whether it be a home freezer or air conditioner, the *amount of liquid will be limited to a certain amount*. This simply means that you will always put just so much Freon in a cap tube type system, and that amount will always be just enough for the liquid to boil in the

Study these two illustrations.

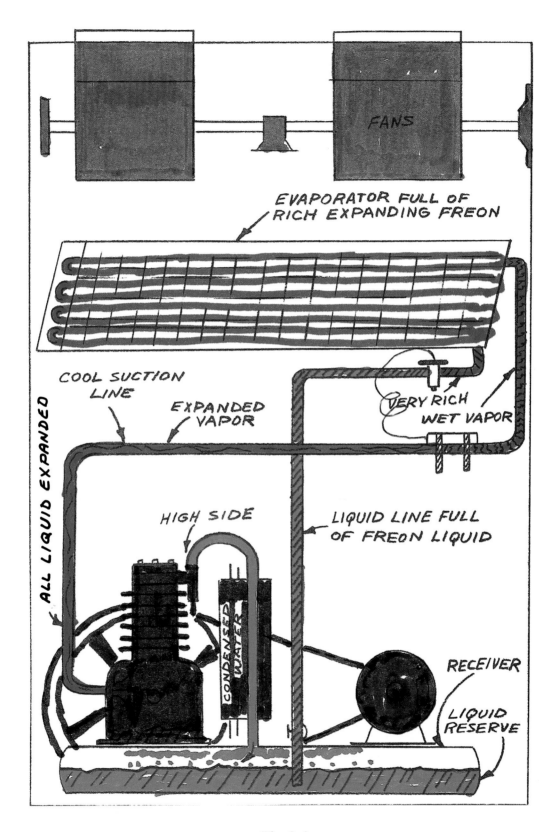

Fig. 3-4

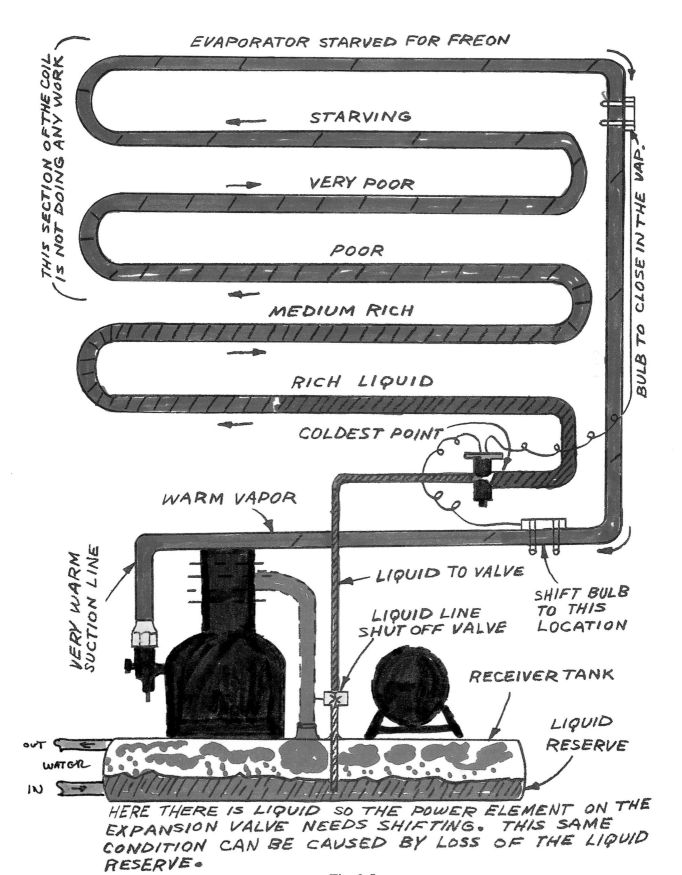

EVAPORATOR STARVED FOR FREON

THIS SECTION OF THE COIL (IS NOT DOING ANY WORK)

STARVING

VERY POOR

POOR

MEDIUM RICH

RICH LIQUID

COLDEST POINT

BULB TO CLOSE IN THE VAP.

WARM VAPOR

VERY WARM SUCTION LINE

LIQUID TO VALVE

LIQUID LINE SHUT OFF VALVE

SHIFT BULB TO THIS LOCATION

RECEIVER TANK

LIQUID RESERVE

OUT

WATER

IN

HERE THERE IS LIQUID SO THE POWER ELEMENT ON THE EXPANSION VALVE NEEDS SHIFTING. THIS SAME CONDITION CAN BE CAUSED BY LOSS OF THE LIQUID RESERVE.

Fig. 3-5

43

evaporator only. Going back to our freezer using a cap tube, not an expansion valve, we will put gas into it until the frost line starts out of the box; and we will then stop and not add any more gas. If you put too much, it would continue on through the evaporator and frost back to the pump. Then you could say in the case of charging any home refrigerator or home freezer using a cap tube, "I will charge the box while it is running; and when it is cold or ready to turn off and has reached the temperature it should maintain, I will go by the frost line. That is, I should have enough gas in the box so that the whole evaporator is frosted and ready to frost out of the box; and here I will stop charging any more gas in the box." If you will memorize the words on this page and the next few pages, then you will have the key to charging all equipment using the cap tube for a metering device.

Once more, the charge in any equipment using a cap tube, no matter what kind, will be limited to just enough to take care of the evaporator and no more. You could say, "I will not put one more drop of Freon in a system with a cap tube that is not necessary." That is, I want liquid to boil in the evaporator and no further than the evaporator; and the only way I can limit the charge is to let the box pull down to the desired temperature. At that point, I will add a small amount of Freon to bring the frost line just out of the evaporator, or I will bleed off a little gas to drive the frost line back closer to the evaporator if it threatens to come on out of the box where it might drip on the floor during the off cycle.

This is not idle instruction; this rule for charging cap tube units is iron bound. Even in the factory where they manufacture cap tube equipment, more especially the freezer and household refrigerator, they follow the same iron bound rule.

Here is the way the factory charges a new refrigerator or freezer. First, they weigh in a charge through a special charging board, which is very accurate. This charging board will measure Freon into a system to less than one-tenth of an ounce of refrigerant. However, this is not accurate enough. For some peculiar reason many refrigerators and freezers do not require as much Freon as others, or they may require just a little more. This is because some pumps pump harder than others. You could likewise say that some cars are just a little faster than others, even though they come off the same assembly line. After the charge is weighed in, the factory men turn the box on and let it run. When the box has cooled down to the desired temperature on the inside—that is, zero in a freezer and approximately storage temperature in a household refrigerator—an inspector inspects the

charge in the unit. He looks in one place only, and that is where the frost line is starting out of the evaporator. If the frost line is coming too far out of the box and there is a possibility that it could cause a drip on the owner's floor, the inspector bleeds off a little of the original charge until the frost line backs up inside of the box. In the case of the frost line not coming out of the evaporator at all so that he can assure himself that the box has a full charge, he may add a little Freon until the frost line starts out of the evaporator. To charge these boxes, you will follow the same procedure and use the same method of determining the right amount of refrigerant. Remember, a box will pull down even with a partial charge. We will go through this together one time and then move on. Here is exactly how you would proceed.

First, we would tap the pig tail or suction line and evacuate the unit with a vacuum pump. See pig tail and line tap procedures as illustrated in **Figure 3-6**. After we have evacuated the system, we attach our drum of Freon to the charging hose and purge the air out of the hose. We break the vacuum by letting in some Freon. You don't need gauges to do this, but they will at all times show you the pressure condition inside of the system. Now we have a little pressure in the system, at least enough to break the vacuum. We plug the unit in and it starts to run. With just a little vapor in the system, just enough to break the vacuum, the unit shows that it is going to pump all right. Now we crack our charging valve in the manifold and give the unit a good slug of vapor from our drum. Now there is enough vapor in the system to condense a few drops in the condenser and these drops begin to feed to the evaporator through the cap tube. Immediately, you can listen at the door of the evaporator (ice cube maker in a household box) and hear the Freon boiling, gurgling, and hissing as it feeds into the evaporator. Take a look at **Figure 3-7**. While you are listening, feel of the evaporator right where the cap tube comes in; and you will feel a cold spot or even a frost spot where the small amount of liquid is boiling just as it leaves the cap tube. If you stopped right here and never put in another shot of gas, the little frosty spot right on the outlet side of the cap tube would remain just a little frosty spot because the few drops of liquid would be all the refrigerant that was available; and this few drops would be going around and around through the evaporator and condenser cycle. So we go ahead and add another slug of Freon, and this spot increases in size until we have at least half of the evaporator frosted up. Again, if we stopped right here, we would always have a half frosted evaporator; and it would probably take

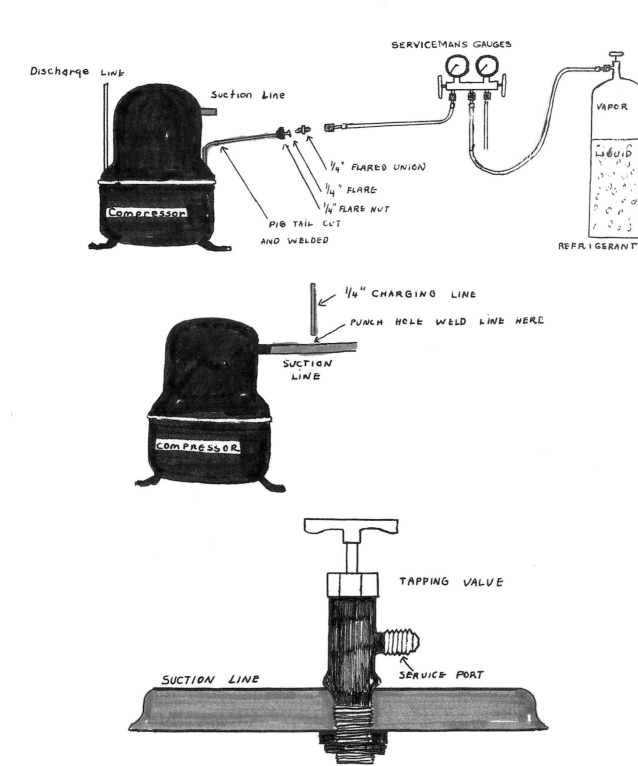

Discharge LINE

Suction Line

SERVICEMANS GAUGES

VAPOR

LIQUID

1/4" FLARED UNION

1/4" FLARE

1/4" FLARE NUT

PIG TAIL CUT AND WELDED

Compressor

REFRIGERANT

1/4" CHARGING LINE

PUNCH HOLE WELD LINE HERE

SUCTION LINE

COMPRESSOR

TAPPING VALUE

SUCTION LINE

SERVICE PORT

Fig. 3-6

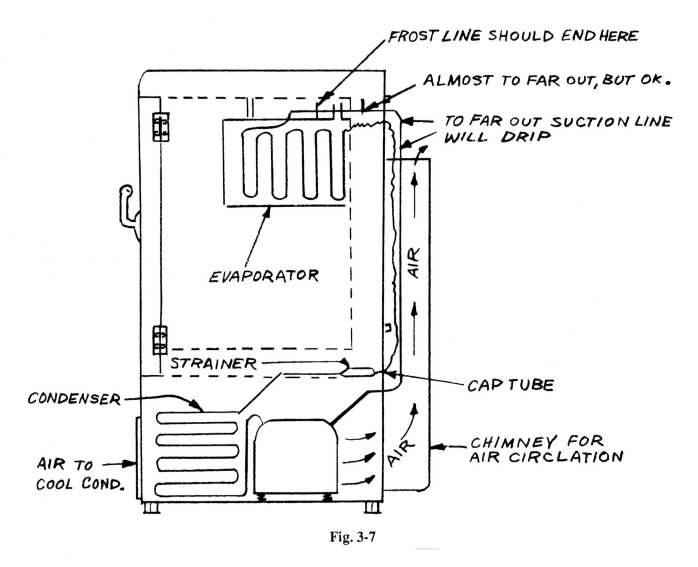

FROST LINE SHOULD END HERE

ALMOST TO FAR OUT, BUT OK.

TO FAR OUT SUCTION LINE WILL DRIP

AIR

EVAPORATOR

STRAINER

CAP TUBE

CONDENSER

CHIMNEY FOR AIR CIRCLATION

AIR TO COOL COND.

AIR

Fig. 3-7

twice as long for this half charge of Freon to pull the box down. Remember, even a partially frosted evaporator would eventually cool the box down. This is not what we want. We want the whole evaporator to work; so we add some more gas through our charging valve. The frost comes on up the evaporator until we now have about three-fourths of the evaporator frosted up. An expert would stop right here and go off and drink coffee. He would leave this box running; and when he came back, the evaporator would still be three-fourths frosted up; but the interior of the box would now be almost down to the storage temperature. Now very carefully give here a little more gas until the whole evaporator is frosted up. Generally, a good practice is to give the box just a little too much and wait until the box is ready to turn off; that is, wait until it is good and cold. Let it turn off once or even twice; it does not matter; but when it comes on and starts its running cycle again, you watch

the frost line on the suction line coming out of the evaporator. If it frosts too far out—that is, comes on out of the box—bleed off a little gas through your charging line, which is now disconnected from the drum. This final operation is called bringing the frost line back into the box; and when this is complete, you have a perfectly charged unit. Remember, many factories follow the same procedure; and they know this is the best and surest method of properly charging a box.

What has been said here applies to a home freezer, also. Let's analyze this operation now. It would be possible for you to put the gas in very quickly and just guess at the charge, and you might hit it right on the nose. Ninety-nine times out of a hundred you will miss, just as the manufacturer misses, even though he has a special measuring device for weighing in the Freon. Do not attempt to charge one of these boxes in minutes. Take your time. It is the best practice to start your charge as we have discussed here and take all day to finish the job. Work on something else while the box is seasoning out for the final top off on the charge. There is not a better way in the world to charge one of these boxes than the simple method we have discussed here, other than charging by weight.

At the risk of tiring you, let's state our rules for charging household refrigerators and freezers, regardless of the type of refrigerant, as long as they are equipped with cap tubes.

1. I will take my time.
2. I will evacuate the system before charging.
3. I will let in some gas gradually and watch the frost on the evaporator.
4. I will continue to add gas until the evaporator is almost frosted up.
5. I will wait a reasonable length of time for the box to cool down.
6. I will then top the charge off with a final shot of gas.
7. If I put in too much gas, I will back the frost line back into the box by bleeding.
8. If I didn't put in enough gas, I will add a slight amount until I do have enough.

It does not matter whether the freezer is an upright or chest model or any other kind. It does not matter what brand or style of household refrigerator or what kind of gas is used in either the freezer or household box. It does not matter that the boxes are old or new. It does not matter that they may have different types of pumps. The only thing that matters is that it is a *cap tube feeder*. You realize, of course, that right here and now with what you know and these instructions, you can handle a lot of charging in a lot of boxes.

Window units are generally equipped with cap tubes; and they are dealt with in exactly the same manner as refrigerators and freezers, only much faster; but frost will not be quite so important. When charging window units, you will depend upon your being able to tell when the suction line flashes back cold. This is the indicator of a full charge. Later we will deal with cap tube air conditioning in greater detail. But before we leave the domestic refrigeration charging method, let's compare again the expansion valve with the cap tube in relation to the frost back on these domestic refrigeration machines. If you remember, the power bulb on the expansion valve will immediately respond to any frost on the suction line where it is attached. It will throttle down and control the frost line by regulating the amount of gas feeding in, rather than the gas being regulated by its quantity as in the case of a cap tube. The cap tube job will always frost back just as far as the amount of gas in the system will permit. It will always come back to this very point unless some gas is lost. In case of loss the frost line would just back up in proportion to the loss. Here is an important point. There can be lots of reserve Freon in a system using an expansion valve. The quantity does not matter since it cannot get on through the evaporator and flood back like an overcharge in a cap-tube job. The power bulb would stop the flood back or frost line. Therefore, it stands to reason that you could have more liquid Freon than was needed in a job where an expansion valve was used. Certainly, you can have more liquid than is needed to just frost or cool the evaporator. In some cases you can have ten times as much as is needed, and this liquid will be stored in the condenser, if it is large enough, or a receiver tank which the condenser coil feeds with the cooled down condensed liquid.

Why have a receiver tank? It's simple. If the system leaked and there was a receiver tank and expansion valve, it could leak for a long time before all the liquid was gone and there was not enough to fed the evaporator. You might ask me what happens when the unit is on the off cycle where there is a large liquid reserve in the receiver. Does it feed on through the valve? The answer is yes, some of it; but the valve tends to close or let only enough liquid through to expand into vapor. This expanded vapor holds a back pressure on the incoming liquid through the valve. And even if it did continue to feed liquid into a cold evaporator on the off cycle, it would just fill the evaporator up; and when the unit started, it would have to clear the evaporator by pumping out the large charge of boiling liquid. See **Figure 3-8**. However, as soon as the evaporator is cleared, the valve would again be feeding right along. Some valves are snap action. They tend to close off when the unit is not running because they not only have their thermal power bulb control built into them, they are so balanced that certain pressures from the condenser and low side must set them into thermal operation.

Thermal operation means thermostatic operation. You do not need to go into this fine engineering feature of valves. It is enough that you understand their normal duties and operation. Let's go back to the frost line where a thermostatic valve is used. If the frost line needed changing, you would simply move the watch dog power element further out on the line or closer in on

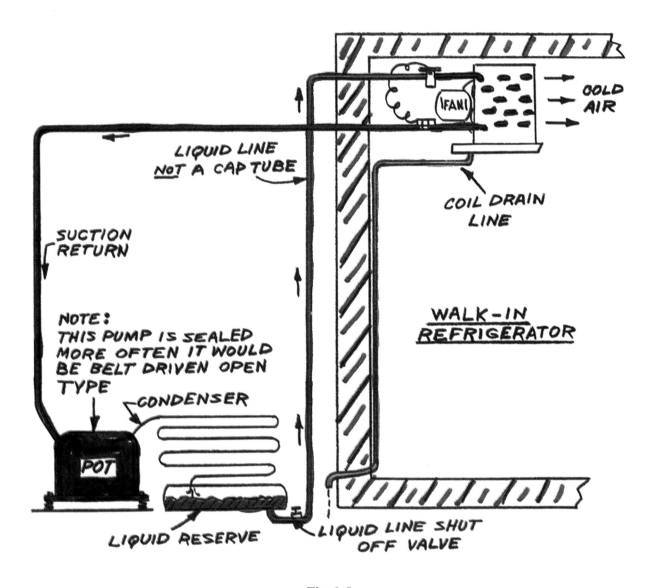

Fig. 3-8

the evaporator.

We have spent a lot of time charging up a cap tube refrigerator and a freezer, and for a good reason. By understanding the charging method involved in the units, we hold the key to charging all units. As a matter of fact, the proper procedure for charging a household refrigerator and freezer is far more difficult than charging almost any other type unit. The big units, both commercial and industrial, are so simple to charge that once you charge one, you will realize why more men prefer to work on big stuff than the little headache jobs like household and window units. Many men who call themselves first class servicemen are not skilled enough to charge the household units. They have never had any

formal training, and many of these men do not understand what you know right now about refrigeration. There are no mysteries connected with refrigeration service work. Do not listen to untrained servicemen who may make statements about certain operations of units that may mislead you. You learn the right way and the correct way and follow your own judgment in keeping with the training you are receiving here. See **Figure 3-9.**

Move on up to the window units now and have a look at the rules for charging these units. They are almost always cap tube feeders. In some cases, there is more than one cap tube. This does not change the basic feature of a cap tube. If two are used, then they are both

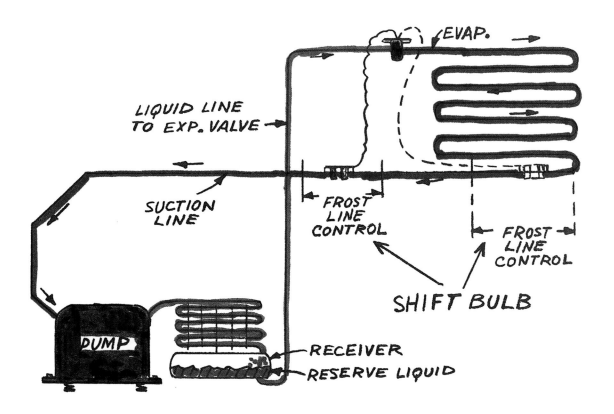

LIQUID LINE
TO EXP. VALVE →

SUCTION
LINE

EVAP.

FROST
LINE
CONTROL

FROST
LINE
CONTROL

SHIFT BULB

PUMP

RECEIVER
RESERVE LIQUID

Fig. 3-9

equal to the size of the hole in one. That is, they are just a multiple of the single tube. This is just a design of an engineer who believes that two tubes will do a better job than one, and they will. You would have a better delivery of liquid to the evaporator if it were more equally distributed to the coil. When more than one cap tube is installed, the cap tubes will enter the evaporator at equally spaced intervals.

Generally speaking, all window units have one cap tube. If there are two or more doing the same job, treat the job as if it had only one cap tube. Now we have stated that the frost line controls in the low-temp domestic field. In a window unit you are not concerned with low temperatures. All that you are concerned with is keeping the evaporator as cold as practical without icing it up when in use. We stated before that this temperature would be about 40°F, and the cap tube would be designed to feed insufficient liquid to boil up a pressure that would always be above freezing. We stated just a few pages back that we charge one of these units just like we do a household refrigerator, that is, by the frost line. The only difference is that we start out with frost, and it disappears when the unit is fully charged. So let's

begin our work on window units with this statement, "The gas charge in a window unit is limited just like other domestic type refrigeration units using cap tubes." Just so much gas should be put into a window unit, and the quantity will vary with the size and make of unit. The key to a full charge is the suction line coming out of the evaporator just like the domestic units, only here we do not have frost; we have a cold sweaty line to tell us that there is gas all through the evaporator and enough that it is spilling over slightly into the suction line coming back to the pot. (See **Figure 3-10**.) If we put too much gas in the window unit, a flood back or spill over will occur. We do not want this to happen because it will cause the pot to hammer, knock or slug when the piston or moving parts try to move liquid rather than vapor. The pump would probably handle the liquid all right, even to the point of the pot itself getting sweaty and cold; but it would be hard on the pump; so we avoid overcharging if possible. How do we prevent this? Suppose we go through the charging job on an ordinary cap tube one-ton window unit. We will evacuate this one, and later on, we will go into the evacuation very carefully. We have a unit that has a hole in one of the

49

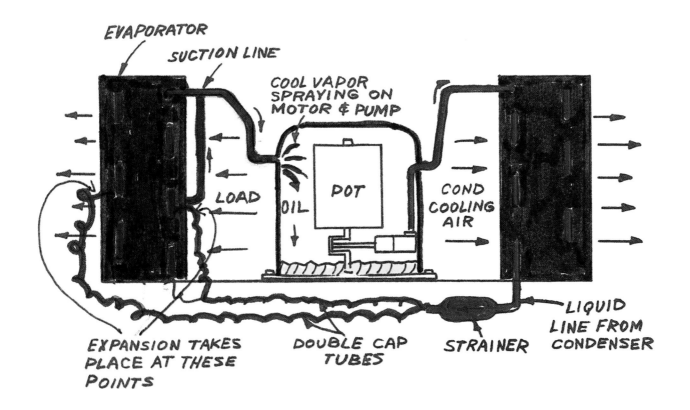

EVAPORATOR

SUCTION LINE

COOL VAPOR SPRAYING ON MOTOR & PUMP

LOAD

OIL

POT

COND COOLING AIR

EXPANSION TAKES PLACE AT THESE POINTS

DOUBLE CAP TUBES

STRAINER

LIQUID LINE FROM CONDENSER

Fig. 3-10

vapor lines, which was rubbing against the side of the frame of the unit. We cut off the pig tail and silver solder the hole in the vapor line. At the same time and while the torch is lit, we solder on a new pig tail and replace the liquid line strainer drier. We attach our compound gauge, sometimes called the vacuum gauge, even though it reads both pressure and vacuum. The pig tail on all window units as well as domestic refrigeration is always on the low side of the pot, that is, the suction side.

After we make our hose connection from the pigtail to the gauge manifold, we attach the center charging line to a vacuum pump. Now, we pull a deep vacuum on the system until our gauge records all that can be pulled, for instance, 29 inches plus. Now we close off the low side of the gauge manifold and move the charging hoses over to a service cylinder of F-22. We purge out the hose from the drum to the manifold by cracking the drum valve while the other end of the hose attached to the manifold is loose. After a purge we tighten up all hose connections and proceed to break the vacuum. Remember, when you closed the gauge manifold while the vacuum pump was running, the vacuum held in the unit. All of the time you were changing the charging hose

from the vacuum pump to your drum of F-22, the vacuum was trapped in the unit. Now with the hose purged and the drum open, we blast some F-22 into the vacuum; and it will go in plenty fast since it is going into a vacuum with the pressure of the drum behind it. What is the pressure in your drum of F-22? The pressure that was equal on your chart to the room temperature where you were storing the drum or working. Let in enough pressure to bring the pressure in the system up to 60 to 80 psi. Test your silver solder joints for leaks. Soap bubbles, oil or a halide torch will do.

Until now we have not run the unit. We have merely soldered the hole, evacuated, and tested the welds with a small holding charge. Now we are ready to start the unit and charge it up. Actually, a fast serviceman can charge a window unit in ten minutes. We start the unit running, and we notice immediately that the low-side gauge is pulling down very rapidly, and why shouldn't it? There is nothing in the unit but the test vapor we put in before we started. Now we crack open the charging manifold valve and let in some F-22, and there will be an immediate response on your low-side gauge. It will start reading higher (more gas). After about one or two

50

shots of Freon vapor out of our drum, there will be enough vapor in the unit to condense into liquid and start expanding and picking up heat as the liquid leaves the cap tube. There will be a flash of frost right where the cap tube leads into the evaporator, and why shouldn't there be? The unit is running at a steady low pressure because it is extremely short on gas. The cap tube is feeding some liquid, true, but not a steady stream like it would if it were a fully charged unit. This little bit of liquid that we have in the unit now condenses into liquid and feeds intermittently into the evaporator. Most of the stuff going into the evaporator at this point is just drops mixed with vapor. It does not yet have enough body nor is it in sufficient quantity to make a steady supply of Freon available in the cap tube. Now we give it more gas, and we immediately note another rise in the steady evaporator pressure.

The pressure is not yet high enough because the unit is still short of gas, but the significant thing to watch at this point is that the system is taking the gas, and the unit is responding by increasing the evaporator pressure with every shot of gas. But the pressure in the evaporator is way below its factory designed pressure, and the evaporator coil will have frost on that part that has liquid refrigerant boiling in it. We add another shot of gas through our manifold, and the pressure comes up again. Suppose now the unit is running steady at 40 psi and is still short of gas. At this pressure on F-22, there is frost on that part of the evaporator coil that has liquid boiling in it. We add some more vapor from our drum which in turn is pumped through the condenser and turns into liquid and continues on through the evaporator. Now our pressure has come up to fifty pounds steady running evaporator pressure. We are getting close to a full charge. There is still frost on part of the coil, which has liquid boiling in it. As a matter of fact, that frost line has been moving further up the evaporator every time we give it a shot of gas. Now we give the unit one more shot of vapor from our drum. The frost suddenly disappears from the evaporator coil, and the pressure on our gauge comes up to 69 psi. Of course, the frost disappears. If you will look on your chart, you will see that this is 40°F.

Now we are ready for the final top off, just like the case of the domestic refrigeration; but we are not going to look for frost because it has all disappeared. We will feel of the suction line, and we will find it nice and cool to the touch with almost a full charge in the unit. However, there might be a few inches of the evaporator right where the suction line comes out that it not doing any work because there is just a slight shortage of refrigerant. So we feel of the suction line and note the degree of coolness and add a little more Freon, and immediately you feel the surge of cold in the suction line where it comes out of the evaporator. This surge is the final amount of liquid you put in to be sure the whole evaporator was doing its job. No matter what the gauge says, give or take a few pounds, the suction line temperature will control. Knowing what the gauge pressure should be simply verifies what your hand on the suction line knows for sure, that there is ample Freon if the line is good and cold. What would happen if you kept on adding more Freon? The answer is nothing, since there is no problem of frost dripping here. You would simply be building up Freon in the condenser which was not needed; and when the unit shut off, this Freon would feed on over to the pump and cause it to be noisy when it started up; or it might even stir up the oil and raise Cain generally until it was pumped back over into the condenser. So here is a rule. An overcharge in a window unit is a nuisance to the pump. An undercharge will cause the coil to ice up, and the unit will not cool like it should, and the ice will build up until it blocks out the air circulation. In the case of a big package unit with a cap tube feeding the evaporator, you would follow the same identical procedure.

Summing up the domestic work, which includes household refrigerators, freezers, and window units, we limit the amount of gas in a household box because we do not want melting frost dripping on the kitchen floor. In the home freezer the same reason applies. We limit the supply of refrigerant to a window unit because we just don't want a lot of pure expanding liquid feeding all the way back to the pot to stir up the oil in the pot and slug the valves in the compressor. Besides, what point would there be in turning the pot into an evaporator? Actually, the manufacturer of these sealed pots wants some cool vapor to spray in on the pump to help keep the motor windings and pump parts cool, but just cool vapor, not liquid. Cool vapor has just the smallest little specks of Freon boiling, and they will do no harm. Here is a drawing of a pot and an explanation of the oil and motor windings. See **Figure 3-11**.

Go back and review your method of charging a home freezer that uses an expansion valve. Compare this with the method of charging the cap tube type. In this section you were led to believe in that first freezer charging operation that the gauges were of the utmost importance and that press-temp was the whole key to charging a unit. This is not the whole story. True, gauges are important because they let you know at any time what is going on inside of the unit. Remember, you cannot see through copper tubing. The press-temp is very important in that you must have some understanding of what the designer was working for when he designed the system. Now you should realize that you do not even need to own a set of manifold gauges in order to charge either a household box, a home freezer, or a window unit. You could simply put a hose on your drum to the charging-in place on the unit and use the valve on the drum and your sense of feel and eyes to complete the job. However, if anything went wrong, you could not tell it by trying to see inside of copper tubing. The gauges would tell you quickly if a pump suddenly stopped pumping or if a line were stopped up. Gauges will be absolutely necessary when we begin to set pressure controls.

∎

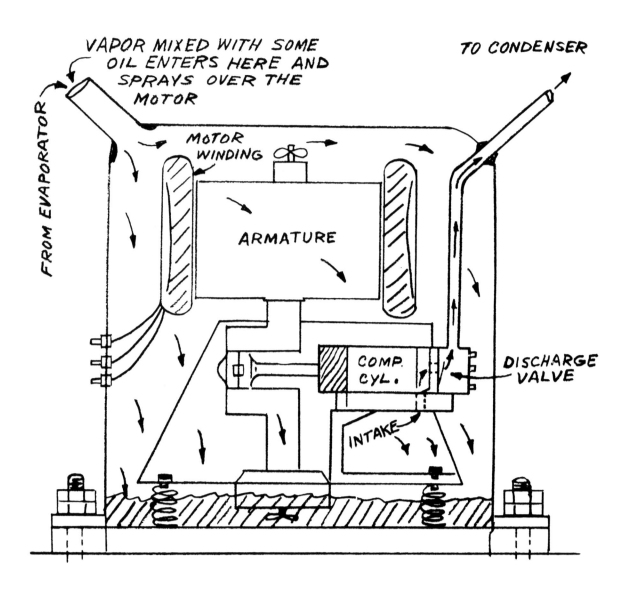

Fig. 3-11

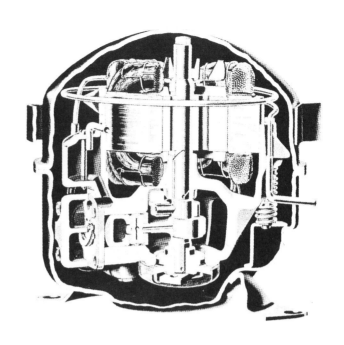

Hermetic

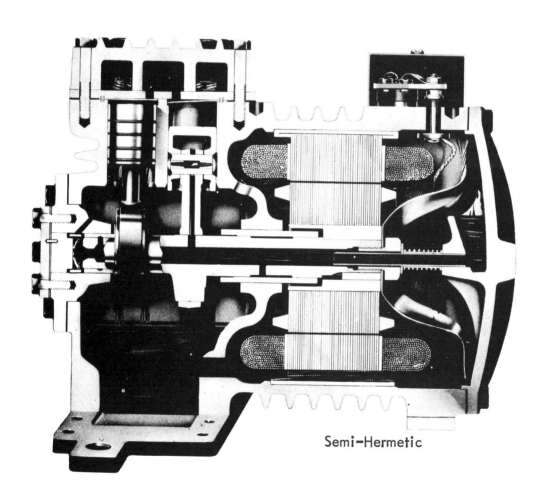

Semi-Hermetic

53

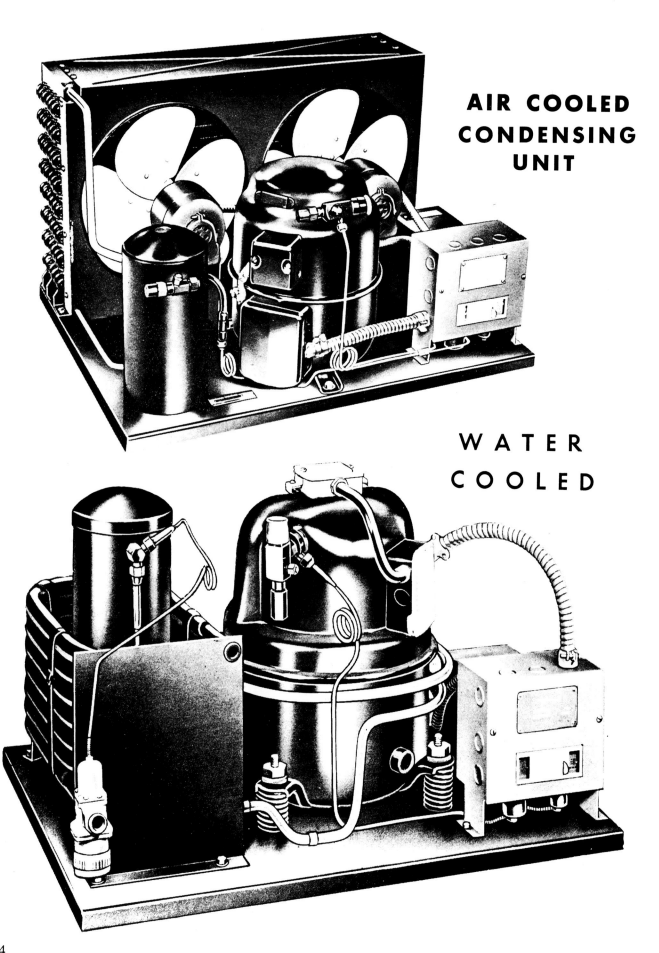

AIR COOLED
CONDENSING
UNIT

WATER
COOLED

54

CHAPTER 4

SERVICING COMMERCIAL EQUIPMENT

You take household boxes, freezers and window units to your shop in order to charge them properly. Never charge domestic equipment in a home unless it cannot be done otherwise.

In the case of central and packaged air conditioners, you service this equipment wherever it is located. You will have plenty of time in your own shop to carefully examine domestic equipment; but on a package unit, you will be expected to proceed, wherever it is located, with whatever repairs are necessary, and in some cases, with the owner standing right beside you. Even so, set those tools down, get out your flashlight and make a thorough inspection of the equipment before you touch a tool. A man accustomed to hiring this service will recognize you as a better serviceman for taking the time to look the equipment over. With your flashlight, look inside and see if you can spot anything out of place. Look the condenser and evaporator over. Locate the disconnect or master switch. See if any belts are broken or slipping. Look the unit over like a detective might look over an automobile. Look for the trouble right in front of you. Don't complicate what may be a very simple job. See if you can determine whether the unit has been worked on recently or has had a lot of work done on it. Read the signs before you disturb the dust, grease or dirt. Talk to the owner if you like. You might find out from him when the unit began to fail or if he had heard any unusual noises.

The owner may have never heard of the refrigeration cycle, but he may know the peculiarities of his own refrigeration equipment; and any information he may be able to give you might help you in your work. If you question the owner in a business-like manner, you will lose no face. While you look the job over, take the time to locate the low-side service valve and the high-side valve. Suppose this particular unit was running and wasn't putting out. The owner has told you that it was gradually getting worse, and today it quit cooling altogether. You confirmed this by running your hand over the discharge air from the evaporator. It was not cool; and of course, you knew the room where the unit was working was in need of air conditioning. You feel of the high side, and it is not hot, just warm. You note that the evaporator blower is running. You also noticed that the unit is running. You step outside and look the water tower over, and it seems to be working all right. That is, there was ample water circulating between the condenser and the water tower. By now, you have looked around until you have a pretty good idea where everything is located, including your service valves in case you need to attach your gauges. You could say right now to yourself, "If everything seems to be running smoothly, yet no refrigeration, then there is obviously a loss of gas." If there is no gas, it had to get out somewhere. Gas does not wear out. It leaks out. With these thoughts in mind, you are now ready to confirm your suspicions. You are ready to put the gauges on and see what is inside of the unit. Your first move would be to stop the unit. Pull the switch. You can always start the unit up again when you are ready. Now take the cap off the low-side service valve stem. This is the first time you have touched the unit with a tool. All you were doing before was looking, feeling, and thinking.

With the unit cut off and the guard caps off the service valves, you are ready to attach your gauges.

Put your service-valve ratchet wrench on the valve

stem and check to see that the service valve is firmly backseated. Now with the valve backseated, unscrew the pipe plug in the gauge connection port which you just made sure was closed by backseating. With 1/8" pipe plug removed, you now screw in a 1/8" pipe to 1/4" flare adapter. See **Figure 4-1**.

We will be mounting gauges on a water cooled package system as shown in **Figure 4-2**. Use a box wrench on the pipe plug and adapter. Do not use pliers on brass fittings or any other nuts or bolt heads. With the adapter screwed in tight, attach the compound gauge hose to the adapter. With the assumption or guess that there is a little pressure in the system, you make sure both valves in your charging manifold are closed. Now unbackseat the low-side service valve where you have the low-side gauge attached. One or two ratchets or turns will do. Now you should immediately have a reading on your vacuum or compound gauge, if there is any vapor left in the system. Take your time. You can say that if you have a little pressure in the unit that is probably Freon vapor left over from the original charge. You also should know that you have trapped some air in your hose that was in there when you unbackseated the valve.

You can purge air out through the high-side hose or loosen the quick coupler on the hose where it enters the manifold and give it a good blow. **(Fig. 4-3)** This will get rid of the air in your hose. A good practice is to have your drum ready and hooked up to the charging hose.

With the drum hooked up to the charging hose, you may purge all the way back to the drum.

You are going to need to check for a leak; and that will require that you put some gas in the unit, at least enough to raise a higher pressure in the system. You may have noticed that you can likewise purge from your Freon drum all the way to the service valve on the unit before you unbackseated it.

Never be afraid to waste a little Freon. Freon will work for you if you will let it. Now we have the gauges on, and we take a look at the pressure on the low side. You may have noticed that we are not even going to bother with hooking up the high-side gauge. Suppose there were a pressure reading of about thirty pounds when we finished purging hoses and with the unit still off. This 30 psi on the low side is just some left-over vapor in the system after the main body of liquid had leaked out. In fact, there is not even a full head of vapor, much less one drop of liquid left that might boil up a pressure that was equal to the room temperature for liquid Freon. Had there been one drop, one pint, or one pound of liquid in the unit, then you would know that there would be a temp-press relationship; and even though the unit has less than the required amount of refrigerant, there would still be enough to boil up the same pressure that was in your full drum.

Let's go through this point very slowly and see if we can explain this more fully. Suppose there were a leak, and it started the week before you got there. When the

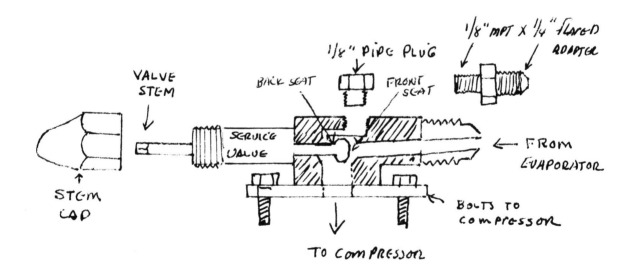

(Fig. 4-1)

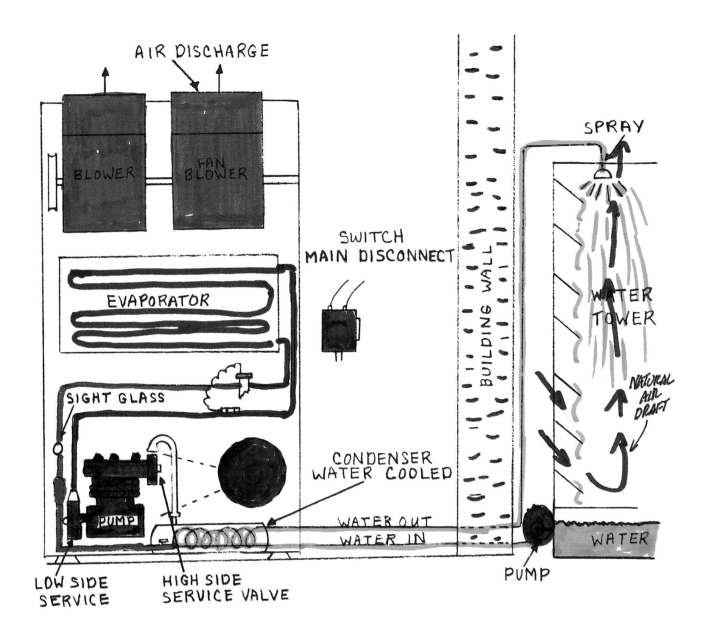

(Fig. 4-2)

leak developed, there were ten pounds of liquid Freon in the receiver and other parts of the system. As long as there was some liquid in the system with the unit stopped, the pressure in the system would be equalized and would read what any drum of Freon would read in the same room. Now the gas is leaking out, and the Freon keeps boiling off more vapor to keep up the pressure that stabilizes liquid Freon. Now suppose there were one drop of liquid Freon left in the system under the pressure of vapor. With one drop in the receiver and the unit off, there would still be a temp-press relationship just as you read it on the chart. Now with this slow leak, the last drop of Freon boils off to make vapor

which is escaping. The instant this last drop changes to vapor to make up for the loss, there will be no more liquid to boil up more vapor. The vapor in the system will leak on out, and the pressure will just keep going on down till all vapor is gone and the reading on the gauge is zero.

However, most commercial systems have low pressure controls that will cut the unit off before the system has a chance to pull a vacuum on itself. Therefore, once the leak is repaired you can add to the existing charge. You could compare this to boiling water in a small container and letting the steam escape at a steady rate through a small pin hole while maintaining a back-

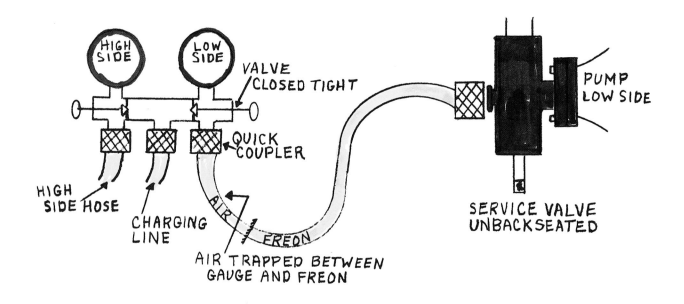

(Fig. 4-3)

NOTE: The serviceman's gauges are reversed as to how you would normally see them. Don't let it confuse you.

pressure on the boiler. As long as there was one drop of water left, you would have steam going on out the pin hole. But the minute the last drop of water disappeared, you could not make any more steam; and the steam that was in the container would just go on out the pin hole until there was no pressure left. This is what happens inside of a unit when it loses its charge. Likewise, you could have two drums of Freon setting in your shop, both sealed up and not being used. One drum, let's say, has twenty-five pounds of liquid Freon in it; and the other has one ounce of liquid Freon in the bottom of the drum. The temp-press would be the same in both drums regardless of the amount involved as long as the smallest amount was still capable of boiling a little to make up for the changes of temperature. The instant a little vapor was released from the short drum and the last drop of liquid disappeared, the temp-press rule would not apply.

Let's go back to our package air conditioner. We have thirty pounds gauge pressure with unit off. Now, if the unit had any liquid at all, your gauge would read the pressure of liquid Freon under pressure that is on your chart. In this case, suppose the room were eighty degrees. Then if there were one drop of liquid Freon in the system under vapor pressure with the low side and high side equalized, the pressure on your gauge would be 143 psi F-22. So there is no doubt that all the liquid is gone, and all that is left in the unit is some vapor that has not as yet leaked out.

Now, you may or may not want to blow out all of this old vapor left in the system and start with clean Freon. Suppose you leave what was in the system for the time being and let some gas blow into your unit from your service cylinder. Bring the pressure up to about fifty psi, a good pressure for finding leaks. Now, cut the charging gauge valve off and back off from the job. You

58

have got to find the leak. Remember, Freon does not wear out. *It leaks out.* Suppose that you saw a suspicious looking oily spot on the fitting coming out of the expansion valve where it hooked to the evaporator. All right, light up your Halide torch and sniff where the oily spot is. Sure enough, the flame turns green. You double check with some soap bubbles around the flare nut on the valve and puddle some soap bubbles around the threads on the flare nut. You see a light bubbling. You have found your leak. Now tighten the nut and check again. All right, you stopped the leak; and the line is tight to the valve again. Now you are ready to charge the system. If you had your vacuum pump with you, you could proceed to blow out everything in the unit until your gauge read zero and you could then hook your vacuum pump to the gauge manifold and pump a deep vacuum on the system.

Now you would break the vacuum with gas from your drum and proceed to charge the unit. Take your time. There is no hurry. Just start the unit and let the gas pump into the system through the low side from your drum. As soon as you have enough vapor in the system to condense into liquid, the liquid will start feeding from the condenser to the expansion valve and on into the evaporator. The more you put into the system, the further over on the evaporator it will expand, that is, get cold. When the suction line starts to cool down, you know that you have at least enough gas to take care of the evaporator. Anymore that you will add, will be the reserve that will fill the receiver. Never fill any container, whether it be your service cylinder, receiver or any tank holding Freon more than half full of liquid.

This is a good safe rule until you know exactly how much you can put into a particular container. The static pressure that will build up very rapidly on Freon in any container if too full, will rupture the container. Now what is a full charge in our package unit? First, we know what a minimum charge is. It is just enough liquid in the system to feed the whole evaporator and keep the suction line cool. If you stopped with the bare minimum, the first small leak that developed would start reducing the boiling liquid in the coil; and the evaporator would begin to warm up until there was only one drop expanding through the valve. Estimate the size of the receiver and act accordingly. You may use a bathroom scale if you are in doubt as to how much you are putting into the unit. One way to tell how much a receiver will hold is to compare it to your service cylinder which will always be designed to hold a certain amount of F-22. Compare it in size and act accordingly. One way to know how much liquid there is in any container, is to heat a line up and down on the end or side of the cylinder. Where the liquid line level is, the thin steel drum will not be hot like the rest of the line where you were heating. **Caution.** Never heat any refrigerant cylinder or any container of refrigerant liquid any hotter than you can hold your hand on the heated area. *This is a must rule. Never break it.* See these two tanks for liquid level in **Figure 4-4.**

Suppose the receiver on this package unit is about as big as a thirty pound Freon service drum. Then you could say that the maximum you could charge safely into the tank would be thirty pounds. This would be too much. About twelve to fifteen pounds would be ample.

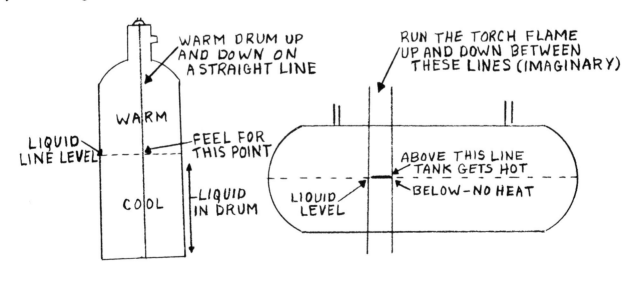

(Fig. 4-4)

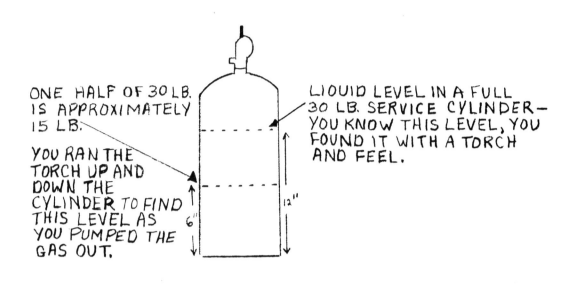

ONE HALF OF 30 LB. IS APPROXIMATELY 15 LB.

YOU RAN THE TORCH UP AND DOWN THE CYLINDER TO FIND THIS LEVEL AS YOU PUMPED THE GAS OUT.

LIQUID LEVEL IN A FULL 30 LB. SERVICE CYLINDER — YOU KNOW THIS LEVEL, YOU FOUND IT WITH A TORCH AND FEEL.

12"

6"

(Fig. 4-5)

Here is a service drum, with an estimated fifteen pounds missing **Fig. 4-5**).

Ordinary good judgment is the rule for charging any type equipment which uses a receiver or reserve tank for holding Freon. With a little practice, you will charge into any unit, just about the correct charge in keeping with a fair estimate of its requirements, and capacity. There is usually always a liquid line sight glass on a commercial unit to tell you when the system has enough gas. It will be located in the liquid line near the unit. When the system is fully charged the sight glass will be clear. When it bubbles the system is low. If you do not have a sight glass, pump the system (explained later) down and install one. You may also charge by listening to the expansion valve. If the system is low the expansion valve will make a hissing sound. Charge until the hissing stops. This applies only to units which use expansion valves and receivers. You already know that a cap tube type refrigerating unit has a limited charge in the system.   Let's go back to our charging. We have cooled the suction line down with a minimum charge and we then added an estimated ten pounds more of Freon into the system. With the minimum charge or the full charge, we can now read our low-side gauge in terms of press-temp in line with the chart and what is going on in the evaporator. You remember that the room temperature was eighty degrees; therefore, we would have a load blowing over the evaporator at eighty degrees.

With the eighty degree load we'll say you want to know if your suction pressure is accurate. Most air con-

ditioning evaporators, with the proper charge, and proper air movement will run approximately 20° colder than the air temperature entering the evaporator. The refrigerant temperature will run approximately another 10°F cooler. So, you have a total of 30°F to work with. For example, we already know the air temperature is 80°F. Take 30°F away from that reading and we have an approximate refrigerant temperature of 50°F. Upon looking at the pressure-temperature chart you can see a corresponding pressure using F-22 would be around 84 psi. Pull the room temperature down to 70°F and deduct 30° for the refrigerant temperature and you would have a corresponding pressure of 69 psi. The pressure has gotten lower because the air temperature is lower. Remember, just because you have been told an R-22 air conditioning system should read 69 psi, don't take it as an iron clad rule. Granted it should read around 69 psi, but there are too many things that will vary the pressure. Worry only if the pressures are extreme such as 55 psi (too low) or 100 psi (too high).

Do not place too much importance in the gauge readings. In the last unit we charged, you could glance at your low-side gauge and say that it should be reading somewhere between sixty to seventy pounds. So if it happened to be a hard pumper and read 60 pounds, don't worry; or if it read 70 pounds and everything seemed normal in your judgment, let it go at that.

Suppose in this case you sold the man twelve pounds of F-22 and you are winding up the job. Shut down the unit for a minute and take your gauges off. Leave the valve tightly backseated, put the 1/8" pipe plug back in and cap the valve stem. While the unit is off and before you leave, give the whole unit a careful leak check. If there are no leaks, close up the panels, start the unit, check the thermostat to see that you are leaving it on the setting that the owner used and write out your bill for twelve pounds of F-22 and labor. Collect your money or have your bill signed and be on your way. *Don't hang around and chew the fat after you get your money. Be strictly business. Time will tell whether you have done your job well.*

Commercial units are the easiest of all units to charge. **Figure 4-6** is an air cooled condensing unit and **Figure 4-7** is a water cooled condensing unit. They are usually located outside, or in a back room, or under the case. You may have plenty of privacy while you are working. You can take your time and may even have time to examine other equipment around, which you may later be called to work upon. You check commercial equipment just like you would a package air conditioner. The only difference is that you will have to go to the box where the evaporator is located to check for leaks and other things that could be wrong. The units are generally equipped with complete service valves on both the high side and low side. There will also be a liquid line shut off valve coming out of the receiver. Every possible service valve is on the equipment for

your convenience. For example, you are called to a meat market to work on a display case. Suppose the complaint is the same as the package unit job we just finished. The unit is running and the box is getting warmer and warmer. You would carefully inspect every part of the equipment with a flashlight. You would look for the obvious things like a broken line or big leak, with oil dripping. You would shut down the equipment and attach your gauges and follow the same procedure as on the package unit. Test for leaks with more pressure from your cylinder. Fix the leak and put the system back on the line and charge it up. Every commercial unit will have a receiver. Sometimes it will be a separate tank, and sometimes it will be the lower section of the water cooled condenser.

The receiver on a commercial condensing unit serves a three-fold purpose. First of all, it is large enough to take care of surges from the pump and condenser where the pump and condenser are overloaded for short periods of time. Secondly, it will store sufficient liquid to keep the expansion valve adequately supplied with liquid, even though there are small leaks which may leak over a long period of time. Finally, the receiver is usually large enough to hold all of the refrigerant in the system when the system is pumped down. Pumping down any system which has a receiver

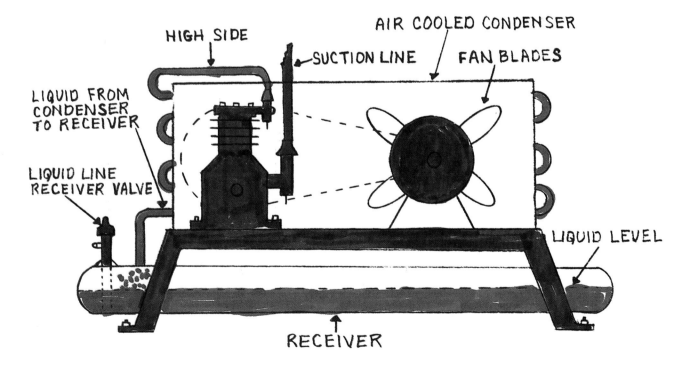

(Fig. 4-6)

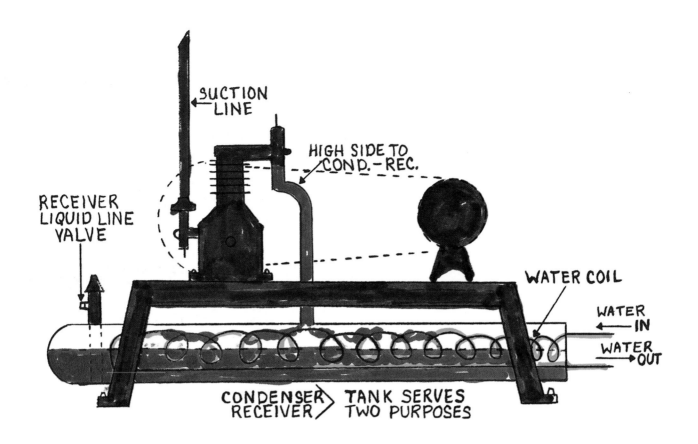

RECEIVER
LIQUID LINE
VALVE

SUCTION
LINE

HIGH SIDE TO
COND.-REC.

WATER COIL

WATER
→ IN

WATER
→ OUT

CONDENSER
RECEIVER ⟩ TANK SERVES
          TWO PURPOSES

(Fig. 4-7)

is the simple process of closing off the liquid line valve feeding the system, thereby blocking any further flow of liquid to the valve and evaporator. You may have to force the pump to run in order to pump every last drop of refrigerant into the receiver, but it can be done. Pumping down a large system will permit the removal of any part between the liquid line shut-off valve and the discharge valve on the pump. At the same time, you will salvage all of the expensive refrigerant contained in a fully charged system. To force a pump to run simply means shorting or jumping across the control device such as a LP control or thermostat. This will keep the unit running until all the gas between the receiver valve and the discharge service valve is pumped into the receiver storage tank. You could then remove the dryer, liquid indicator, evaporator and even the pump if necessary. How would you know when a system is pumped down? When your low-side gauge is down to zero, the inside of the whole system between the receiver shut-off valve and the discharge valve on the pump will be zero or the same as the atmosphere. If you then opened up part of the system, nothing would go in and nothing would come out. The system is neutralized pressure

wise. Pump down is an ordinary every day practice by a serviceman repairing equipment that can be pumped down. Window units cannot be pumped down, because there are no service valves which can be used to trap the gas where you want it, that is, in the receiver. A package air conditioner may be pumped down if it has a receiver. Any unit using a receiver, together with a liquid shut-off valve and service valve can be pumped down.

Going through this together, suppose we want to take down the evaporator coil in **Figure 4-8** and have it steam cleaned. First, we close the receiver outlet valve, then let, or force the unit to run until our low side gauge reads zero or just above. We would then pull the switch on the unit, front seat the low side service valve to keep air out of the compressor, and pull the evaporator coil. The charge of refrigerant would be trapped in the receiver and the condenser. When the evaporator coil is reinstalled, tighten all line fittings and nuts except the suction line nut that screws on the low side service valve. We now purge through the low side by cracking the receiver outlet valve. Air in the system will be pushed down the suction line and out around the loose nut. Tighten the loose suction line nut while gas is still

62

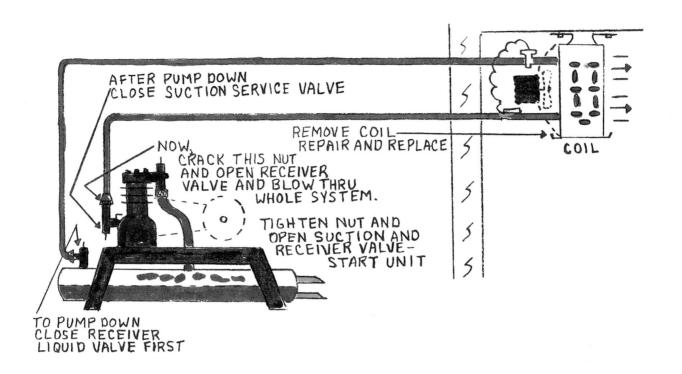

AFTER PUMP DOWN
CLOSE SUCTION SERVICE VALVE

REMOVE COIL
REPAIR AND REPLACE

COIL

NOW,
CRACK THIS NUT
AND OPEN RECEIVER
VALVE AND BLOW THRU
WHOLE SYSTEM.

TIGHTEN NUT AND
OPEN SUCTION AND
RECEIVER VALVE—
START UNIT

TO PUMP DOWN
CLOSE RECEIVER
LIQUID VALVE FIRST

(Fig. 4-8)

blowing through the system. Open all valves to normal operating position, and turn switches on. You are now back on the line.

You could charge a unit through the low-side and pump into the receiver from your service cylinder without ever opening the receiver valve until you were ready to put the unit on the line.

Let's stop and briefly review the charging rules for all systems. Remember any unit with a cap tube will have a limited charge—just so much gas—and the charge will be determined by the frost line on the suction line or the cool state of the suction line. In home boxes, both freezers and refrigerators, the frost line controls. In window units using cap tubes the cool suction line controls, and frost is important to watch up to the point where it disappears. In package units using cap tubes, the above rule applies; and where package units are equipped with an expansion valve, the rule of a cool suction line and good judgment as to what the receiver will hold determines the amount you sell the customer. In commercial units, which are almost always equipped with expansion valves, the same applies. Also use the liquid line sight glass to determine a full charge.

At this point, it is time again to tell you that Freon is a wonderful crazy liquid which boils at a very low temperature. The only thing that tends to upset a serviceman is the fact that it is not like water which boils at 212 degrees. Freon boils at twenty degrees below zero. Freon is a friendly liquid. Don't ever be scared of it. True, it is expensive. But you will never use enough to break you. You have been using Freon for years and may not have been aware of it. At one time insect spray cans were charged with liquid Freon to keep up pressure to carry the fly poison out in a Freon vapor spray. You have bought Freon unknowingly for many different purposes.

There is an important phase in commercial refrigeration which must be understood before we continue. Before now, you have always thought of frost as being an indicator of a cold evaporator. This is still true. Frost is the natural result of an evaporator getting cold enough to freeze the moisture or humidity in the air carrying the heat to the evaporator. However, frost is actually an insulation, if you have too much of it on an evaporator. A commercial evaporator ordinarily runs below freezing, and any frost on the coil becomes a coating of insulation, which slows down the flow of heat into the evaporator. Because of the insulating nature of frost, we have to defrost certain coils. We have to get rid of the coat of insulation which can be a nuisance, when it gets too thick. Frost is not too important where the evaporator is very large, for instance, in a freezer which may have the whole tank as an evaporator. However, in the case of a very small coil frosting up, it would soon build up so thick there would be no work done. What has all this to do with commercial refrigeration? All right, let's start at the beginning by studying the development of commercial refrigeration in this country. We will then better understand why frost is a problem in commercial refrigeration.

Many years ago, the average commercial installation used the ordinary condensing pump, which we are now familiar with, and the same refrigeration cycle. But the evaporator was a real old fashioned monster. No one had thought to build a blower type coil. These original evaporators were little more than long rows of pipes suspended from the ceiling of the box. Later, they added fins to these pipes, but they were still huge and bulky. The old installation consisted of the condensing unit, liquid and suction lines, and evaporator coil. The box would be inside of the store much like it is today, and the unit would be in a back storeroom or outside of the building. Inside the box the whole ceiling would be covered with the evaporator coil. This was before the invention of the expansion valve, and most of these evaporators were controlled by boiler float systems. The pump worked on these huge coils and the air circulation within the box was nothing more than cool air settling like water to the bottom of the box, and the warmer air would rise to the top where the evaporator was located. This created a gentle circulation within the box. After so long an operation, these monster coils would be iced up; and the proprietor would have to pull the switch on the box and defrost the coil with a garden hose or just wait until it melted off. Contrast this to a modern day commercial box which has an evaporator which may only be eighteen inches in diameter. This modern evaporator will have a high speed fan recirculating the air in the box over the evaporator. This coil is called a forced evaporator. That is, air is forced through the cold finned surface of the coil. You could compare this modern evaporator with the small heater in an automobile. Modern auto heaters are forced heaters. Here is the problem that was encountered when manufacturers started reducing the size of the evaporator. Since the coil was very small, any frost or ice accumulating on the coil would block the flow of air being forced through it. When this happened, there was no more refrigeration, other than that which was derived from blowing air over a small chunk of ice.

The press-temp chart you have was the manufac-turer's life saver. He reasoned that if this press-temp relationship were constant and never lied, then all he had to do was to use pressure to control the temperature of the box rather than the old fashioned thermostat. He built a low-pressure control and hooked it on the low side of the pump. This pressure control was set much like the pressure control on an ordinary garage air compressor, that is, to be cut off at a certain point and cut back in at another point when the low side pressure began to rise. The manufacturer knew that the low side pressure would always be the same in the evaporator as well as the low side of the pump, less the pressure drop across the line. With the unit off, there wasn't even any pressure drop to consider. So he hooked the control to the pump on the low side and waited until the pump pulled the box down to the desired temperature. He immediately adjusted the control until it shut off. Now with the box off, the pressure in the evaporator and the suction line and the low side of the pump, became the pressure equal to the temperature of the frosty cold evaporator.

Suppose he had pulled the box down to 36 degrees and then caused the pump to shut off on the pressure control. Immediately, the pressure in the evaporator would be equal to the temperature of the evaporator. Now, if you will remember that the evaporator leads the load, (superheat) you could look on your chart and say that if the valve has about twenty degrees superheat, then the evaporator is actually twenty degrees colder than the box temperature of thirty-six degrees. That is, the coil is leading the load by twenty degrees. Or, thirty-six degrees F-12 equals 33.4 psi, and thirty-six degrees minus twenty degrees equals sixteen degrees; and sixteen degrees equals 18.4 psi for F-12. Now the instant the unit shuts off on the pressure control, the low-side pressure at the pump where the LP control is attached will read 18.5 psi, which was the cut off pressure. Likewise, the minute the unit shuts off, the coil will start to warm up; and this 18.5 pressure will begin to rise. The evaporator is bound to warm up to the thirty-six degrees which is the temperature in the box. The evaporator will reach this temperature very quickly with the air inside of the box being blown over the coil. All right, as the evaporator starts to warm up, the original pressure of 18.4 starts to rise. You could tell right now just how high this pressure must rise during the off cycle in order to clear the frost off the coil. Look at your chart for thirty-two degrees (where frost starts to melt). You will see that the pressure for thirty-two degrees plus is 30.1 psi for F-12. Maybe the temperature of the air in the box by

64

this time is thirty-seven or even thirty-eight degrees. At any rate, the air blowing over the coil will soon have the coil the same temperature as the air in the refrigerated box. Suppose it does warm the coil up to the same temperature as the box, that is, about thirty-six to thirty-seven degrees. Now the pressure is 33 psi or better and the frost is melting off the evaporator very rapidly, since the melting point began as soon as the metal of the evaporator got thirty-two degrees. Now, the manufacturer, in order to be absolutely sure that there is going to be no ice or frost on the coil, sets the pressure control to come on at 36 psi. He can't miss. He has a pressure control that will cut off the unit when the pressure gets down to 18.4 psi, and the unit comes back on when the pressure in the low side reaches 36 psi. In fact many of these LP controls are connected to the service valve right where you connect your low-side gauges. That is, there are two pipe plugs. One has been removed and used for the connection to the LP control, and the other is for your service gauge adapter. More often, the manufacturer of the pump has a special fitting some-where on the pump which will feed the low side pressure to the control just as if it were reading to your gauge. See **Figure 4-9**.

The importance of gauges may be apparent to you since it will be necessary to set this control on a defrost cycle. The only accurate way to set it is to attach your gauges on the low side when you adjust this control. All of these controls are made so that they cut off at a certain pressure; and when the pressure comes back up again, they turn the unit back on. Your press-temp chart will come in very handy here. It will help in setting your pressure control. **(Fig. 4-10)** The job of the LP control is to break the circuit to the motor driving the pump and turn it back on again when the pressure comes up again. By this I mean, nothing takes place but the stopping of the condensing unit when the control cuts it off and the turning back on of the condensing unit when the control comes on again. The blower fan on the evaporator stays on all the time. It never shuts off unless you turn it off for some reason, perhaps to work on the coil, or when the motor burns out.

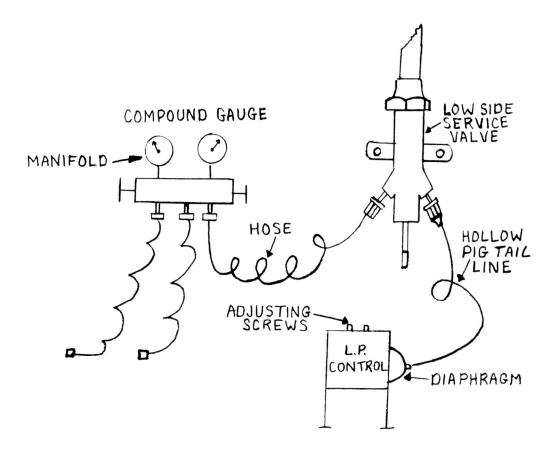

(Fig. 4-9)

Setting a pressure control together, let's assume that we have an ordinary one horse power pump and an air cooled condenser out in back of a grocery store. The liquid and suction line run into the store and lead into a big reach-in dairy box with a blower coil for an evaporator.

Now, suppose the complaint on this job was that there was no refrigeration. When you arrived, you took your flashlight and while looking the system over, you found that there was solid ice on the evaporator coil. In fact, there was so much ice that it was pushing the metal of the coil around. It was so thick, the fan blade on the blower was hitting the ice. You found the box cool, but not nearly cool enough, since the fan was blowing air over ice; and even so, most of the air flow was completely blocked off. You would shut the pump down by pulling the switch. You would open the doors on the reach-in, and with the fan blowing, the ice would slowly melt off the evaporator. You may have to use a torch to speed up the process. With a clean evaporator you are ready to adjust the pressure control so that this will not happen again. You attach your gauges, start the unit up, and note that it has plenty of gas. Apparently, there is nothing wrong except the LP control was way off its proper setting. You could determine what its old setting was by simply closing in (front seating) on your suction line service valve until the unit begins to pump down. As soon as your gauge pressure starts falling, and the cut off point is reached, the unit will stop. Now, open the suction line service valve slightly and let the pressure come back into the pump again. You can watch your gauge carefully and get the reading where it is going to start again. This is much like setting the cut-in and cut-off switch on a water well system using a storage tank. Suppose you found that the unit pulled down to 12 psi, and then cut off and came back up to 28 psi, and cut back in again. Right there you could look on your chart and see that on the low side, including the evaporator, the pressure did not come up high enough to clear the ice off the coil. See that 28 psi F-12 equals thirty degrees, and ice melts at thirty-two degrees. As soon as you change the setting so that the unit comes up to between thirty-three and thirty-six psi before it comes on again, you can look on your chart and see that this will be above freezing. With the blower moving air over the evaporator, the evaporator will defrost before the unit starts again. The pressure on the low side where you have your gauges and the LP control connected will at all times be the same low side pressure that is in the evaporator coil.

Where do you want the unit to cut out? The answer is wherever it happens to be when the job is pulled down. For instance, suppose the owner of the box wants the box to stay at about thirty-six degrees. You could say to yourself, "The evaporator coil refrigerant temperature will be about 30° colder than the air temperature. So subtract 30° from 36° and you have a near 6° refrigerant temperature." Using the pressure-temperature chart, convert 6° to pressure. That would be about 12 psi, so the old setting was all right. This is *not* the best way to set the cutout point of a pressure control. It is simply the way the designer of the equipment expected it to work; and in many cases it will work just as he wanted it to, at those exact pressures. However, as has been stated, some pumps are harder pumpers than others; and pressure drop and superheat will vary; so this method of determining the pressure setting is not perfect. It is just a guide. The best practical method is the simplest. That is to screw the control down colder than you want the box ever to go pressure-wise and put your thermometer inside of the box. When the box gets down to the thirty-six degrees which the proprietor wants, you would simply go to the pressure control and back it up until the unit turns off at this temperature, regardless of the pressure, which might vary from the chart pressure by several pounds. But you will know that whatever the pressure, the box will always pull down to thirty-six degrees; and at that pressure on the low side (whatever

Low Pressure Control

High Pressure Control

(Fig. 4-10)

it is) the unit will cut off; and it will not come back on until the pressure is exactly at that point which has been set for the control to kick on. This cut-back in pressure should always be high enough that there will be no ice left on the coil.

Don't try to master the pressure control and its relationship to the defrost cycle in one lesson. I will lay down a hard and fast rule now, and we will constantly be referring again and again to our job of setting the common pressure control. Here is the rule; memorize it: "I will always set my cut-in pressure so that it is above freezing. This pressure setting will take care of 99 percent of the LP controls in the world. For F-12 it should be set at 34 to 36 psi gauge pressure at the service valve where my gauges are attached. I will never set the cut-out pressure until the box is pulled down to the desired temperature on my thermometer. While the unit is running at this temperature, I will back up on the cut-out point until the unit cuts off; and the pressure when it cuts off does not matter as long as that pressure gets the job done." The 34 to 36 psi is standard the world over for the cut-in point for F-12 simply because the temp-press chart is always true for a particular refrigerant, and the LP control will be more accurate than any other type control, if we assume the purity of the Freon. You could even set the cut-in point on a new pressure control in your shop by using your gauges and the pressure in a service cylinder. You could do this before you hooked the control up. But the cut-out point would be where the load was pulled down and no other.

Don't worry about these pressure controls for many of you may never do commercial work. But it is nice to know and understand. In time- you will understand the advantages and accuracy of the pressure control over the thermostat, which is a crude control device compared to pressure control and your temp-press chart.

Many commercial units in addition to LP controls have a HP control. This high pressure control is attached to the high side as a safety device to kill the unit in the event of extreme high pressure on the high side. Package units are almost always thermostatically controlled with a thermostat in the air stream. An LP control would serve no useful purpose on a package air conditioner because frost is no problem. The coil never frosts, unless the unit is running out of gas or the filter is stopped up. Many package units have, in addition to thermostats, high pressure controls to protect the condenser and receiver against high pressure in case the condenser cooling agent should fail.

■

SATURATED VAPOR PRESSURE TEMPERATURE CHART
FOR FREON-12 AND FREON-22

Temp. °F	Gauge Pressure (PSI.)		Temp. °F	Gauge Pressure (PSI.)	
	Freon-12	Freon-22		Freon-12	Freon-22
—40	10.92*	0.61	60	57.71	102.5
—38	9.91*	1.42	62	60.07	106.3
—36	8.87*	2.27	64	62.50	110.2
—34	7.80*	3.15	66	64.97	114.2
—32	6.66*	4.07	68	67.54	118.3
—30	5.45*	5.02	70	70.12	122.5
—28	4.23*	6.01	72	72.80	126.8
—26	2.93*	7.03	74	75.50	131.2
—24	1.63*	8.09	76	78.30	135.7
—22	.24*	9.18	78	81.15	140.3
—20	.58	10.31	80	84.06	145.0
—18	1.31	11.48	82	87.00	149.8
—16	2.07	12.61	84	90.1	154.7
—14	2.85	13.94	86	93.2	159.8
—12	3.67	15.24	88	96.4	164.9
—10	4.50	16.59	90	99.6	170.1
—8	5.38	17.99	92	103.0	175.4
—6	6.28	19.44	94	106.3	180.9
—4	7.21	20.94	96	109.8	186.5
—2	8.17	22.49	98	113.3	192.1
0	9.17	24.09	100	116.9	197.9
2	10.19	25.73	102	120.6	203.8
4	11.26	27.44	104	124.3	209.9
6	12.35	29.21	106	128.1	216.0
8	13.48	31.04	108	132.1	222.3
10	14.65	32.93	110	136.0	228.7
12	15.86	34.88	112	140.1	235.2
14	17.10	36.89	114	144.2	241.9
16	18.38	38.96	116	148.4	248.7
18	19.70	41.09	118	152.7	255.6
20	21.05	43.28	120	157.1	262.6
22	22.45	45.53	122	161.5	269.7
24	23.88	47.85	124	166.1	276.9
26	25.37	50.24	126	170.7	284.1
28	26.89	52.70	128	175.4	291.4
30	28.46	55.23	130	180.2	298.8
32	30.07	57.83	132	185.1	306.3
34	31.72	60.51	134	190.1	314.0
36	33.43	63.27	136	195.2	321.9
38	35.18	66.11	138	200.3	329.9
40	36.98	69.02	140	205.6	338.0
42	38.81	71.99	142	210.3	346.3
44	40.70	75.04	144	216.1	355.0
46	42.65	78.18	146	222.0	364.3
48	44.65	81.40	148	227.4	374.1
50	46.69	84.70	150	232.3	384.3
52	48.79	88.10	152	238.8	392.3
54	50.93	91.5	154	244.0	401.3
56	53.14	95.1	156	251.0	411.3
58	55.40	98.8	158	257.2	421.8
			160	263.2	433.3

* Inches Vacuum.

FREON-12 CONDENSER CHART

Typical Head Pressures and Air OFF Temperatures at Different
Suction Pressures and Air ON Temperatures

Suction Pressure (PSI. Gauge)	Coil Temp.	Air Temperatures On Condenser																	
		75°		80°		85°		90°		95°		100°		105°		110°		115°	
		Head Press.	Air Off Temp.	Head Press.	Air Off Temp.	Head Press.	Air Off Temp.	Head Press.	Air Off Temp.	Head Press.	Air Off Temp.	Head Press.	Air Off Temp.	Head Press.	Air Off Temp.	Head Press.	Air Off Temp.	Head Press.	Air Off Temp.
30	32	121	89	130	94	138	99	147	104	156	108	166	113	175	117	186	122	197	127
32	34	124	90	132	95	140	99	150	104	159	109	169	114	178	118	189	123	200	127
33	36	126	91	134	96	142	100	152	105	161	109	171	114	180	118	192	123	203	128
35	38	128	92	136	97	144	101	154	106	164	110	174	115	184	119	195	124	207	129
37	40	131	92	139	97	147	102	157	107	167	111	177	116	187	120	198	125	209	129
39	42	133	93	141	98	150	102	160	107	170	111	180	116	189	120	200	125	211	130
41	44	136	94	144	99	153	103	163	108	173	112	183	117	192	121	203	126	213	130
43	46	138	94	146	99	155	103	165	108	175	112	185	117	194	121	205	126	216	130
45	48	140	95	148	100	157	104	167	109	177	113	187	118	197	122	208	127	218	131
47	50	142	95	151	100	160	104	170	109	179	113	189	118	199	122	210	127	221	132
49	52	145	96	154	101	163	105	173	110	182	114	192	119	202	123	213	128	224	132
51	54	148	97	157	101	165	105	175	110	185	115	195	120	205	124	216	128	226	132
53	56	150	97	159	102	168	106	178	111	188	115	198	120	208	124	218	129	228	133

FREON-22 CONDENSER CHART

Typical Head Pressures and Air OFF Temperatures at Different
Suction Pressures and Air ON Temperatures

Suction Pressure (PSI. Gauge)	Coil Temp.	Air Temperatures On Condenser																	
		75°		80°		85°		90°		95°		100°		105°		110°		115°	
		Head Press.	Air Off Temp.	Head Press.	Air Off Temp.	Head Press.	Air Off Temp.	Head Press.	Air Off Temp.	Head Press.	Air Off Temp.	Head Press.	Air Off Temp.	Head Press.	Air Off Temp.	Head Press.	Air Off Temp.	Head Press.	Air Off Temp.
58	32	205	92	219	97	232	101	246	106	260	110	275	115	290	119	307	124	324	128
60	34	209	92	222	97	235	101	249	106	263	111	278	116	294	120	311	125	328	129
63	36	212	93	226	98	239	102	253	107	267	111	283	116	298	120	314	125	330	129
66	38	216	94	232	98	245	102	258	107	274	112	290	117	305	121	319	126	333	130
69	40	221	95	235	99	249	103	263	108	278	112	292	117	307	122	323	126	339	130
72	42	226	96	239	100	253	104	267	109	281	113	295	118	310	123	326	127	342	131
75	44	230	96	242	100	257	104	272	109	286	114	300	118	316	123	332	127	348	131
78	46	234	97	247	101	262	105	276	110	290	114	304	119	320	124	336	128	352	132
82	48	237	97	252	101	266	105	280	110	294	114	309	119	325	124	340	128	355	132
85	50	242	98	257	102	271	106	285	111	299	115	313	120	329	125	344	129	359	133
88	52	249	99	262	103	276	107	290	112	304	116	317	120	333	125	348	129	363	133
91	54	254	100	266	104	281	108	295	112	309	117	322	121	338	126	353	130	368	134
95	56	259	100	270	104	285	108	300	113	314	118	327	122	342	126	356	130	370	134

APPROXIMATE SETTINGS ON THE
LOW PRESSURE CONTROL (R-12)

TYPE OF EQUIPMENT	CUT-IN SETTING	CUT-OUT SETTING
Walk-in Meat Cooler	38 psi	12 psi
Walk-in Dairy Cooler	38 psi	14 psi
Walk-in Produce Cooler	38-40 psi	16 psi
Walk-in Florist Cooler	40-42 psi	18 psi
Open Meat Display Case	26 psi	9 psi
Closed Meat Display Case (Gravity Air Evaporator)	24 psi	4 psi
Closed Meat Display Case (Forced Air Evaporator)	32 psi	12 psi
Open Dairy Display Case	30 psi	9 psi
Open Produce Display Case	38 psi	16 psi
Medium Temperature Reach-in	38 psi	12 psi
Beverage Box (Soft Drinks, Beer, etc.)	38 psi	12 psi
Air Conditioner (central) (Safety cut out only)	120 psi	54 psi

TYPE OF CASE	REFRIGERANT 12		REFRIGERANT 22		REFRIGERANT 502		APPROX. TEMP.
	SUCTION PRESSURE	HEAD PRESSURE	SUCTION PRESSURE	HEAD PRESSURE	SUCTION PRESSURE	HEAD PRESSURE	
Walk-in Meat Cooler	12 psi	150-175 psi					34°F
Walk-in Dairy Cooler	12-14 psi	150-175 psi					36°F
Walk-in Produce Cooler	18-20 psi	150-175 psi					45°F
Walk-in Freezer	0 psi	130-150 psi	5 psi	225-250 psi	15 psi	280 psi	0°F
Open Type Freezer	0 psi	130-150 psi	5 psi	225-250 psi	15 psi	280 psi	0°F
Ice Cream Case	5" Vacuum	120-150 psi	0 psi	200-225 psi	5 psi	260 psi	-15°F
Open Meat Display Case	6-9 psi	140-160 psi					32°F
Closed Meat Display Case (Forced Air Evaporator)	12 psi	150-175 psi					34°F
Closed Meat Case (Gravity Air Evaporator)	6 psi	130-150 psi					34°F
Open Produce Display	16-18 psi	150-175 psi					45°F
Open Dairy Display	9 psi	150-175 psi					36°F
Beverage Box	12 psi	150-175 psi					34°F
Reach-in Cooler	12 psi	150-175 psi					34°F

These pressures are approximate and will vary with outside temperatures. If the outside temperature is low, the operating pressures will be lower than normal. If the outside temperatures are unusually high, the operating pressure will be higher than normal. Varying the controlled temperatures will also vary the operating pressures.

TROUBLE SHOOTING GUIDE

COMPRESSOR OPERATES BUT WILL NOT COOL

CAUSE	REMEDY	PRESSURE REACTION
1. System low on gas.	Repair leak and recharge.	Head pressure and suction pressure low.
2. Air in system.	Pump all refrigerant into high side of system. Let condenser cool to room temperature. Bleed air from discharge service valve.	Head pressure high. Suction pressure higher than normal.
3. Moisture.	Replace drier.	Suction pressure low. Head pressure low.
4. Restriction in low side (drier, expansion valve, etc.).	Replace drier if restricted. Check strainer in expansion valve.	Suction and head pressure low.
5. Bad valves in compressor.	Replace valve plate if compressor head is exposed.	Suction pressure high. Head pressure low.
6. Evaporator coils iced over.	Remove ice and determine why evaporator iced over.	Suction pressure low. Head pressure low.
7. Evaporator fan motor off.	Determine why fan motor is off. Replace if needed.	Suction pressure low. Head pressure low.
8. High head pressure.	Clean condenser. Compressor location may be too hot. Ventilate equipment room.	
9. Restricted or partially restricted expansion valve.	Replace expansion valve.	Suction and head pressure low.
10. High heat load.	Determine why head load is high.	Suction and head pressure high.
11. Condenser fan motor off.	Determine why fan motor is off.	Head pressure extremely high.
12. Compressor too small or wrong application.	Take model and serial number of compressor. Check manufacturer's specifications.	Suction pressure high.

CAUSE	REMEDY	PRESSURE REACTION

COMPRESSOR LOSES OIL AND WILL NOT RETURN

CAUSE	REMEDY	PRESSURE REACTION
1. Low head pressure.	Use mean to maintain normal head pressure.	
2. Compressor valves bad in compressor.	Replace valve plate and oil will return after a period of operation.	Suction pressure high. Head pressure low.
3. System operating low on refrigerant.	Repair leak, recharge, and oil will return after a period of operation.	Suction pressure low.
4. Evaporator iced over.	Remove ice. Determine why ice formed. Oil will start returning to compressor after ice is removed.	Suction pressure and head pressure low.
5. Improperly sized suction and liquid line.	Check specifications.	Suction pressure low if lines are undersized.
6. Oil leak.	Repair oil leak and add oil to proper charge.	No affect on operating pressures.

SUCTION LINE ICES BACK TO COMPRESSOR

CAUSE	REMEDY	PRESSURE REACTION
1. Iced evaporator.	Remove ice and correct cause of evaporator icing.	Suction pressure low.
2. Expansion valve overfeeding.	Size expansion valve properly or replace defective valve.	Suction pressure higher than normal.

COMPRESSOR SHORT CYCLES

CAUSE	REMEDY	PRESSURE REACTION
1. Control differential set too close.	Widen differential between cut-in and cut-out.	Suction pressure drops, head pressures rise then control brings compressor on. Suction pressure rise. Head pressure drops when control opens.
2. Discharge valves in compressor leaking.	Replace valve plate.	Pressure rises rapidly on off cycle.

CHAPTER 5

BASIC PRINCIPLES
OF ELECTRICITY

You are not supposed to be an electrician, and you are not expected to do electrical work outside of the controls and motor starting devices on your equipment. Stay out of the other man's trade. Unless you are already doing electrical contracting, stick strictly to your own trade. When you have electrical work or trouble in the electrical service in a place of business or home, call the customer's electrician or your own. By doing this, you will gain a good reputation among the other crafts and the electricians in your community. They will send you new customers in turn. Again, stay out of the electrician's field of work. Not only can it involve you in work that you may not be equipped to handle, but you may run into legal restrictions, which would make it very difficult for you to collect your money if there were the slightest argument. You will find that most top refrigeration servicemen draw the line very quickly as to how far they will go on electrical repairs. In many cases, these men will work no further than where the electrical service leaves the unit. You will never lose any face by asking the owner who his electrician is and getting him out on the job if you need him. If the owner has no electrician, suggest your own man. Remember, the electrician that you call will probably do the same for you if one of his customers needs a refrigeration serviceman.

You may use your own judgment in this matter, but I strongly advise you to stick strictly to your refrigeration equipment. The owner of this equipment will respect you for using good judgment where special work has to be done.

All right, if you are going to stay out of the electrical contracting business, how much electricity do you have to know in order to carry on your own work? The answer is, very little. In most cases, you should know at least enough to know when to call an electrician.

Here is the extent of electrical knowledge you should know:
1. What kind of voltage runs the equipment, 120 V, 240 V, and/or 3 phase.
2. Whether the voltage is getting to the equipment or not.
3. Whether the voltage is of sufficient strength.
4. Whether there is a fuse blown or not.
5. Where the unit disconnect switch is located.
6. What a line starter is and its function.

In addition to a general knowledge of the above, you will be required to be an expert on the following devices:
1. High pressure, low pressure and thermostatic controls.
2. Electric motor starting devices.
3. Capacitors, both run and start.
4. Relays.
5. Thermal overloads and heaters.

To know and understand these devices, it will be necessary that you learn a few simple tests and have an understanding of the volt, amp and ohm meter. The volt-amp combination meter will be used to check voltage and amperage and help solve electrical problems.

Beginning with our first list of the general knowledge of electricity, we will review some basic electricity. You will find that you already know most of this, but I want to be sure that you have it straight in your mind. Those of you who are electricians and do electrical work, just bear with us, because we are going to start at the very bottom of elementary electricity as though none of us knew any more than the other. Keep in mind at all times, that we are not studying to be electrical engineers, nor are we going to study electricity outside of the area in which we are going to be working as servicemen. We will confine ourselves to service work that is absolutely essential that a competent serviceman understand in order to trouble shoot refrigeration equipment.

Let's begin by taking a service call in a home. We arrive in a truck, and get our tools and start up to the house. Glance up at the electrical service leading from the pole to the house. You see two wires. What can you say to yourself about the electrical service available to this house? You can say that there are only 120 volts in the house. There can't be anything but 120 volt service to the house. In one wire there is 120 volts and the other wire is a ground. The ground wire is part of the whole electrical system in the house. The switch box, conduit and weather-head are all tied to this ground wire; and if you had not noticed, the ground wire on the pole belonging to the utility company is grounded to a rod driven into the ground by every other pole back to the power house. These rods are driven into the ground. There is a ground rod driven into the ground right at the base of the meter loop on the house, also.

To operate any appliance or device of 120 volts, all you need is one hot wire equal to 120 volts. The other wire is a ground, and it stays grounded at all times. The circuit is the whole electrical system in the house. When a circuit is closed, it means that the hot wire (120 V) has made contact with the ground wire somewhere, perhaps inside of a light bulb or an iron. (See **Figures 5-1 and 5-2**.) When this circuit is closed, the resistance starts electrically; and heat results in the iron and light by white heat of the bulb element. The switches in the house break the circuit between the ground and the hot wire.

Now, suppose instead of two wires you had seen three wires leading to the building. You could say right then and there, that there were 240 volts and 120 volts available to the house or building. There would be 120 volts in two of the wires, and still you would have the old ground. Now, the ground would still work with one of the 120 volt lines to make 120 volt service to light bulbs or other appliances. There would also be available 240 volts, usually for big appliances like ranges, window units or electric heaters. Now, remember the 120 meets with the ground in an appliance, and there you have the marriage or circuit closed for 120 volts. You do not need a ground in order to close a circuit on two 120 volt lines making 240 volts in an appliance. These two 120 volt lines are kissing cousins. All they have to do is meet and

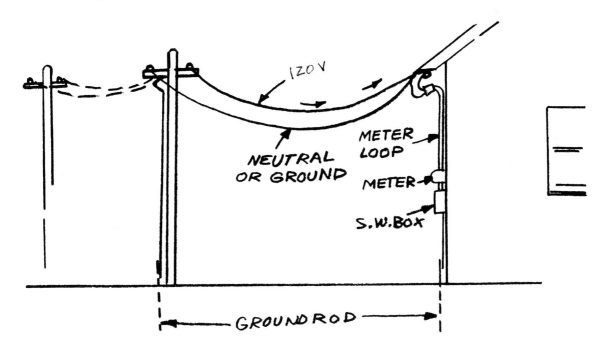

(Fig. 5-1)

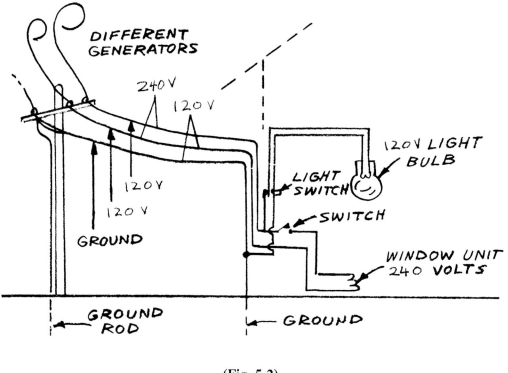

(Fig. 5-2)

wham! You have 240 volts. When I say kissing cousins, I mean that they are not from the same family or generator, or at least they are not out of the same phase of a generator. When they meet, they burn, just like one line meeting a ground and burning. The third wire you have seen running along with the two hot 120 volt wires comprising 240 volts is simply (in the case of a 240 volt appliance) a safety device. The national code and many of the underwriter specifications require that where 240 volt motors are used on appliances, they must be grounded. Look at the sketch of a polarized cord, plug, and the ground or third wire in **Figure 5-3**.

You could get the same effect by grounding your window unit or range to a water line in the house, but it is much neater and cleaner to run this ground wire in the cord itself. The reason the plug on all of these three-wire cords is crazy looking, is so that the ground will always go in the right hole of the receptacle. I repeat. You do not need a ground to run a 240 volt motor. The ground is a safety device to keep the operator from being shocked in case of a short. In the case of a 120 volt motor, the ground serves two purposes. First, it is part of the 120 volt circuit. Second, it is a natural ground to the motor or appliance.

All window units have this third wire if they are 240 volt units, although this third wire, the ground, just bolts

to the frame of the unit to make sure that the whole unit is grounded. Never do they take one side of the line, as it is called, to make a 120 volt circuit for something other than the main motor. In the case of electric ranges, they do use one of the 120 volt lines together with a ground to operate a 120 volt attachment to their range, like a clock or light. Here is a simple schematic, where the 240 volt service is tapped for 120 volts.

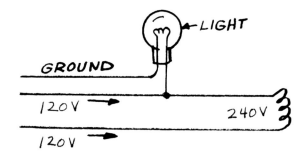

Suppose you had seen four wires coming from the pole to the house. You could say that there was probably three-phase voltage available to the building or house. In the case of a big home or place of business, you should not be surprised to find these four wires leading to the meter loop.

76

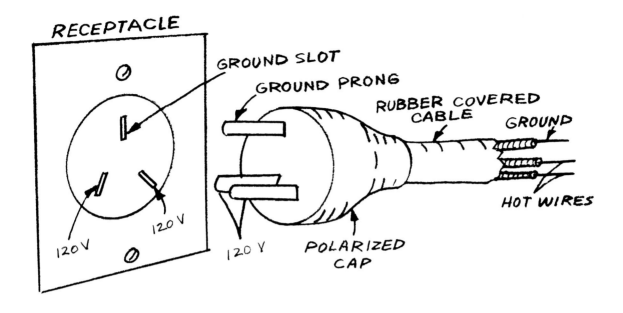

RECEPTACLE

GROUND SLOT

GROUND PRONG

RUBBER COVERED CABLE

GROUND

HOT WIRES

POLARIZED CAP

120 V 120 V 120 V 120 V

(Fig. 5-3)

You can say that this house has 120 volt, 240 volt, and three-phase service. Out of these four wires, you get all of these combinations of voltage. There has been a trend to install more and more three-phase in residential buildings in recent years, and many of the central combination jobs are three-phase installations. Take a look at the four wires coming to this house and see the combinations that can be made up from the service available. See **Figure 5-4**.

Note that one of these wires is 208 volts or sometimes called the stinger or high leg. You cannot tap this line together with a ground and get 120. It will be 208 to ground and would burn out a 120 V light bulb if used as a 120 volt combination. See **Figure 5-5**.

Figure 5-6 shows an improper take off on a stinger. **Figure 5-7** shows a correct three phase tap for 120 and 240 volts.

You will never find higher voltage than 240 three-phase except in an industrial system. Remember that the modern three-phase you will work with consists of two 120 volt wires and one 208 volt wire, the stinger. No ground is required to run a three-phase motor. Where there is a ground on three-phase equipment, it is used as a safety device to ground the equipment. The same is true with 240 volt service to a unit or any appliance. All of these different possible combinations of service available by the utilities are used in your work. The licensed electrician does not go further than the meter loop on the house. He stays in his own backyard and

calls the electric company to handle any repairs or trouble shooting on the service running from the pole to the house. Likewise, you do not tamper with the electrical system within a building unless it is inside your equipment. You simply want to know what you are working with and nothing more. We can lay it down as a hard and fast rule that you are never likely to work on any more combinations of service than have been mentioned here, unless it is an industrial job. Industrial work might go to 480 volts three-phase. Just remember that 480 volts three-phase can operate a thousand horsepower motor, so you are not likely to run into any higher voltage than 120, 240 and 240 three-phase. All electrical service running to a building has a meter loop and a service entrance switch in conjunction with the meter. The meter may be on the outside of the house or garage wall and the switch can be in back of the meter on the inside of the garage wall. The wires from the meter run through a nipple joining the meter and box through the wall. See **Figure 5-8**.

You are probably asking yourself, "Why does one three phase circuit show 120 V on each of the three hot legs and come up with 240 volts and 208 volts on another?" It has to do with the type of transformer on the utility pole. The utility company installs the transformer they think is most suitable for the job. Your only concern is that the voltage is correct and to expect it to be the way it is, as long as there is no harm to any of the equipment. I am going to tell you this now and then we

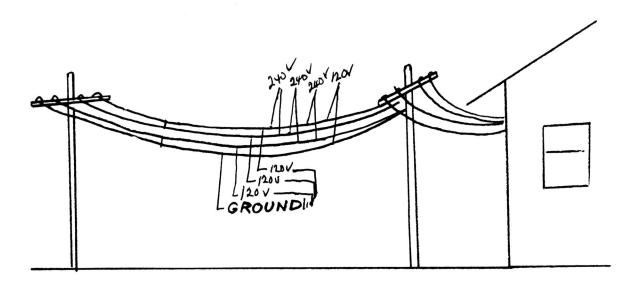

(Fig. 5-4)

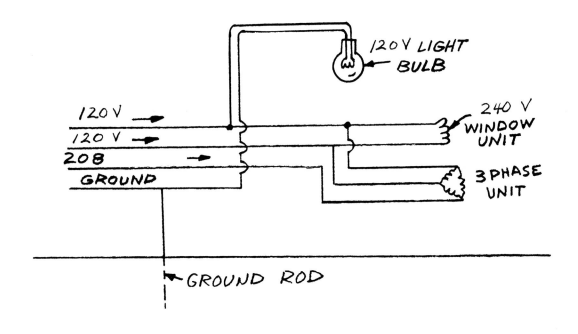

(Fig. 5-5)

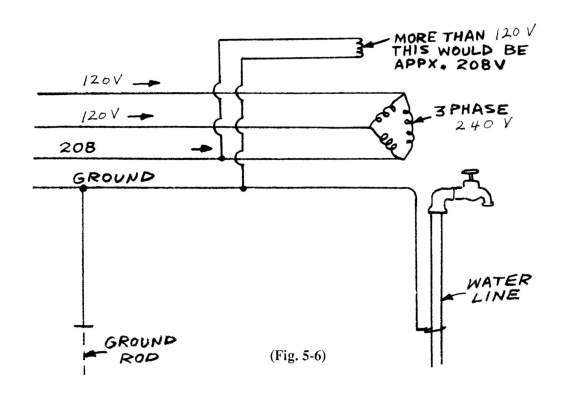

(Fig. 5-6)

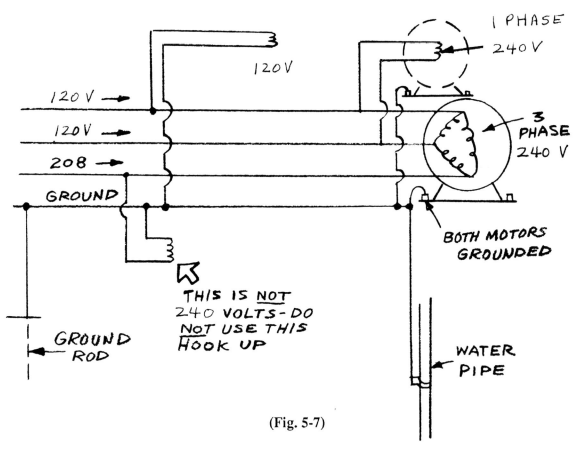

(Fig. 5-7)

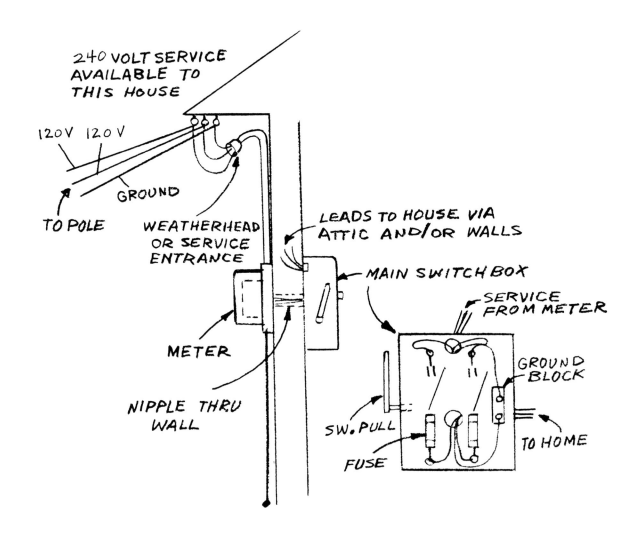

240 VOLT SERVICE AVAILABLE TO THIS HOUSE

120 V 120 V

TO POLE

GROUND

WEATHERHEAD OR SERVICE ENTRANCE

METER

NIPPLE THRU WALL

LEADS TO HOUSE VIA ATTIC AND/OR WALLS

MAIN SWITCH BOX

SERVICE FROM METER

GROUND BLOCK

TO HOME

SW. PULL

FUSE

(Fig. 5-8)

will drop it. You have Wye-type, delta, open delta, closed delta and Wye-star power transformers. All of them will give you various electrical characteristics. All you need to know is the voltage as it should be coming to your equipment, that is all. Notice that the drawing in **Figure 5-9** has three phase 208 V. This voltage is more popular with large building lighting services. However, refrigeration and air conditioning equipment can be run on this circuit. If a motor is specifically rated for 240

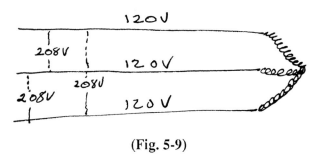

120 V

208 V

120 V

208 V

208 V

120 V

(Fig. 5-9)

volts, then do not put it on a 208 volt circuit. The applied voltage must be within 10% of rated voltage of the motor.

Don't be alarmed if you see twisted wires running from the pole to the meter loop. There has been a trend by the utility companies to use a bare steel cable from pole to house, which serves as a ground and at the same time carries the weight of the conductors. That is, the insulated conductors are wound around the bare ground supporting cable.

When the juice leaves the entrance switch, it may scatter all over the building and run into a half dozen more switch boxes. We call these latter subswitches, or disconnects to the appliances such as air conditioners, ranges and heaters. **Figure 5-10** is a simple sketch of a three-phase installation and the various possibilities of its use with several disconnect boxes.

You will note that the stinger reads 208 to ground.

80

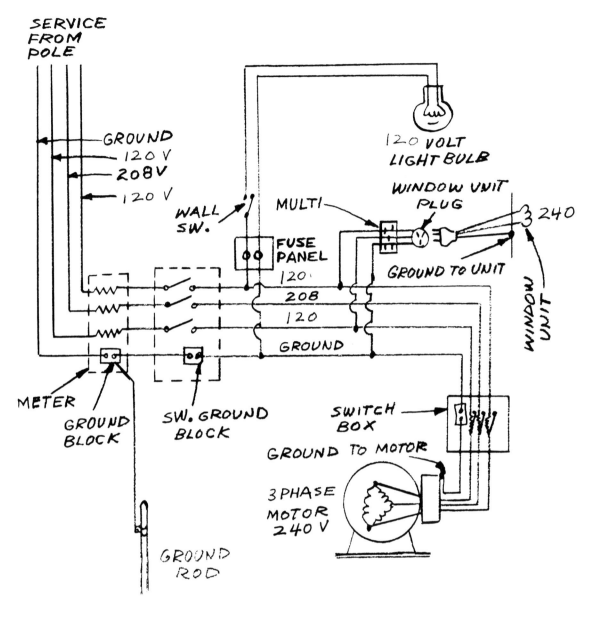

SERVICE
FROM
POLE

GROUND
120 V
208V
120 V

120 VOLT
LIGHT BULB

WINDOW UNIT
PLUG

240

WALL
SW.

MULTI

FUSE
PANEL
120
208
120
GROUND

GROUND TO UNIT

WINDOW UNIT

METER

GROUND
BLOCK

SW. GROUND
BLOCK

SWITCH
BOX

GROUND TO MOTOR

3 PHASE
MOTOR
240 V

GROUND
ROD

(Fig. 5-10)

Now, if you were reading across any two of the leads in the box, you would always read 240 volts on all combinations. For instance, 120 and 120 will read 240; and the 120 and stinger (208) will read 240. This is the phenomena of the delta transformer connections used by the power company. Don't worry about this. Just be sure the voltage is as it is supposed to be to the motors and so forth.

Now, let's review the service available to buildings or homes. Two wires equal 120 volts only, and the wires will be 120 volts and a ground wire. Three wires equal two 120 volt wires and one ground, and you can have both 120 volts and 240 voltage. You do not need a

ground to any 240 volt motor. The two different 120 volt lines collide, and you have 240 volts. One leg or hot wire, together with a ground wire equals 120 volts. Three-phase is carried in three wires and the fourth wire is the ground. You will have the possibility of using 120 V, 240 V, and/or three-phase. Just forget about any other three-phase, like 480; this is real big stuff. The service from the utility pole, whether it be two-wire 120 volts or four-wire three-phase, enters the meter loop through a weather head or service entrance fitting. The lines then run down the conduit to the meter box or base and out of the meter box to the main switch box. They go out of the main switch box to whatever place electricity is needed. It may go in several different directions and may feed several different types of appliances or motors. See **Figure 5-11**.

The big old service entrance switch is on its way out; yet the new boxes are still switches, and the subswitches may be multi-breakers or heater overloading type protection devices instead of a series of plug fuses. In most cases, the disconnect box will be a simple hand type switch with a lever to pull in order to break the circuit.

It is important that you know the stinger leg (208 volts) is in the middle of the three-phase circuit. See **Figure 5-12**. However, should you be testing for a burned out fuse using a light bulb, you could blow it out if you were using a 120 bulb. For this reason, whenever you use a volt-ammeter (**Fig. 5-13**) or light bulb on three-phase, remember that there is the possibility of encountering 240 V. Each separate wire of three-phase will read its voltage to ground.

Where window units are concerned, there may only be a wall outlet. Where package units or commercial condensing units are concerned, there will be another switch for the individual piece of equipment, or at least a cord that can be pulled. This switch or cord is installed so that the unit can be disconnected without cutting off everything in the building. From this switch you will

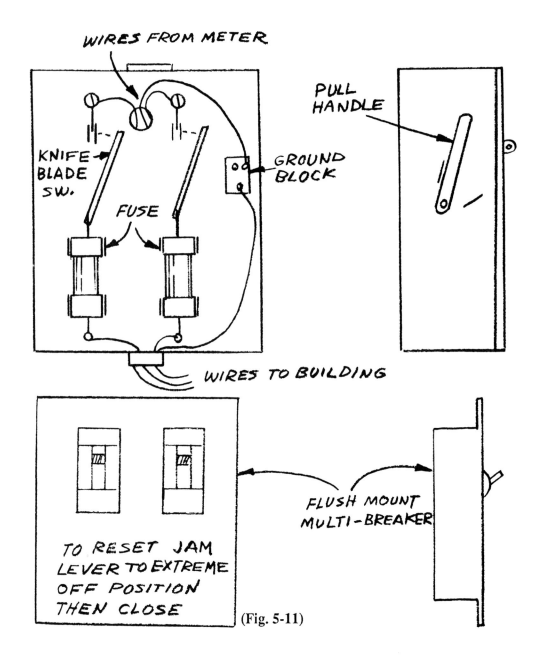

WIRES FROM METER

PULL HANDLE

KNIFE BLADE SW.

GROUND BLOCK

FUSE

WIRES TO BUILDING

TO RESET JAM LEVER TO EXTREME OFF POSITION THEN CLOSE

FLUSH MOUNT MULTI-BREAKER

(Fig. 5-11)

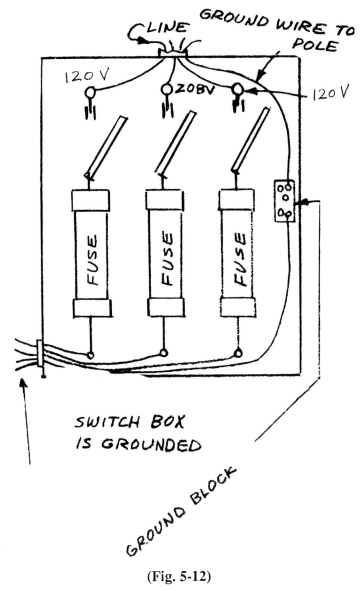

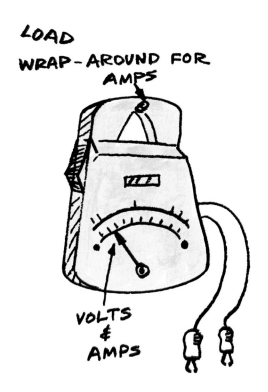

where they attach to the cap prongs. If the cap looks clean and does not show arcing or burned places and the screws are tight, you are ready to make a quick test to see if voltage is coming to the receptacle.

One simple method is to use two 120 volt light bulbs in series in a pig tailed weather proof socket. **(Fig. 5-14)** Stick the two bare wire ends of the pig tail in the receptacle slots; and if the light bulb burns with full brightness, everything is all right to this point. If you have a dull light you will know you only have 120 volts. You may also use the leads off your volt meter to do the same thing and read the voltage on the meter scale. Try each hot leg of the receptacle to the ground. You will read 120 V, or you will have the two 120 volt bulbs half lit.

Look at the test using both lamp and amp-meters as shown in **Figure 5-15**.

Suppose your bulbs will not burn, or you do not get a reading on the meter. Ask where the switch box or breaker panel is for the house. You may look for it if you have no one to ask. Look in the garage or out by the meter loop. In a modern home you will find the circuit breaker or multi-breaker in the kitchen, hallway, garage

(Fig. 5-12)

have wires leading down to your unit or to a line starter. A line starter is sometimes called a magnetic starter. Line starters are generally used on units of 1 hp or larger. From this line starter, you will have service to your pump motor and fans. The line starter is operated by the controls. We will devote more time to a line starter later on. Right now, let's learn a simple test for determining whether there is something wrong between the box by the unit and the main switch box. Likewise, we can learn to test the receptacle in the wall into which a window unit plugs into. Making the window unit test first, we will proceed to find out if there is ample voltage available to a unit that has gone dead. First, we will pull the three-pronged cap out of the wall receptacle. Take a look at the cap to see if the wires are all attached to the screws in the cap. Look for burned off ends of the wires,

(Fig. 5-13)

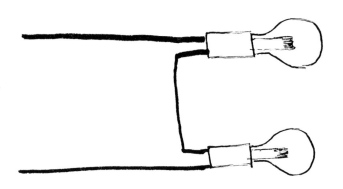

or out by the meter base. One place or another there will be a fuse panel or switch box, whether it be the main switch box at the meter loop or a sub-switch box or panel with several multi-breakers somewhere about the premises. (Fig. 5-16)

Many times a multi-breaker (a resetting device) will have kicked out and just needs resetting. Sometimes an extremely hot day may simply overload the circuit, and it may not kick out again after you reset it.

After you reset the switches that apparently feed the unit, check again with your bulb or meter. If you now have the required voltage, then proceed to find out why the fuse or breaker is out. Right now we just want to

Series hook up. Lights burn bright on 240 V and dull on 120 V.

(Fig. 5-14)

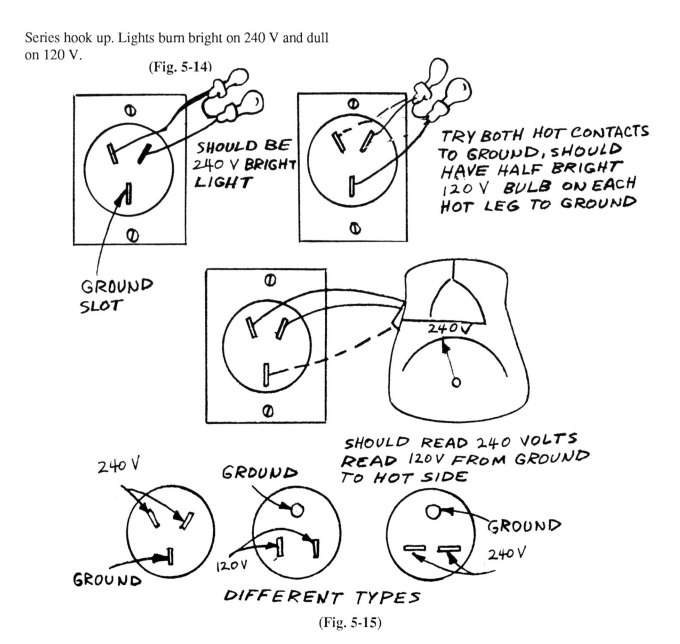

DIFFERENT TYPES

(Fig. 5-15)

84

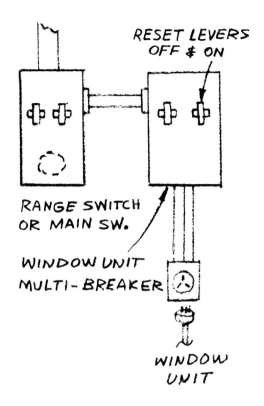

RESET LEVERS
OFF & ON

RANGE SWITCH
OR MAIN SW.

WINDOW UNIT
MULTI-BREAKER

WINDOW
UNIT

(Fig. 5-16)

check services available to the equipment. Later, we will find out why breakers kick out. Suppose you have a commercial or package unit with a large switch box mounted near the unit, where you can easily check for supply voltage. This type of box will have cartridge fuses or plug fuses, and you can make a quick visual check on a plug fuse. Just look in the glass window to see if the fuse link is parted. Where cartridge fuses are used, you can never make a mistake by replacing them with new fuses. If you do not have enough new fuses on hand for a complete change, take the old ones out by pulling them and test them for continuity. Don't complicate the job by trying to test fuses in the box. There may be feed back. The fuse will appear to be good when it is burned out. Here is a simple foolproof test for fuses out of a box. Make up a test cord with a lamp, plug it in somewhere, and test each fuse for continuity. See **Figure 5-17**.

Many servicemen waste valuable time fooling around trying to determine whether a fuse is good, by testing it in the box. Don't waste time. Pull it out and make absolutely sure whether it is good or bad. There are several good fuse pullers made out of insulated material which will cost you only a few dollars. Do not use pliers or screwdrivers around fuse boxes unless you are absolutely sure that you know which side of the fuse is hot and which side is dead. That is, don't short the circuits or shock yourself fighting a fuse. I quit fighting fuses years ago. I just pull them out and test them or replace them. Suppose you found that all fuses were good in the box by the unit, but you are still suspicious that there is no juice to the unit, or that one side of the line is out. In this case, take your light bulb and pigtail socket combination and touch one pigtail end to the prongs of the switch box (**Fig. 5-18**) that seems to be attached to the service leading from the main meter box. You may be able to see where these wires are screwed to the switch mechanism. If so, just touch there. Now, touch the other pigtail to the box itself, the conduit around the box, or any good ground. If the circuit is good, that is, if there is voltage available in that wire, the bulb or your meter will indicate it. Suppose you found a dead wire running to one of the fuse connections. Then you could move out to the main switch box and make the same test there, even though it might cut the service off to the whole building for a few moments. Remem-

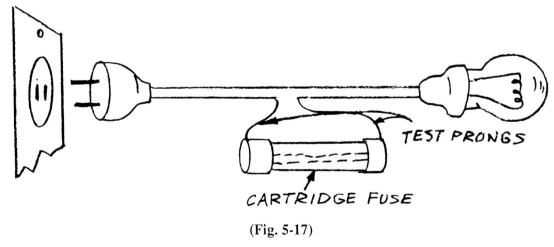

TEST PRONGS

CARTRIDGE FUSE

(Fig. 5-17)

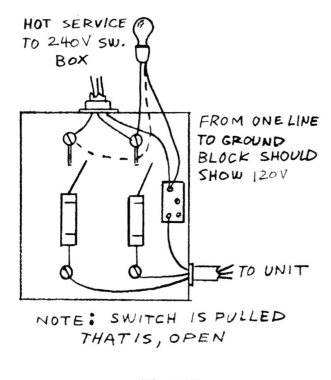

HOT SERVICE
TO 240V SW.
BOX

FROM ONE LINE
TO GROUND
BLOCK SHOULD
SHOW 120V

TO UNIT

NOTE: SWITCH IS PULLED
THAT IS, OPEN

(Fig. 5-18)

ber, you do not have to do any of this. If you have discovered that there is no service to your box by the unit, you would be wise to call in an electrician right then and there. If the owner says go ahead and check the main box, you may use your own judgment. I would suggest that where the electrical system is very simple and plain like in a home or small place of business, that you test all the way back to the main switch box. In many cases the owner will know his own electrical boxes pretty well, and he will show you where the main box or any in-between boxes are located. In the case of a big installation, like a supermarket or place where there is a maze of wiring or many different commercial units operating out of big feeder circuits, do not waste time any further than the immediate switch box by your unit. If your test shows that voltage is not coming through to this box, call the electrician. You should know what kind of voltage is supposed to feed this box, and one way to determine this is to read the voltage rating on the unit that you are servicing. Another way is to look into the box by the unit, and if you see three fuses, you may be fairly sure that this is three-phase, and you should read hot to ground on three wires coming into the box. Namely, 120, 120 and 208. In the case of 240 single phase service, you will see two fuses and a ground block, and there will be three wires coming into the box. The two fused wires should read 120 volts each

before they enter the fuses in the box. In the case of 120, you will find one fuse and a ground block. The fused lead will read 120 and the ground is nothing but a ground. See **Figure 5-19**.

For the record, most window units of one hp or more, generally will be 240 V units. Verify this by looking for the rating on the unit. Look at the cord cap and see if it is three pronged. Window units under one hp are generally 120, but a few under one hp are 240. Read the rating on the unit. Look at the cord. The cord will be the old two-pronged ordinary cap if 120 volts. (Note: All modern units are equipped with three pronged 120 V cords.) You are not likely to ever find a three phase window unit. The three-phase equipment will be packaged and commercial units. This type of equipment will also be 240 voltage. See the rating on the motors or unit. Look into the switch box and see the service available, or you may test all systems with a volt-meter. Generally, all service is either on or off; and the tests that you make are to determine whether it is on or off. Where low voltage problems are concerned, don't worry yourself. If you see that the electrical service in a building is heavily overloaded, let the owner deal with an electrician. Your job is not to repair the electrical system. Your job is to see that it is there or not. If ordinary good judgment tells you that the whole building is overloaded, like if all the lamps dim when you try to start the equipment, tell the owner to get an electrician to find the trouble.

We have examined the service that you may be expected to be familiar with in your work as a serviceman. Do not give a thought to high voltage or special service installations. The service we have discussed here is probably all the electrical you may be called upon to look at in ten years as a refrigeration serviceman.

It will be time to say right here that you can be the best refrigeration serviceman in the business and never need know how to wire the most simple floor lamp. It is not what you know that is important. It is how you apply whatever you know that counts. Let's apply what we have been studying in this chapter. Starting with a call to a home to fix a household refrigerator, you get into the kitchen and find the box and set your tools down. You open the box and find that it has defrosted and is apparently not running at all. The light inside of the box is off also. You can back away from the box and say to yourself right now that this box either has something radically wrong with it, or there is no electricity coming to it. The first thing to do is to see if someone kicked the

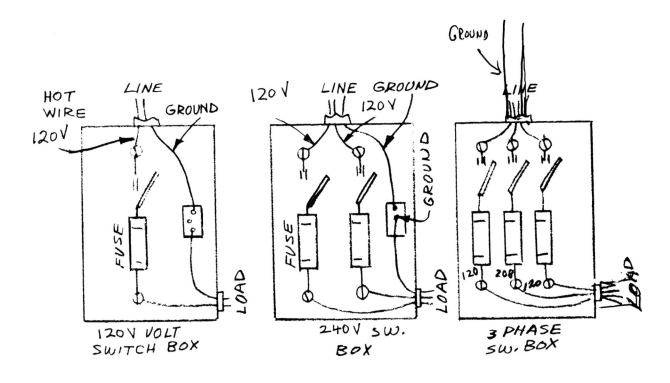

HOT
WIRE
120V

LINE

GROUND

120V

LINE GROUND

120V

GROUND

LINE

FUSE

LOAD

120V VOLT
SWITCH BOX

FUSE

GROUND

LOAD

240V SW.
BOX

120 208 120

LOAD

3 PHASE
SW. BOX

MANUAL SWITCHES

(Fig. 5-19)

plug out of the wall. This happens more often than you can believe. The refrigerator is simply unplugged. Suppose you look behind the box and the plug is in. You wiggle it and still the light and unit seem dead. Now, pull the plug and apply some plain electrical know how. Stick your test light pigtails into the receptacle and see if it is hot. If your test light will not light up, there is no use blaming the refrigerator or worrying about what is wrong.

Go check the fuses or multi-breakers and see if you can find a blown fuse or tripped breaker switch. Now, if you do find a blown fuse and when it is replaced it blows again, then you can say that there may be something radically wrong in the house wiring or the refrigerator. The pot or pump may be burned out, or the wiring harness in the refrigerator may be shorting against something. At least you have established that there is something wrong in the refrigerator which will have to be checked out. We will check out these shorts later on. Right now, you have made the simple test of determining whether or not power is available to the unit. Suppose when you wiggled the plug, the unit started running. Then you can say that the prongs on the cap are making poor contact, or that they are corroded

and there is poor contact. Replace the plug and the wall outlet. Suppose you find a fuse blown. You replace it and the box seems to run fine afterwards. In this case, give the box plenty of time. If the fuse does not blow again, look and see if the lady of the house could have overloaded the circuit with an iron or some appliance. If so, collect your service call and be on your way. Many freak things can happen to cause a breaker to kick or a fuse to go out. Don't condemn the box until you are sure of your power supply. Sometimes, if you do not happen to have any instrument or test light with you, go and get a floor lamp, plug it into the wall receptacle and see if it lights properly. Use your noodle.

Let's take a window unit now. You get the call; when you arrive, you find that the unit is stone dead. You turn the thermostat and switch the off-and-on switch off and on. Nothing happens. Everything seems dead as a door nail. Look at the cord plugged into the wall. Just like the household box, maybe the kids stumbled over it; and it is pulled slightly out of the receptacle. All right, suppose it is tight when you wiggle it; and when you examine the prongs after pulling it out, they all look clean and coppery. None of the leads are loose from the screws in the cap. Everything is in order.

You are now looking at the empty receptacle. Stab your 240 volt lamp test socket pigtails into the two matching slits. You may immediately get a lit up bulb, a bright light equal to the light in a 240 volt bulb. If so, then you can start looking to the unit for your trouble. But suppose you don't get any light in the two 120 volt female slots equals to 240 volts. A good test now to make is to try each of the 120 volt slots with the off pole, the ground slot. Suppose you get a half dim light in your 240 volt test lamp. Now, you try the other and get no light at all. Now you know that one leg, or one 120 volt line, is dead somewhere back in the house or in the fuse panel.

Suppose you find a blown fuse. You replace it and the unit immediately starts to perform. Stay long enough to make sure that the fuse is not going to blow again immediately. Collect your bill, if it holds up. Suppose it blows a fuse as fast as you replace them. Take the unit to your shop for repairs. Never attempt repairs on a window unit in a home unless you are selling window units and are scared that if you pick it up they won't let you bring it back. It won't cost anymore to repair a unit in your shop. You have a service charge on the unit already. You made a few simple tests, like you would make sure you had gas in your car before you worked on the carburetor.

Now, take a call on a packaged five-ton air conditioner. You set your tools down, get out your flashlight and find the unit won't run. You immediately look for the switch that was installed by the unit. It may be on the wall, or it may be just inside of the unit itself. Look and find the switch. Open the door of the switch box and look inside. You may be able to see a burned fuse. Suppose they all look the same. Pull the switch, and pull the fuses. Check them for continuity. Do not rely on testing the outlet side of a fuse with the switch on, even though you may own a volt-meter. In many instances the voltage in one line that has a good fuse will back up in another line to the end of a blown fuse, and you will read the fuse as good. Don't waste time. Be sure. Pull the fuses. Suppose all fuses are good at this box. While the fuses are out, place your test light on the wires leading into the box and make a circuit between the leads running into the box and the box itself.

Generally, all boxes are grounded. At least they are supposed to be grounded. You can make sure by examining the box. With this test, you will determine whether or not there is voltage from the leads to ground. Suppose one lead to the head of the fuse clamp is dead. Put your fuses back and go further up the line. You may have to go to the main switch box. Pull the fuses there if it is permissible and replace them. Be cautious around the main service entrance. Suppose that all of the fuses are good but one. Replace the fuse and you are ready to go back and start the unit by throwing the switch by the unit that you left pulled. If the unit starts properly, you can check the equipment over for places to oil and clean. Collect your money for the service call and be on your way. You would follow the same procedure on any commercial unit. Summarizing the problems involved here, you never make work for yourself. Learn to make quick thorough checks on the service available to the unit before you cut or begin to take anything loose. There is nothing more embarrassing than to hunt around on a unit for an hour and suddenly notice that the plug is out of the wall.

The first thing a good serviceman does on any unit that is apparently dead is to check the electrical service to the unit. Of course, there are many reasons why a unit may be dead and still have power available. You will know what they all are in time. The main thing at this point is to remember that you must first look for the simple things. The common sense approach is the best. If you don't have fancy instruments, use a floor lamp on 120 volts. If you do not have a 240 bulb or an amp-meter, use a 120 volt bulb on one-half of a 240 service. Presently, there is a very good combination amp-volt meter on the market that sells for about fifty dollars. This instrument will do as good a job, for the use that you have for it, as a $400 instrument. Don't tie up a lot of money in instruments which you may not need once a week.

As a refrigeration serviceman, you may need never know what ohm's law is; but you will need an ohm meter. You will also need a volt meter and amp meter. All of these meters may be purchased in one meter (a volt-amp-ohm meter). For under $100.00 you can own one. Plan on getting one as soon as possible. There are selectable scales on the meter to check either volts, amps or resistance in ohms. There are adaptable leads that attach to the meter for testing. Upon purchasing the meter, read instructions. For checking amps the snap around clamp is put around one of the line wires gong to the appliance or the motor. Never put two wires inside the clamp because you will get an erroneous reading.

You can expect to find a dead unit, because of power failure in at least one out of ten calls and sometimes more often. These calls are not big money calls, but they will add up if you will take them.

We have not discussed the many reasons a fuse will blow or a breaker kick out. Generally, this happens when the motor goes bad or the motor starting mechanism is on the blink. Shorts in the actual wiring of a refrigeration system are rare and nothing to worry about. Throughout this chapter, we have just assumed that the electrical failure was caused by a minor mishap like the wall plug getting loose or a fuse just dying of old age, or an overloaded circuit. More often, the case is that the motor and other electrical devices have caused the failure.

Look your own house wiring over. Go take a look at the wires leading to the house. Buy yourself a weatherproof pig tail socket with a 240 V light arrangement. Make some simple tests around the house. Get acquainted with the electrical service as you see it.

Use your own judgment on commercial equipment both condensing units and/or package air conditioners.

Questions that need answering:

What is a feed back? Take an ordinary 240 volt switch box and test the outlet side of the fuses going to the unit. **(Fig. 5-20)**

Now assume that the fuse on the right is blown, only you do not know it. You test with light or meter, and you find that you have voltage to ground. **(Fig. 5-21)**

Here is why you are showing voltage on the bottom side of a blown fuse. The electricity is feeding all the way around the system and right back to the end of the dead fuse. The way to avoid feed back is to take the fuses out and test them. **(Fig. 5-22)**

What is a ground block in a switch box? The ground block is just a plate welded or bolted to the switch box

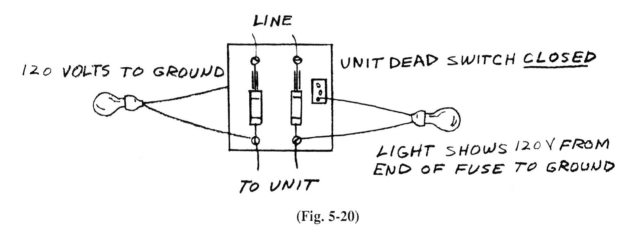

(Fig. 5-20)

I want those of you who have a few tools to start taking service calls as soon as you finish the electrical. Be on the look out for good used equipment that can be bought for a song. Don't pay over ten to twenty dollars for household boxes, twenty to thirty for window units.

itself to make sure that the switch box is grounded to the ground wires that come into the box to the block and go out of the box from the block. Everything metal on any electrical system, except the copper voltage conductors, is or should be grounded some place. ∎

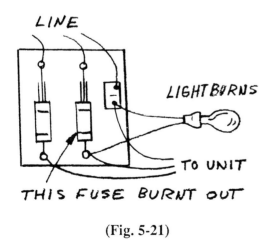

(Fig. 5-21)

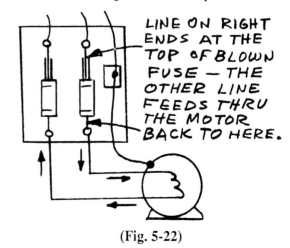

(Fig. 5-22)

AMPROBE

Here are a few of the many applications of AMPROBE

MEASURING LINE VOLTAGE
connect leads to line. Reading should be within 10% of rated voltage line.

FUSE CHECKING
measure voltage between load side of fuse and line side of the next fuse.

MEASURING RESISTANCE
apply test prods to leads of resistor. Read actual value on ohms scale. Compare with plate rating.

MEASURING LOAD CURRENT
snap instrument around wire and check reading against fuse or plate rating.

CIRCUIT TRACING
facilitates circuit tracing . . . merely use AMPROBE as ohmmeter to determine condition of circuit.

FINDING HOT SIDE OF LINE
a voltage reading can be obtained between hot side of outlet and ground.

The original pocket size
snap-around volt-ammeter-ohmmeter

AMPROBE RS-3

AMPROBE JR.

CHAPTER 6

ELECTRIC MOTORS

The voltages you will work with will be 120, 240 single phase and 208-240 three-phase. You know that it requires only two wires to have 120 volts to a unit. You know that only three wires are required for a 240 volt circuit. The third wire that is always present where 240 voltage is used is a safety wire to a ground. The two hot wires to 240 are 120 volts each. They are equal to 240 when the circuit is closed. You know that three-phase voltage is simply two 120 volt wires and an additional stinger or high leg of 208. All three wires of a three-phase circuit are hot. The fourth wire, if present, will be the ground.

Household refrigerators and freezers are always 120 volt units. Window units of one hp and better are generally 240. Window units under one hp are generally 120 volts. Package and central air conditioning may be 240 or three-phase. Rarely do you find this type of equipment with a 120 volt motor unless it is a midget size unit. Commercial units are generally 240 single phase or 240 three phase. Some small units are 120. You may say as a general rule that the size or hp of the equipment will dictate the voltage. Single-phase equipment, which means both 120 and 240 volt equipment, is confined to small sized units and motors no larger than five hp. Most single-phase motors are so wound that they will operate on both 120 and 240 volts. To change a 120 volt motor to 240 volts would require a different hook up from motor leads to line. The serviceman will generally find the change-over schematic on the motor lead terminal box. You are not likely to be called upon to make this change-over unless you are installing new equipment or changing out a motor. Where new equipment is involved the directions for hooking up to 120 or 240 will be on or about the motor.

Here is an outline of the voltage requirements for most units manufactured in this country.

1. Domestic refrigeration
   A. Household refrigerators .... 120 volts
   B. Home freezers .... 120 volts
2. Window units
   A. Under one hp and over .... 120 volts
   B. One hp and over .... 240 volts
3. Central air conditioning systems
   A. Up to three hp .... 240 volts
   B. Above three hp .... Three phase
4. Package air conditioners
   A. Floor console models under one hp ...... 120 volts
   B. Floor console and package
      above one hp .... 240 volts
   C. Above three and five hp .... Three phase
5. Commercial condensing units
   A. Under one hp .... 240 & 120 volts
   B. Above one hp, generally .... 240 volts
   C. Above three hp, generally .... Three phase

Remember you may have three-phase motors on any size equipment. However, where electrical consumption would be slight, it would be pointless to have a three-phase installation. Single-phase motors may be built as large as five hp. Above five hp single-phase motors are not practical, and you will never find a single-phase motor larger than five hp.

The reason why certain voltages are used in certain types of equipment is not particularly important. The reason is simply that the manufacturer of certain equipment has adapted the motors to suit the voltage that is most likely to be available in the areas or place where the equipment is to be marketed and installed. For instance, a household refrigerator could run on three phase as well as single phase 120 volts. But who has three-phase outlets in his kitchen? Besides, there would be no advantage to having three-phase winding in so small a motor. The builder of window units hopes that the sale of a unit will not be held up because the buyer does not have 240 already installed. However, most houses today do have 240. So that is no problem. But there is a technical problem involved in the manufacturing of window units above one hp. A 120 volt motor above one hp draws very high amperage, and there would be a terrific drain on the house wiring system if a unit of one hp or larger operated on 120 volts. For this reason, the builder sticks to 240 where one hp or larger window units are concerned. Not- withstanding, 240 volts has the quality of higher voltage that will give the motor the added starting power to get big window unit pumps to rolling. Actually, the only advantages of 240 motors over 120 motors are less amp pull and higher starting torque. You will never find a window unit hermetic pot operating on dual voltage. It will either be straight 120 or straight 240. You can't change from one to other. This applies to all full hermetics.

It costs a little less to build a motor to run on 240 than it does to build a motor to run on 120 for the same horsepower. This has nothing to do with the 120 volts cost of operation. It costs just as much to run a 120 volt motor. Your customer may ask you questions concerning the cost of operation of certain equipment which you service. Be prepared to give him straight answers. Remember, it costs a certain amount per horsepower. It does not matter that the motor runs on 120, 240 or three phase. So much horsepower consumes so many watts. The consumer buys watts. It does not matter that the voltage is three-phase voltage or 120 volts. The only advantage cost wise of 240 or three phase over 120 is that the original motor will cost less. That is, you can buy a three phase five hp motor for much less than the cost of a 120, 240, single phase five hp motor. There will be less maintenance on the motors using higher voltage, and in the long run this can mean an appreciable saving. The higher voltage motors will take up less space since they are slightly smaller. Even though smaller, the higher voltage motors will start or get off with much more power than a low voltage motor.

Many years ago the refrigeration mechanic had to be a motor rewind man. This is no longer the case. Today there are many motor rewind shops in the larger cities; and even in the smallest of towns, you can generally find a motor repair shop. Here is the policy of most refrigeration servicemen all over the U.S.A. They do not do even minor repair to open-type motors; and never, under any circumstances, do they attempt to rewind a conventional motor. The motor rewind shops will do the job for you at a discount, and you are permitted to mark up the cost of motor repair when you bill your customer for the completed job. It isn't even a good practice to do minor repair to electric motors such as rebushing or cleaning. You can job this work out to a motor shop that is set up for doing this work, and you will have their guarantee that the job is satisfactory. Generally the motor shop will have a set price for minor repair. This minor repair will include such work as rebushing, cleaning and replacing starting switches and some capacitors. The motor will be turned over to you in good condition; and should the job not hold up, the motor shop should stand back of its work. The type of motor that you should take to a motor shop is every type that can be disconnected or detached from a unit and is portable enough to throw in the back of your truck and take to a motor shop. Your time is too valuable to spend with a greasy old motor which may require parts that you would have to order from a supply house or buy from the motor shop anyway.

When we refer to conventional motors or open type motors, we mean the type that is bolted onto a unit much like a washing machine motor. It will have a shaft and pulley and will generally drive the pump through belts.

Rewinds are standard price among the motor shops, and the price won't vary much from shop to shop. You should make anywhere from 60% to 100% on a rewind (60% on your cost above $300.00 and 100% on lesser cost) plus your service. Yes, you guarantee the job to your customer because you have the motor shop's guarantee. The guarantee applies only to faulty workmanship, not to burn-outs that are caused by faulty or over worked refrigeration equipment.

You will most likely find open-type motors on large package equipment or commercial equipment. Open-type motors will drive the fan on package units, even though the pump may be hermetic. To take an open-type motor out is a simple operation. You pull the switch and disconnect the motor leads where they attach to the line running to the motor. If you are in the slightest doubt as to how the leads go back, draw a picture of the hookup; or better yet, tag the leads You will find bolts holding the motor to its mounting bracket or base. After you have taken the motor loose, put the nuts back on the bolts. It is good practice to pull the pulley before you take it to a motor shop. If you do not have a small wheel puller, let the motor shop pull it; and you take it with you. Don't let the pulley get lost in the motor shop. In the case of a complete rewind, if the motor runs on 240 volts, it is a good idea to tell the motor man what kind of voltage it will be running on and have him tie the leads coming out of the motor for this particular voltage. If the motor should turn in one particular direction, the motor man will show you how to reverse it. The motor man wants your business, and he should help you in every way possible. Don't be afraid to ask him for help. However, you are not likely to have to say a word, just pay for the repair and proceed to put the motor back in and get the job going again.

Let's examine the open-type motor and see some of the troubles that a serviceman may encounter when servicing refrigeration equipment.

You get a call to a grocery store. You locate the unit.

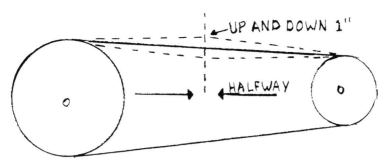

(Fig. 6-1)

You find that it is a belt-driven unit and won't run. You smell the strong odor of burnt wiring, and you look into the motor and find it cooked to a crisp. Pull the switch, disconnect the leads, take out the motor. There is nothing to it. What caused it to burn out is the problem. Many things can cause a motor to fail. The first thing to consider is why didn't the fuse blow or the overload device kick out? You must consider the job as a whole. Maybe the motor was so oily and dirty that it broke down from too much oil. Some owners believe that if a little oil is good, then a whole lot is better. Over-oiling may have been the trouble. Maybe the condenser was so dirty that the pump had to buck such a high head pressure that it just loaded down and burnt out. If the proper fuses had been installed, this would not have happened. You may find that there are 30 amp fuses in the box to protect a motor drawing only five amps. Constant loads on a motor rated at five amps would cause it to break down, with say a heavy load amp draw of 8. The fuse cannot blow at 8 amps. Your judgment would be as good as mine as to what caused the failure, and maybe the motor man can make a good guess if he finds certain breakdowns within the motor itself. Bad bearings can cause a burn out. Poor wiring and many other factors can cause a motor to burn out. After a motor is burnt out, there is nothing for you to do but have it rewound and put it back in service and determine the cause of the burn-out. Many owners of motor driven equipment do nothing toward prolonging the life of the motor. They may have their automobile serviced every thousand miles and let a ten-thousand dollar air conditioner sit and run for months without ever looking back of the front panel. When it fails though, they wonder why, as though they expected it to sit there and run for ten years. Had they known, most open-type belt driven units need regular service. The belts should be adjusted for slack. The motor should be oiled at regular intervals, and the pulley should be checked for looseness. Even when you have made other repairs to a system, you should make these checks on your own as part of your service call. Here is the rule for checking the tightness of a vee belt

drive. With the unit switch pulled, tighten the belt or belts until the span half way between the motor and the pump will give up and down between your two fingers no more and no less than about one inch. (**Fig. 6-1**)

Every experienced serviceman uses this check system for tightening belts. The amount of oil to put into the end bearings of a motor will have to be determined by you. See the size of the motor and use oil sparingly. To check the pulley, pull the switch and roll the pulley around until the Allen set screw hole is exposed and tighten the Allen set screw with an Allen wrench. Remember, sometimes there will be two Allen screws in the same hole. One of the screws actually tightens down against the key and shaft, and the other is a lock on the first one. (**Fig. 6-2**)

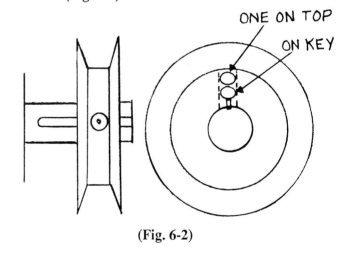

(Fig. 6-2)

The above operations of oiling, belt tightening, and checking the pulley are routine work for a serviceman. Now the next most important thing to know about the open-type motor is to know when it is bad or needs to be taken out. Bad bearings, or bushings as they are called in the trade, cause motor failure; and the windings may still be good. Bad bearings can be detected by the grinding or vibration of the motor when it is running, especially when it is starting. A loose pulley may sound the same way. The front bearing, the one nearest the belts, generally gives the most trouble. The cure is to take the

93

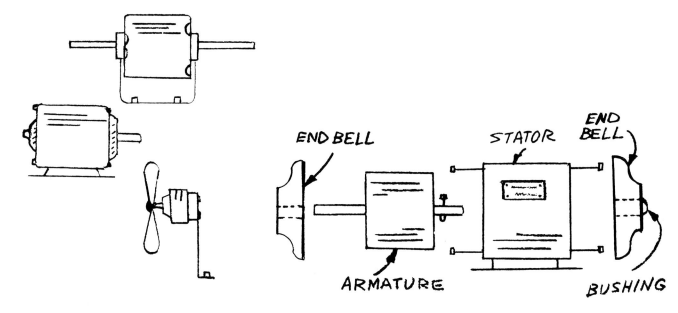

END BELL   STATOR   END BELL

ARMATURE   BUSHING

(Fig. 6-3)

motor out and have it rebushed; and when you install it again, make sure the belts have the proper tension. You may even replace the belts if they are worn and ragged. You will make more friends if you will take out every motor that is suspicious and have it cleaned and checked by the motor shop. You are getting paid for being thorough, so be thorough. Even though the motor does not require a visit to a motor shop, the motor man may catch something that was ready to cause a major, breakdown or burn out. You would not go for years without having your car greased just because it seemed to run all right.

In your last chapter a great deal of emphasis was placed on the importance of checking and knowing whether there was voltage to the motors or not. Don't take motors out that do not need removing simply because you failed to check the supply of voltage. You may bring the motor back from the motor shop and find you are right back where you started.

Summing up the servicing of open type motors, we could say right here that to service an open-type motor does not require the services of an expert serviceman. You could send a helper out to a motor job and give him a pair of side cutters, a box wrench and set of Allen wrenches, and tell him to pull the switch, take the motor loose, loosen the Allen screw, pry off the pulley and fan, take the motor down to Joe's Motor Shop.

You can't take a hermetic motor down to Joe's Motor Shop. You are the motor man where these units are concerned.

We work with single phase and three phase AC motors, no others. You know a three-phase motor will require three-phase current. It will either be a three-phase motor, or it won't be a three-phase motor. There

is no middle ground or special types. In this country we do not use the term double or dual phase. We class all 240 volt motors as single-phase motors for no other reason than the fact that most of these motors are wound to run either on 120 or 240 V. There are some two phase motors which are specially wound for 240 volts only. Just keep in mind that the term "single phase" means both 120 as well as 240 motors.

Therefore, for all purposes, we now have only single phase and three phase AC motors in our business. How many single phase motors outside of the three phase motors are we concerned with? The answer is about three or four. Out of this three or four, there is only one that is really important. This motor is the backbone or workhorse of the air conditioning industry. It is called the split-phase capacitor-start induction-run motor. We will consider the three phase motor, another workhorse, later on. Right now let's classify and describe the single phase motor family right down the line to our workhorse. Here is the single phase motor family which we may encounter in our line of work.

Sketches of the shaded pole motor are shown in **Figure 6-3**.

This motor is used on small fans. This motor has no starting power whatsoever. Once the juice is turned on, this motor just barely can get rolling. These motors are generally very small fractional horsepower motors. Once these motors reach their running speed they deliver their rated horsepower. For instance a 1/20 hp shaded pole motor will deliver 1/20 hp when it reaches its normal running speed. Up to this point it is so weak is can hardly get rolling. The shaded pole motor has no starting device. The motor is designed to start because of the shade on one or more poles. Shade means that one

94

pole or coil is slightly off in degree relationship to the other coils. A shaded pole motor has four coils or poles. These four coils are the running windings. The armature is a solid piece, and the end bells hold the bearings which support the armature. These shaded pole motors are very weak starters and are generally used on fans where there is no starting load. The fan may be a condenser fan or an evaporator fan or both.

The next motor in the single phase family is the ordinary old washing machine motor. It is a weak starter, but still stronger than a shaded pole motor. See **Figure 6-4**.

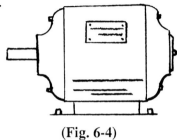

(Fig. 6-4)

This motor is practically worthless in the air conditioning business. You will find one used now and then in a window unit as the fan motor where a pretty rugged fan motor is needed. It is stronger than the shaded pole motor and much heavier. The washing machine motor, instead of having a shaded pole among its four poles or running windings, has a separate set of starting windings installed in between the running winding poles. See **Figure 6-5**.

The starting winding here is used to get the armature rolling; and once the motor has reached two thirds of its rated running speed, the starting winding is disconnected from the circuit. A small centrifugal switch is used to disconnect the starting winding. See **Figure 6-6**. This switch is mounted on the inside of the end bell

away from the shaft end of the motor. The lever that opens the switch is mounted on the armature shaft inside of the motor; and when the motor reaches a certain speed such as 1000 r.p.m. for a 1500 r.p.m. motor, the weight on the arm of the lever swings out against the spring or governor and kicks the switch open for as long as the motor continues to run. The designers or electrical engineers incorporated this starting winding into the motor to give it added power to get started when a washing machine pulley and belt were connected to the drive shaft. No single-phase motor will start of its own accord on the running winding alone. True, you could have a shaded pole, that is, one of the running windings slightly off center; but if you did have such a starting arrangement, the motor would be very weak. Because of this inability to start, the single phase motor must have a starting winding.

With no starting winding in a single-phase motor, the motor would just set there and hum until it burned up. If you were around every time it needed to start and reached down and gave the pulley a whirl, the motor would run without any further help. Many of you probably have owned a motor that had to have some help to get started. You may not have known it at that time, but the trouble was that there was something wrong with the starting device and/or the starting winding. Step by step let's see if we can nail this down.

You will note in the drawings of the single-phase washing machine motor that the only moving parts are the armature and the centrifugal switch which kicks the starting winding out when the motor reaches its running speed. Actually, if there were some way to get this motor rolling, nothing would be needed but the solid armature, the running windings and the housing to hold the bearings and stator. All single-phase motors have

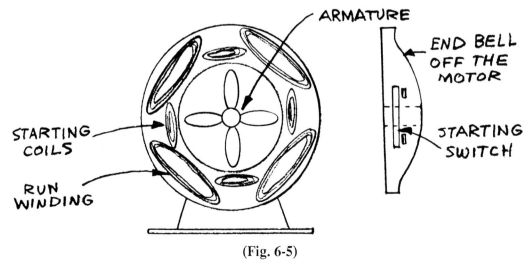

(Fig. 6-5)

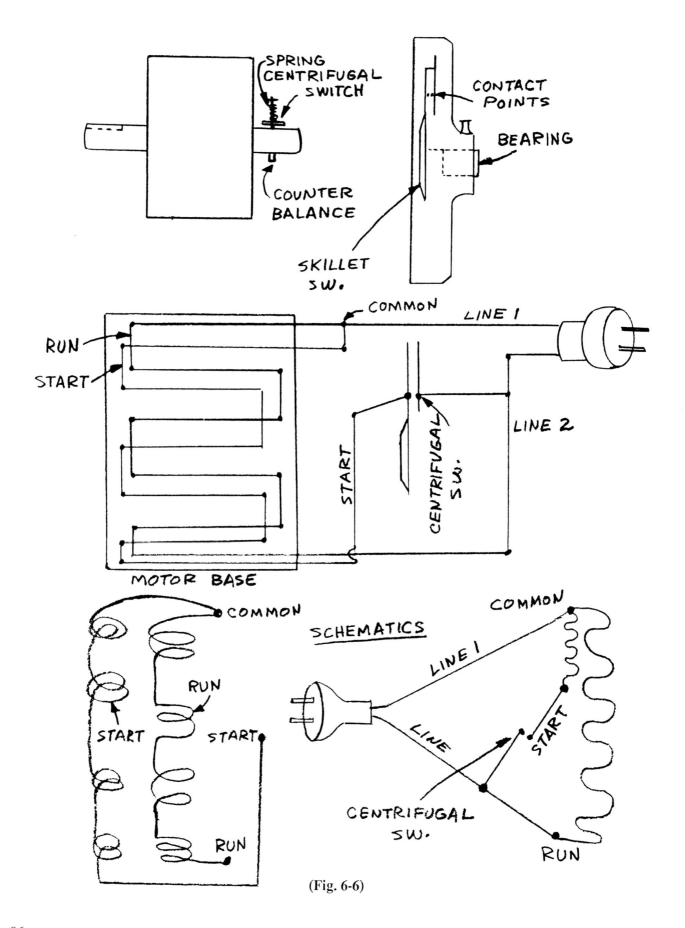

SPRING CENTRIFUGAL SWITCH

CONTACT POINTS

BEARING

COUNTER BALANCE

SKILLET SW.

COMMON LINE 1

RUN

START

START

CENTRIFUGAL SW.

LINE 2

MOTOR BASE

COMMON

RUN

START

START

RUN

SCHEMATICS

COMMON

LINE 1

LINE

START

CENTRIFUGAL SW.

RUN

(Fig. 6-6)

solid armatures; and for that matter, so do three-phase motors. Likewise, both are induction motors once they are running. In summing up the motors we have discussed so far, we can say that they all have one thing in common; that is, they need something to help them to start; and the most practical thing is another winding. The three-phase motor does not require any starting device. It is a natural motor capable of starting under a load and delivering its power right from the start.

The capacitor-start motor is the work horse of the industry. The capacitor-start split-phase motor is a washing machine motor with a capacitor added to the starting winding. The capacitor gives the extra power on the start that is so necessary where a motor has to start against a pump pressure. The capacitor is a modern invention by the electrical motor industry. It serves one purpose and one purpose only. When the current is turned on (with the starting switch closed) the current

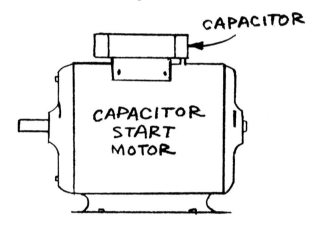

CAPACITOR

CAPACITOR
START
MOTOR

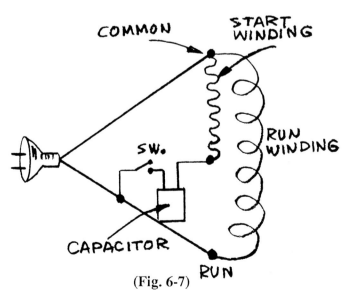

COMMON

START WINDING

SW.

RUN WINDING

CAPACITOR

RUN

(Fig. 6-7)

goes through the starting winding, as well as the capacitor. The capacitor, being in the line, builds up a charge of high voltage. This charge can be several thousand volts. More than 60 times a second, this charge of high voltage is unloaded into the starting winding. Something has to give; and no matter how much load is on the motor pulley, the motor will attempt to pull it or get it rolling enough that the running winding can take over. Take a look at this layout of the hook-up with a capacitor added to the circuit. See **Figure 6-7**.

The minute the motor reaches its running speed, the running winding takes over; and the capacitor as well as the starting winding is disengaged. If you will remember that the starter on your car is not expected to start the engine with the car in gear; you will appreciate the starting load that a motor on a refrigeration pump has to start under. There is no neutral. The motor has to get the pump rolling with pressure on the first stroke of the piston. This has been the headache of the air conditioning industry for years and years; that is, how to build a motor that will start under this load, and at the same time, build it cheaply enough that it can be sold with the unit at a fair price. It's certain that a washing machine motor will not get a pump rolling with a load with only the assistance of the starting winding, which is very weak by itself. The capacitor was a windfall for sure. Today it is used on motors whenever there is the slightest starting load. The repulsion induction motor was powerful but bulky, complicated and expensive to build. The capacitor was the answer.

There is nothing mysterious about the capacitor's function. It simply unloads such high voltage into the starting winding that it tends to jar the armature into action, and it begins to roll under unbelievable loads. When the capacitor was perfected, it opened up all kinds of possibilities for small air conditioning units which had been plagued with high starting load conditions. The whole problem that we have been discussing here hinges on getting the motor started under load conditions. The one place where this load was the greatest was in an air conditioner. The air conditioner manufacturer took this capacitor-start motor and sealed it up in a pot with the pump. The hermetic pot, or welded up container for the pump and motor, was built for one purpose and one purpose only. That was to get rid of the seal leak around the crank shaft of the conventional open-type pump. Now the manufacturer can warrant the unit without the fear of the old seal leak that was the sore spot in all refrigeration pumps. True, we still have some open-type pumps; but the seals are expected to

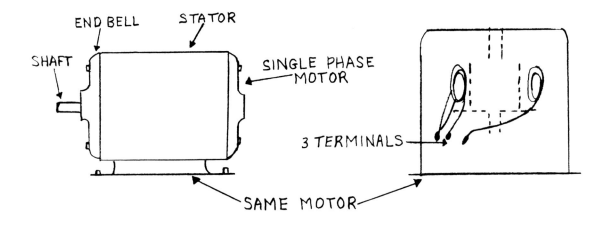

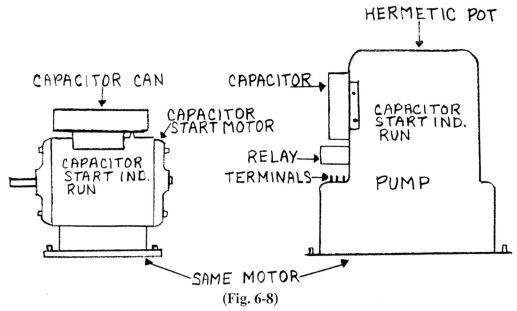

(Fig. 6-8)

develop leaks; and they are serviceable. But in the case of the small household refrigerator and freezer as well as the window unit, the manufacturer wanted to give a warranty without the old trouble spot seal forever leaking. This brought on the birth of the hermetic. Now with that problem solved, the manufacturer had another problem. He could build a pot to contain a motor which was the same as the washing machine motor. But it had a very low starting torque and had to be specially designed mechanically to reduce the load when the motor started to run. Your refrigerator pump may be one of these split-phase, plain starting type pots. Look and see. If it does not have a capacitor on the pump, it is a plain washing machine type motor.

Now your next question should be this. What happened to the throw-out switch that had to disconnect the starting winding after the pump began to run? My answer would be that the manufacturer had to get rid of the centrifugal switch the minute he welded the motor up inside of the steel dome. In the first place, all motor switches will spark; and it certainly would not be a healthy condition to have a sparking switch inside of a hot oily steel dome which could be a bomb if properly set off, notwithstanding the fact that you could not service a mechanical device inside of a welded-up pot. Anything inside of the pot that might need servicing would defeat the purpose of the pot to begin with. So the electrical industry came up with the starting relay. This relay took the place of the mechanical switch. With this relay the points can burn without any visible effect on the workings inside of the pot. Likewise, the relay can be serviced without disturbing the hermetic weld on the pot. Let's sum this up by saying that the capacitor start motor was simply shifted over into a welded-up pot. The starting device was arranged so that the motor would start with a capacitor, and the relay on the outside of the pot would disconnect the starting winding when the motor reached 2/3 of its normal speed.

Many household refrigerators have small pumps of 1/6 and 1/9 horsepower that are the no-help type start-

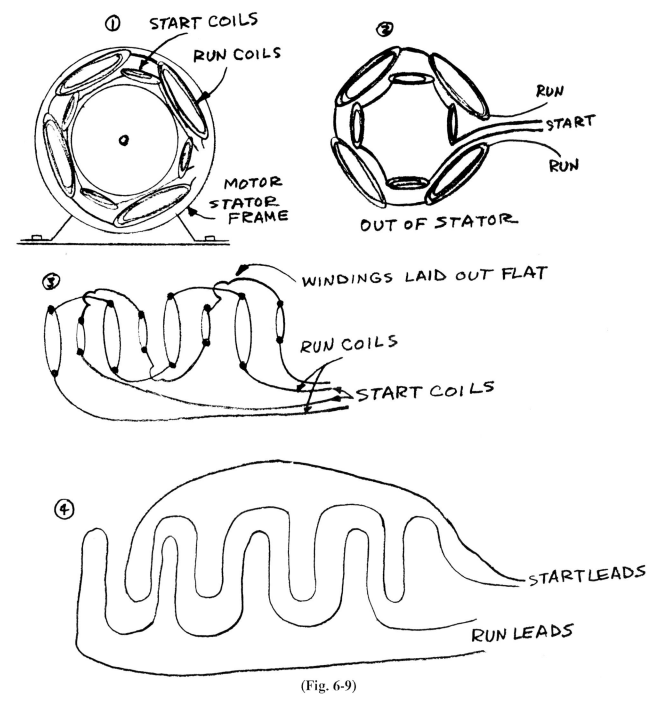

① START COILS
RUN COILS
MOTOR STATOR FRAME

② RUN
START
RUN
OUT OF STATOR

③ WINDINGS LAID OUT FLAT
RUN COILS
START COILS

④ STARTLEADS
RUN LEADS

(Fig. 6-9)

ers. The manufacturer has put the old washing machine motor over into a sealed pot with a relay replacing the centrifugal switch. The motors are still the same, even though one has been welded up in a pot and the other is still the old conventional motor.

See this comparison in **Figure 6-8.**

Since we are mainly concerned with the starting mechanism of hermetic motors (we job out open-type motor repair), let's develop out a schematic so that we can speed up the study of the relay and capacitor on the hermetic.

In order to understand the hermetic schematic, let's develop the hook-up of split phase motors step by step up to the sealed pot. Start with number one and study right through. **(Fig. 6-9)**

You see in these drawings that the terminals coming out of the sealed pot are the ends of the motor windings. You are not concerned with the actual coils of wire in a motor unless they are burned out or damaged. You are concerned with the leads coming out of the pot and the manner in which you check them for continuity and ground. Likewise, you want to be able to start the

99

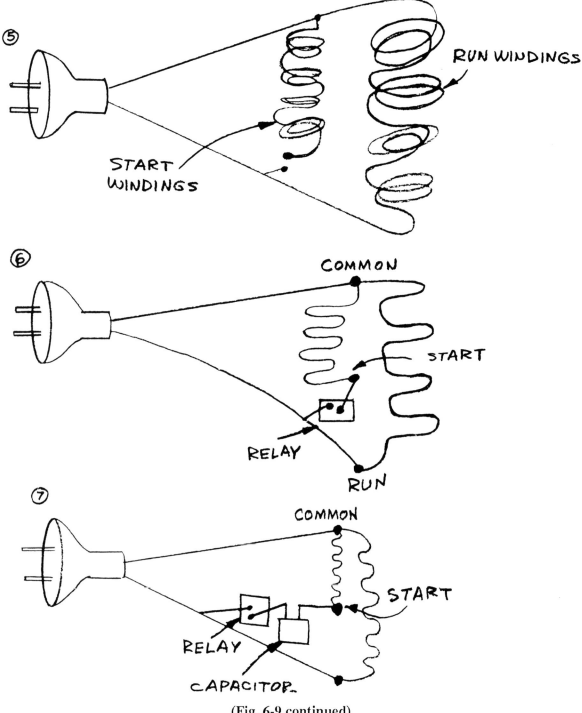

⑤ RUN WINDINGS

START WINDINGS

⑥ COMMON

START

RELAY

RUN

⑦ COMMON

START

RELAY

CAPACITOR

(Fig. 6-9 continued)

pot of your own accord in order to know whether it will run or pump. You also noticed that we are getting rid of the complicated turns, or coils of wire, representing the windings; and we are just using a wavy line to describe the inside windings.

A well trained serviceman does not even think in terms of coils inside the pot. He thinks only in terms of what the terminals coming out of the pot represent. See **Figure 6-10** for terminal arrangements.

You will be a better trouble shooter if you will just think of the motor inside of the pot as having these wavy lines which represent the many coils of copper wire inside of the motor. You will note that in all of the drawings so far, there are four coils of wire for the run winding and four coils of wire for the start winding. The ends of these sets of coils are the motor leads. See **Figure 6-11**.

The manufacturer could have placed four terminal bolts in his pot and had the four inside leads from the coils attached to the four terminals. But he decided, why

100

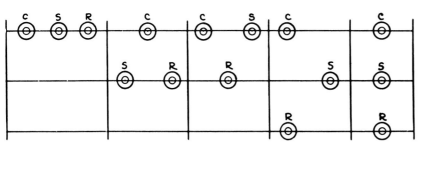

(Fig. 6-10)

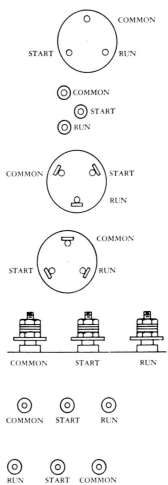

drill four holes in the pot and put in insulated terminal bolts when three would do just as well since one end of the running winding would be tied to one end of the starting winding. So he drilled three holes for bolts and soldered the motor leads to the end of the terminal bolts on the inside. You will note that he has tied two ends to one terminal. See **Figure 6-12**.

By doing this he has eliminated one more hole and terminal in the pot. These terminals are always subject to leaking; therefore, the less terminals there are, the better chance of no leak.

Here is a summary of the main body of this chapter. We job out open-type motor repair. We know that all single-phase motors have to have a starting device. This starting device is either a shaded pole on the running winding or a separate set of starting windings. Where separate starting windings are used, we must have a disconnect switch or starting relay to cut this set of windings out of the circuit. We know that there are two types of hermetic pots, the single phase no-help type and the single phase capacitor-start type. We understand now that the single phase starting winding type of motor has four leads coming out of the windings. It has one lead on the end of each set of windings. We know that there is no sense in having four terminals in a pot since one end of the run and start windings are to be joined anyway. Therefore the manufacturer has soldered the ends of these two windings onto one terminal, and it serves as the common. That is, one terminal is common to both the running winding and the starting winding. One end of the starting winding is called the "start," and the other end of the running windings is called the "run." The term is Common Start and Run. That is, the three terminals on any pot which is single phase are called the common start and run. Where they

are located or their arrangement is not important here. When we troubleshoot the hermetic wiring, we will be concerned with the arrangement or order of these terminals.

Finally, we said that many household refrigerators were the plain old single-phase motors sealed up in a pot with a relay for a disconnect switch. The starting relay takes the place of the centrifugal switch. We said that this pot was the washing machine motor welded up in a tank. We then stated that the improved washing machine motor was the same motor with a capacitor in the starting line. This motor then becomes the capacitor start motor. The capacitor motor is a high starting torque motor and the best motor for the window unit equipment. With this information in mind, we will move to the window unit and central system and every other type

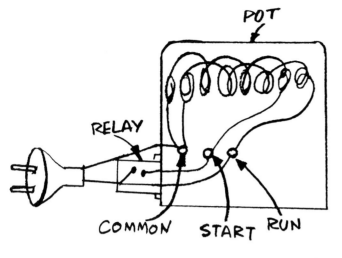

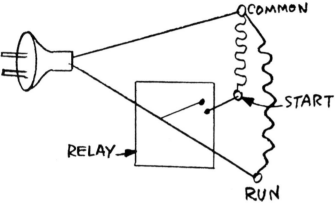

Questions That Need Answering

What can be said about the three-phase hermetic pot? The three-phase hermetic pot needs no starting devices, and the terminals are all used as plain run winding ends. That is, there are three sets of separate coils of three each in a three-phase motor, and no help is needed to start the armature to rolling in this pump. The three-phase motor is a natural starting motor. Once the circuit is closed, it is ready to go as a high starting torque motor with no assistance from switches, capacitors or special windings.

Can a capacitor start motor have more than one capacitor? Yes. This is up to the designer of the particular motor. It may have one or many. However, the starting principle remains the same.

How is the pump crankshaft hooked to the motor in a hermetic pot? The crankshaft of the pump is the end of the motor shaft. Practically all motors in hermetics are in a vertical position with the pump underneath the armature. It makes no difference though if they are horizontal.

How do the terminals come through the side of a steel pot without grounding out to the pot? They are insulated with fiber washers and Neoprene sleeves, and the bolt terminal itself does not touch the metal pot in any place. The leads from the motor windings are attached to the terminal bolt on the inside of the pot.

Do you ever open up a pot for any reason? Never. This is a factory operation. ■

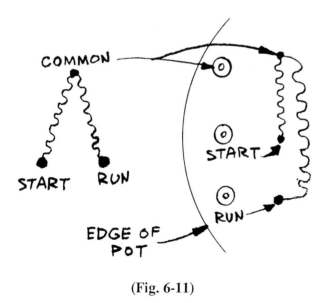

(Fig. 6-11)

of unit using a hermetic pump. Remember, hermetic pumps are now being built up to three and five horsepower; and in the case of the three-phase hermetic pots, they may be almost any size.

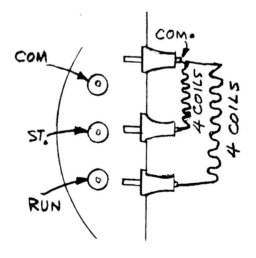

(Fig. 6-12)

CHAPTER 7

ELECTRICAL TROUBLE SHOOTING OF HERMETIC COMPRESSORS

I am going to prove to you that the electrical problems involved in working on hermetic units can be as simple as ABC. First, I want you to do a little mental arithmetic with me. If I told you to multiply 25 times 29 you would put down 29 and put the 25 underneath. You would say to yourself mentally 5 times 9 is 45, put down the 5 and carry the 4 and so on. Then you would add up the total and give me the answer. You would be going step by step. You would be doing it mentally except for putting down the sums of the digits. Don't you agree? All right, then let's go through the same procedure mentally on a hermetic motor. Take the simplest type, the no-help single-phase pot. You have the problem, a sealed-up hermetic pot with three terminals sticking out. You want to run it or see at least if it will run. Your first mental thought should be: *This pot has a common terminal.* I will attach one end of a cord to this common terminal, and I know it will be connected to one end of both the run and start windings, even though I cannot see inside of the pot. Next I will attach the other side of the cord to the run terminal; and if I plug the cord in, the pot will not start. It will just hum. Now I have one terminal left, *the start winding terminal.* I know that this terminal is connected to the other end of the start winding opposite the end that is tied onto the common. I know that if I put a jumper across the run terminal to this start terminal and then plug the cord in, the pot should start and continue to run, even after I have taken the jumper away from the start to run. Now the problem is solved. Either the pot will start and run, or it won't run.

When you made your first mental calculation on the problem; that is, 5 times 9, you knew for certain that it was 45. You can say the same about your first move, and it is just as certain as the multiplication. You can say that

one terminal is a common…common to both the start and run winding. There are no exceptions, it's just as positive as multiplication. When you carried your four over in the next step of multiplication, it was no surer than when you carried the other side of your cord over to the run terminal. When you did this, you had the run winding square in the circuit. You plugged in. Nothing but a hum came out of the pot. Then you knew that it would not start because you know that every single-phase alternating current motor made has to have a helper to get started. You knew that the start terminal was the end of the start winding helper. The only thing left to do was to get a circuit going through this start winding just like the circuit through the run windings. So you put a jumper across from the run terminal to the start terminal, and the motor started. You pulled the jumper away, and the motor continued to run.
Note: On a no-help start compressor the pressures must be equalized between the high and low side. If they are not, the compressor will not start when jumped. You would erroneously condemn the compressor.

I have compared this to a mathematical problem because there is no doubt in your mind that 25 times 29 is 725, and you would be willing to stake your life on it. Positive, indisputable, true, and without any catches, 25 times 29 equals 725. It is just as true that every 120 and 240 volt single-phase motor sealed in a pot will have a common start and run terminal. When these combinations of terminals are properly approached, they will add up to the motor starting, if the motor is good. There are no tricks, no gimmicks, no mysteries. It's just like good mathematics, honest, true and indisputable. Don't try to complicate the terminals on single-phase pots. If they are three in number, you can be sure that they are one common, one start and one run. It does not matter

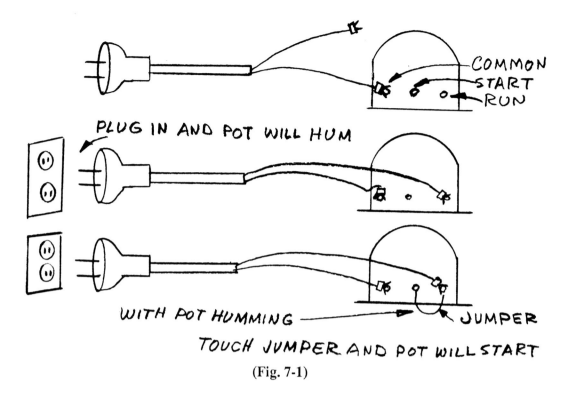

COMMON
START
RUN

PLUG IN AND POT WILL HUM

WITH POT HUMMING ———→ JUMPER

TOUCH JUMPER AND POT WILL START

(Fig. 7-1)

whether you are using your own cord to start this pot or the regular wires in the refrigerator. Step by step, here is what you were doing. (Fig. 7-1)

In these drawings you have disconnected all of the refrigerator wiring in order to make this conclusive test. Take a look at the simple wiring diagram of this refrigerator. We will add the thermostat here.

Never confuse a thermostat with a thermostatic expansion valve. They are two entirely different things. The thermostat is a switch, and the thermostatic expansion valve is a valve. Both, however, have the power element which is charged and responds to temperature. One causes a toggle switch to open or close, and the other causes a needle to modulate or open and close. Notice that the thermostat in this last drawing does nothing more than break the line circuit much like the wall switch in your house breaks the circuit through the light bulb.

Now on this same box take all of the wires off the terminals. Bend them back out of the way and go in with your cord. See this method of testing a pot in **Figure 7-2**.

Why should you want to make such a test on this? For one thing you have eliminated the thermostat and the relay from the wiring when you disconnected the regular running gear. That is, you have bypassed all of the electrical devices on the refrigerator in order to establish whether the pot was good or not.

The thermostat is used to cut the box off when it

reaches the desired temperature. If you did not have a thermostat, the box would freeze up everything in storage: eggs, celery, and other things that are injured by freezing. The relay is used to kick the juice through the starting winding for a few seconds in order to get the pump armature to rolling; that is, jar it into starting. Once it is rolling, the run winding will do the job. Actually, when you disconnect all of the wires off of a pot and use your clip cord and a jumper, you are doing manually what is ordinarily done automatically. Your jumper takes the place of the relay. *I repeat, your jumper takes the place of the relay.* Since your clip cord is tied directly to the terminals from a wall socket bypassing the thermostat; the unit will never shut off as long as you have your clip cord tied on. The thermostat has no bearing on the system now. It does not matter how cleverly the manufacturer has hidden the wires in a refrigerator or how they are arranged. It does not matter that he has put one hundred connections around about the system; the pot will still have the old common start and run; and the pot will operate directly from your clip cord. Where you have a capacitor start pot, instead of a piece of wire or a screwdriver for a jumper, you would use a capacitor. See this quick method of kicking off a capacitor start pot where a clip cord is used in **Figure 7-3**.

Here instead of a jumper you are using a capacitor (Fig. 7-3) with a couple of pig tails for the jumper. The juice from the run side of the line will go through the

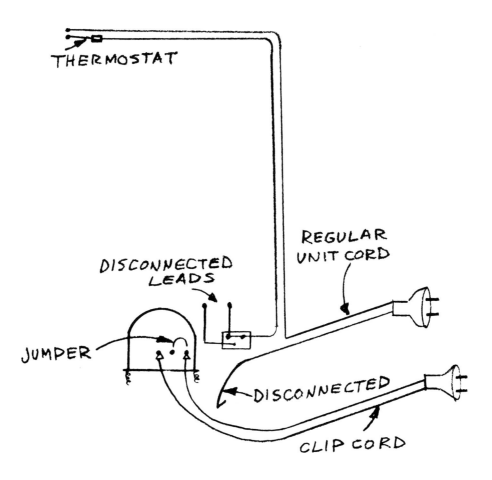

THERMOSTAT

REGULAR UNIT CORD

DISCONNECTED LEADS

JUMPER

DISCONNECTED

CLIP CORD

(Fig. 7-2)

capacitor and unload into the start through the start terminal as soon as it is touched with the end of the capacitor pig tail. You might be interested to know that even though a pot is designed to start with a capacitor, it may start a few times with nothing but a wire jumper instead of the capacitor. Likewise, a plain old single-phase no-help design pot might start better if instead of a plain jumper you used a capacitor, even though the pot was not designed to ever use a capacitor.

Everything that has been discussed here relative to starting household hermetic pots applies to window unit pots also. You are the boss, once a pot is stripped of its electrical harness. You can make it run without worrying about whether the rest of the electrical apparatus is good or bad. Once you have established that the pot is in good working order, then you have whipped half the problem of servicing sealed systems.

Summing up this manual starting of a pot, we can say that your act has taken the place of the relay which ordinarily does the job of jumping from the run to the start in order to energize the starting winding. Relays are a constant source of trouble to hermetic pots. The relay is not perfect and is subject to wear and tear since it has a tough job to do. You never repair a relay. They cannot be repaired. They are delicate little electrical devices which are either good, bad or getting so bad that they are in need of replacing. Would you try to repair the points on your automobile? Never. You would replace them without a second thought. Likewise, you replace relays when they are bad or even suspected of being bad. Generally, the points get burnt just like the points of the distributor on an automobile. When this happens, even though the relay as a whole may be in good shape, the relay will have to be changed out.

It is not particularly important that you know exactly how a relay works since you are not going to design or manufacture relays. However, you will be a better serviceman for understanding its purpose.

Remember, it does not matter who makes the relay; it will have the same job to do. Basically, there are three

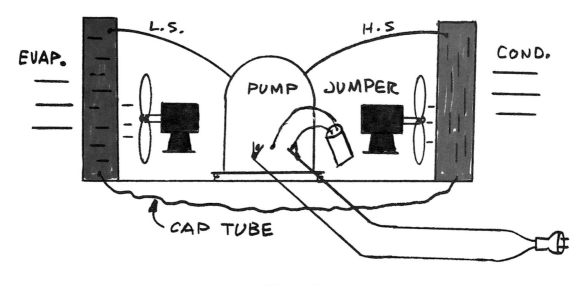

(Fig. 7-3)

different types of relays. The one most common to the household refrigerator is the plain amperage relay. (**Fig. 7-4**) This relay will be found on freezers as well as household refrigerators. For that matter you are going to find a relay on or about every hermetic pot, no matter where the hermetic pot is installed. You are going to find that the relay has one job to do, no matter what the pot is used for. That is, the relay is to start the motor and nothing else. There is one exception to this rule. In recent years there has been a discovery that a capacitor can be used to start a pot and then act as a resistor in the line without being disconnected. This is a very recent trend, and we will cover it fully when we have the relay down pat. Remember, for all purposes the relay is hard to beat and is universally used to get the starting job done. If you will study this first relay and know it perfectly, you will have the key to understanding all relays, regardless of what magic the engineers have cooked up electrically in order to get the starting job done.

Look at the basic relay (**Fig. 7-5**). Leaving out the thermostat, take a look at this bare electrical starting device called a relay.

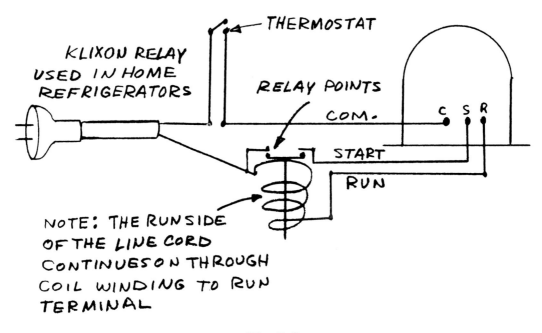

(Fig. 7-4)

106

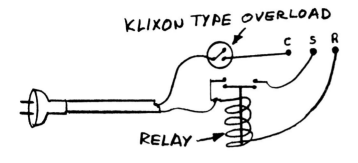

KLIXON TYPE OVERLOAD

RELAY

(Fig. 7-5)

Step by step, here is what happens. Plug the cap into the wall outlet, and juice flows down to the unit. One side of the line is tied directly to the common (you know this common is tied to both the start and run windings). The other side of the line goes through the coil and continues on to the run terminal.

Even if the relay points were burnt completely off, you can see that you would still have a circuit through the running windings. Now the instant the circuit is made on the running windings, the T-shaped bar is rammed magnetically up against the two points which close the circuit to the start windings also. When the unit has reached two-thirds of its normal running speed, the amperage will fall off; and the coil in the relay will no longer be able to hold the T-shaped bar up against the starting contacts. Immediately this happens, the bar falls down and the starting winding is disconnected from the circuit. The motor will continue to run on its running winding. It will be timely to mention right now that all single-phase starting windings are made up of very fine wire. If the start is not disconnected immediately, when the pot starts to run, then these fine starting coils would cook or burn up.

The minute any single-phase motor starts, there is

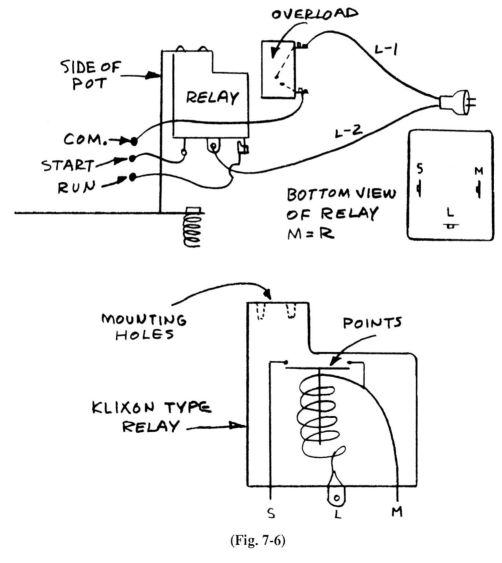

(Fig. 7-6)

107

a back pressure or high amperage pull on the line feeding the motor. This increased amperage on the line, and in the coil of the relay, causes a magnetic action on the relay armature or shaft. This shaft will move; and the points, being attached to it, will in turn break or open. This amperage or back pressure lasts for only a few tenths of a second, but still it is long enough to cause the relay to move. In fact, it works much like a solenoid except that it does not require constant voltage. The relay only requires the back pressure impulse of the line when the motor gets its first charge of voltage. Likewise, this surge of amperage can cause a hot wire to stretch and close or open points.

All right, we have the picture now. But how do we hook up a relay or identify the connections on the relay? Suppose you have our simple household box relay, and it is totally enclosed in a plastic box. All you see are some screw holes with letters reading like this.

Now see the inside in relation to these screw connections (**Fig. 7-6**).

In the above relay the manufacturer has called the connections on the relay the *L, S,* and *M. L* means *line* from the service cord to the relay. *S* means the *start* connection from relay to start terminal on the pot. *M* means the *main* winding connections. That is, the continuation of the line to the run terminal on the pot.

You see that this relay hooked up to the pot is going to do with an electromagnetic coil what you were doing

with a jumper and a clip cord.

This is the basic relay. It is basic to every relay used in our business. Some relays are more refined and fancier, but the operation is the same. We have potential relays, hot wire relays, voltage relays and magnetic coil relays. However, it does not matter what clever design is incorporated into the relay by an electrical engineer. The job is the same, to close the circuit to the start winding for a moment in order to help the motor to start under a load.

But let's go back to our original household schematic. We will add another small circuit to complete the hook up. See the light arrangement in **Figure 7-7**.

This is the basic hook up of practically every box in the country.

In many cases you will never be called on to work on anything but the relay in one of these boxes. In fact, you won't work on the relay. You will just replace it when you find it has burnt out or gone bad. Take another look at the schematic of a household refrigerator without even the outline of the box, just the bare electrical schematic. (**Fig. 7-8**)

It is important that you understand the simple electrical schematic. It is like having a map of a strange country.

Take a call on a household box. The complaint is that the box won't run. You set down your tools and open the door on the refrigerator. You find the light burning, so there is no reason to check the wall plug. Now if the light is burning, you know there is electricity to the circuits within the hook up. But where is it broken? It may be that the windings in the pot are burnt out, and no circuit can be made because of these open windings. More simple yet, it may be that the thermostat has gone bad and won't close the circuit so the box will start. Take the easiest thing first; Suppose the only tools you have with you are a screwdriver and a short electrical lead cord with clips on the other end.

Now let's check this box out by using good old common sense as our guide. First, if the light burns, we have juice available to the box. We know this for sure. Now the pot is dead; and it may be dead for a good reason, like a bad thermostat. You don't repair thermostats. You replace them. How are you going to check the thermostat? Take it loose and expose the two wires running to it; that is, one wire in and one wire out. (**Fig. 7-9**)

Since the only job that a thermostat has to do is to break open the contact points when the evaporator gets cold, and close the points when it warms up, why not

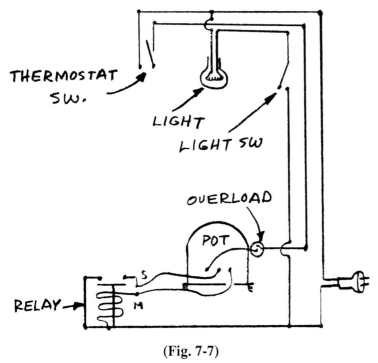

(Fig. 7-7)

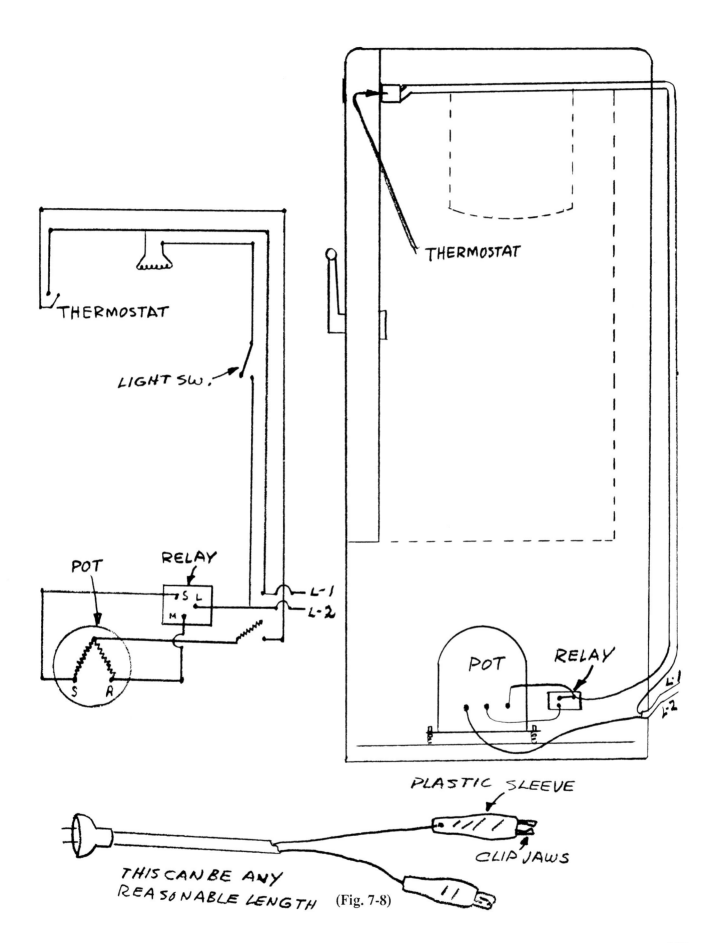

THERMOSTAT

LIGHT SW.

POT

RELAY

S L

M

S R

L-1

L-2

THERMOSTAT

POT

RELAY

L-1

L-2

PLASTIC SLEEVE

CLIP JAWS

THIS CAN BE ANY
REASONABLE LENGTH   (Fig. 7-8)

109

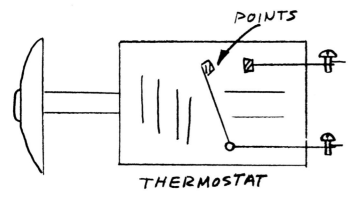

POINTS

THERMOSTAT

(Fig. 7-9)

pull the cord out of the wall, tie both wires together where they are attached to the thermostat switch and bypass the thermostat altogether. That is, the unit will be the equivalent of being on all the time. (**Fig. 7-10**)

We call this putting on a jumper or shorting out the control. Now you plug the refrigerator cord in; and if the pot starts and runs properly, you can say that the trouble is probably in the thermostat. Thermostats go bad quite frequently. But suppose after you have jumped the thermostat, the unit won't start or run. Now we have to check the relay. The quickest way to check a relay is to open it up carefully, look inside and see if it is burned up (some cannot be opened). Smell of it and it will smell burnt if the little electromagnet coils are burnt out. Suppose the relay smells and looks good, but you are still suspicious that it may be bad. Then you should by-

BOTH WIRES HOOKED TOGETHER

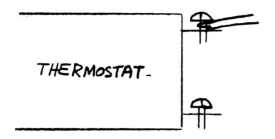

THERMOSTAT.

BOTH METHODS HERE BYPASS THE THERMOSTAT BREAKER POINTS

(Fig. 7-10)

pass it by carefully disconnecting the leads on the pot. Mark them if you care to, and connect your clip cord. Plug your clip cord in and you will hear the run windings humming. Now reach in with your screwdriver and jump from the run to the start terminal. If it starts, then you can say you have found the trouble by the process of elimination. The trouble is a defective relay. But if the pot will not start when you jump across the start to the run, then you can say that the pot is defective. When this is the case, there is nothing to do but to change out the pot. Many things can cause a pot not to run. The most common reason is burnt out windings. Sometimes a locked motor will cause a pot to hum but not start. Another simple test for a good pot without even taking the time to run it is to put a test light across the terminals and see if there are any open windings. However, actually running the pot on a clip cord is the best test.

See the test in **Figure 7-11**.

Everything being discussed here applies to window units as well as any other hermetic units. Don't waste time messing around with complicated relays. Go directly to the pot and see if it is good or bad. Now suppose you find that the pot will start with your test rig. Then you know that you have a bad relay. Of course, the heater overload could be ruined and open, even though it should be closed. This rarely happens. Relays often go bad. Whenever you hear any household box or freezer trying to start and can't, you are listening to a relay in a death struggle. That is, it is almost gone and may work once in a while. When this happens you will hear the tattletale *Hummm, click, Hummm, click, hum and click, hum and click*. Maybe, after a dozen tries, the relay will get the pot off. However, when it needs to start again, after it has turned off, the same old trouble will start. *Hum and click, hum and click*. The click is the little heater overload kicking out. This overload will reset again after it cools down. This is the same as a fuse breaking the circuit except that it is self-setting. The relay does not cut out unless it has a built-in heater or overload device. Generally they are separate devices. We now have a pretty good idea of how to start a pot and how to jump around the devices on a refrigerator in order to eliminate the possible trouble spots. The next problem we have to deal with is how do we determine which terminal is which. That is, if we know how to use a clip cord and jumper on the pot itself, how do we determine which terminals we hook our clip cord to? In most cases where you will have to make this decision, the answer will be right in front of you.

For instance, we know that on every hermetic

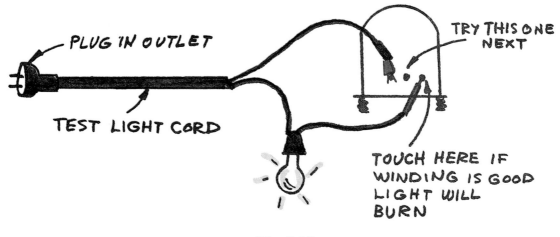

PLUG IN OUTLET

TEST LIGHT CORD

TRY THIS ONE NEXT

TOUCH HERE IF WINDING IS GOOD LIGHT WILL BURN

(Fig. 7-11)

which has a relay, there will be the start contact point. If you will examine the relay, you will find this screw connection marked *S*. If it does not have an *S*, but uses the number system like "1-2-3-4," then you can open up the relay. Find the contact points and follow down from the points to the connection which takes off from the relay whenever this is hooked to the pot. That terminal on the pot will be the start terminal. In many cases you will find that the overload will also tell you which is the common. You know the principle that the pot starts on, so all you have to do is associate the connections on the relay to these particular terminals. Suppose you have a pot stripped with nothing on it to indicate the common, start and run. Then you can use the ohm meter to determine which terminals are which.

In **Figure 7-12** you see three terminals which are not identified. Suppose the terminals are marked 1-2-3 from left to right. Check and get an ohm reading across 1&2, 1&3, and 2&3. The highest reading tells you that you are on the start and run terminals. In this example you are reading 10 ohms. You don't know which terminal is start and run, but you do know that terminal "1" is common. Terminal 1&2 will be the next highest ohm reading and 1&3 the lowest. So, start and run when ohmed across reads the highest. Common to start the next highest and common to run the lowest. The sum of C&S and C&R added together totals the S to R reading. For example: C to S = 6 ohms, C to R = 4. Then 6 + 4 = 10, the total found across S to R.

You may never be called upon to hook up a relay where it has already been detached from the unit. However, you must know how to hook your relay to the pot terminals even though you have no schematic or old relay to go by. One method of handling a bad relay is to draw a schematic or hook up when you remove the old one. Sure, an expert can trace out all of the connections, but an expert does not waste time tracing out wires. He will draw a diagram of the old hook up or he will take each wire loose from the old relay just as he installs it, moving one wire at a time over to the new relay. Don't be too proud to make a drawing. You can save a lot of time by drawing a schematic before you disconnect anything. Here is something to remember when hooking up a strange relay with no help from a schematic. If you will remember that there is always a marriage of the run and the start within the relay, you will have the main guess work whipped. This marriage is the joining of the run and start connection through the points of the relay. The line or main wire going to the relay will continue on to the main or run terminal of the pot and so the offspring of this marriage will be the line running to the start terminal.

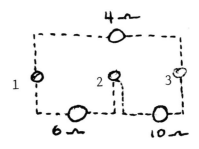

(Fig. 7-12)

111

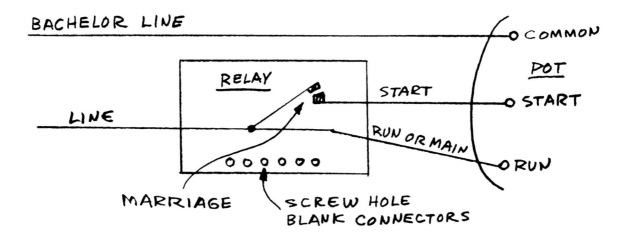

(Fig. 7-13)

See this hook up and the imaginary marriage (**Fig. 7-13**).

There can never be any other combination inside of a relay other than this marriage of the start and run. The common will always be a lonely bachelor. The common line will come to the pot free of all branch connections. The common may be broken here and there by the thermostat or the overload device. But it will never have any connection with the relay or the starting device. The common terminal already has a permanent end of the start winding tied to it inside of the pot.

Now, we can strip the household refrigerator and

see the electrical in two or three different angles. Here is the simplest of all refrigerator schematics (**Fig. 7-14**).

You will note that in all of these drawings the same old basic method of starting the pot is used. In the window unit it does not matter that there are two speed fans, push button controls, green lights, or bug killers; the same basic electrical starting devices are present. Once the household box or window unit is stripped of all its refinements or fancy wiring, the bare terminals are the common, start, and run. The unit will require a relay and it may or may not have a capacitor. (**Fig. 7-15**)

Whether you are checking out pots or by-passing relays, the best tool you can own for this work will be

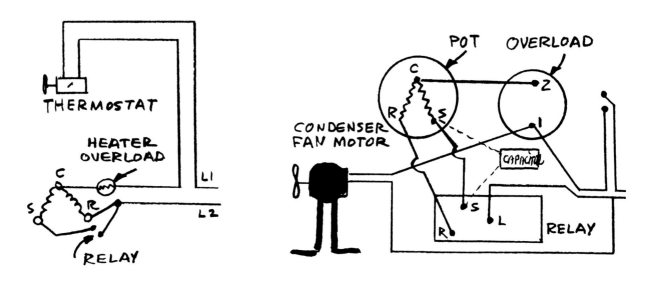

(Fig. 7-14)

(Fig. 7-15)

a clip cord. If you are in doubt about a pot, don't waste time. Take the connections loose from the terminals. Mark them or tag them so you will put them back in the right order in case the pot is good. Connect your clip cord to the common and run and jump from the run to start with a wire or capacitor. Try to start the pot. If it runs good, then you can look to the relay or the thermostat for your trouble. The quickest test here would be to by-pass the thermostat by jumping across the thermostat connections so that it will be on all of the time. If the pot still refuses to run, then look to the relay. Here you will have to use your own judgment. If you have eliminated everything else, it will have to be the relay. Don't be too sure, always watch out for broken wires or loose connections which might indicate a bad relay when the relay may be perfectly good. Many good shops change out relays whether they need it or not, like changing the points on an automobile ignition system whether they need changing or not. **DO NOT FIGHT RELAYS; CHANGE THEM.**

Questions That Need Answering

Why must the starting winding be disconnected every time the unit starts? The start winding is fine wire and has many turns in each coil. If it were left on the line for more than forty seconds to a minute, it would cook. Likewise, if a motor does not start and the run is in the line and humming, it too will burn out if it is not disconnected by the overload device. The start winding will burn if the unit does not start, and it will burn if the unit starts and it is not disconnected.

How do you take a terminal connection loose on a pot? You hold a backup on the lock nut under the nut holding the wire connection tight. Never let the terminal bolt turn when you are taking loose the top nut. If this happens you may break the leads off the end of the bolt on the inside of the pot, and the pot will be ruined.

What is a clip cord or a starting cord? This is an ordinary piece of two-wire, size 12, rubber covered cord. You should make up one about six to ten feet in length. On the ends you solder or bolt the alligator clip. ∎

CHAPTER 8

MOTOR STARTING DEVICES

You know that you are not expected to do even minor repair to an open-type motor. You take these motors to the motor repair shop. The motor shop man may be able to loan you a replacement motor until your motor is ready. If not, the customer will just have to wait or buy a new motor.

You are the motor man where hermetics are concerned. You are not expected to open up hermetic pots or rewind any hermetic motor. In fact, you can service hermetic pots for years and never look inside of a hermetic pot or go any further than the terminals on the outside of the pot. However, you are expected to repair or replace all defective hermetic motor starting devices. You will be expected to know whether a pot is good or bad. If you have to make this decision, make it with the knowledge that you are absolutely sure. There is nothing more aggravating than to condemn a pot and later find that it is perfectly good. If this happens, you have failed to test it properly. Take your time and be sure.

The motor starting devices you will be concerned with are the relay and capacitors. You will, of course, be expected to understand the overload device. But rarely does an overload give any trouble. Overloads, line starters, thermostats, pressure controls, and ordinary off-and-on switches are all a part of the electrical devices on refrigerating equipment. These electrical devices are no more complicated than those which you use every day in the modern home, such as hot water tank controls, washing machine controls, iron thermostats and the usual appliance off-and-on controls. The two or three-speed switch on your kitchen mixer is probably twice as complicated as the switches on a window unit air conditioner.

In reviewing the hermetic pot, we know that we have a single-phase motor in the pot. We know that it has a starting winding and that the ends of the windings are tied onto the terminals. We know that we have replaced the centrifugal switch, common to open motors, with the relay. We know that the electrical engineers came up with the starting capacitor to help the ordinary motor to start with more zip and power. We know that this starting capacitor is hooked in so that it is disconnected from the winding when the relay disconnects the starting winding. I have not mentioned another type of capacitor which is being universally used but is not absolutely necessary in order to operate a hermetic pot. However, if the motor designer engineered his pot to use this special capacitor, it will have to be used. This special capacitor is called the running capacitor. Many hermetics use this running capacitor in conjunction with the starting capacitor. Take a look at **Figure 8-1** and note the hook up where the ordinary relay and starting capacitor are the principle motor starting devices. Compare the relay capacitor start pot to the relay capacitor start, capacitor run pot.

This running capacitor is hooked in between the run and start terminal, and it stays in the circuit all of the time. I repeat, *even though the relay has cut the starting winding out of the circuit and the starting capacitor is out, the running capacitor is still in the circuit.* This running capacitor can always be identified by its hook-up relation to the other starting devices, and it will always be in a metal can since it is an oil-filled transformer or capacitor. You can compare this extra capacitor to the supercharger on a gas engine. The engine will run without the supercharger but not as well as it would run with a supercharger. The electrical engineers have discovered that this oil-filled running capacitor, when used on a hermetic, will actually reduce the amperage load pulled by the motor. But probably more important, the use of the running capacitor permits some savings in the actual construction of the single-phase motor power-

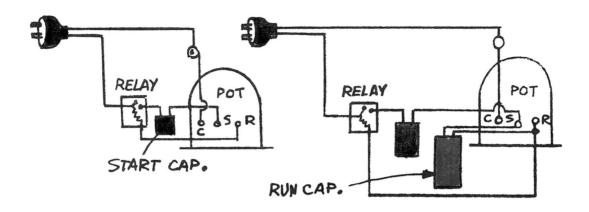

(Fig. 8-1)

wise. There is some question as to whether the actual savings are not off-set by the wear and tear on the running capacitor. Capacitors do blow out and leak their oil. *Do not confuse the running capacitor with the starting capacitor.* One is a dry electrolytic starting device, and the other is an oil-filled running device which stays in the circuit all of the time. What happens electrically within the windings of a single-phase motor when the running capacitor stays in the circuit between the run and start windings is far too complicated to explain here. It is enough that you know when one is bad or good and whether it is the proper size and doing its work. Compare the start capacitor here with the running capacitor in appearance.

Let's briefly review the development of the hermetic pot motor. Originally the engineers had a single-phase no-help pot with run and start windings. Many of these pots are still in use. Then the engineers added the start capacitor between the start windings and the relay. The relay was necessary after the motor was welded up inside of the can. After this system of starting a hermetic pot had been perfected, the electrical engineers discovered that another special device could be added; that is, the running capacitor. They added this capacitor and hooked it between the run and start. The engineers designed this special oil-filled capacitor so there was no necessity to disconnect it from the circuit. So even though the relay cut out the start capacitor and the start windings from the circuit, the running capacitor stayed in the circuit. The manufacturers called this their capacitor start, capacitor run hermetic pot. That is, the pot started by the relay and start capacitor jolting the armature; and still the run capacitor was always in the circuit. This was a pretty good combination and is still used by some manufacturers who favor the capacitor

start, capacitor run type pot over the plain capacitor start, single-phase pot.

Just about the time the engineers had perfected the running capacitor to be used in conjunction with the relay and start capacitor, they made another discovery. They discovered that the oil-filled run capacitor could be made to do double service. That is, act as a start capacitor as well as a run capacitor. With this type capacitor on a pot, there was no longer any need for a start capacitor or a relay; and both were eliminated. Here is the new look in pots. Note that this pot is called the *Permanent Split Capacitor Type.* (**Fig. 8-2**)

This PSC motor is very popular with the air conditioning manufacturers using hermetic pots. This is understandable since it eliminated both the start relay as well as the start capacitor. This new run and start capacitor kicks the pot off just like the start capacitor with relay. After the pot reaches its running speed, the new capacitor becomes a run capacitor with the characteristics of a resistor in the circuit through the start windings. The old capacitor start, capacitor run motor with relay has not really changed; it is still the same old single-phase motor; but now it has the new look. In fact, a relay and start capacitor can be added to one of the modern PSC hermetic pumps, and it will function just as it did before with start and run capacitor with relay. We have not discussed this new PSC pot motor until now for a good reason. Had we gone into this new type combination start and run capacitor which eliminates the relay, you might have gotten the impression that the relay was no longer needed nor served any useful purpose in this trade. This is not the case. The relay is very important to the hermetic pot; and even though the new pot uses this combination run and start capacitor which will do everything for the motor, the manufac-

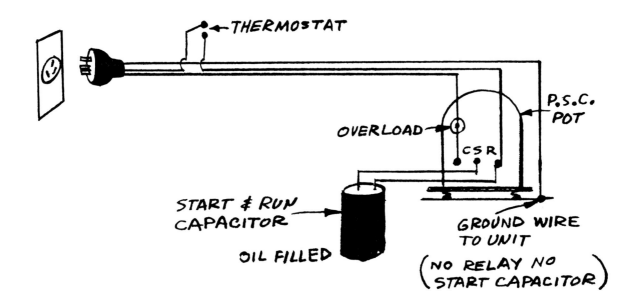

(Fig. 8-2)

turer still reserves the right to go back to the relay in order to get this motor kicked off if the new combination capacitor will not get the job done. You can see that it would be extremely important that you understand the relay if the manufacturer himself has the relay for an ace in the hole in case of the PSC failure. The PSC jobs are wired exactly like the old straight run capacitor hook-ups. That is, between the run and start. To identify a PSC pot, you only have to look for the relay. If it is missing, then it will have to be a PSC pot. Likewise, the start capacitor will be missing since a start capacitor has to be disconnected just like the start winding; and you would never find a plain start capacitor without a relay.

Take a look at this pure PSC hermetic pot wiring diagram:

This special run-start capacitor may be large or in multiples. You will find these PSC pots used in the central system condensing units also.

Just what takes place inside of this oil-filled capacitor used on the PSC pot is of no consequence to us since we will never repair one or do more than test it to see if it is good. The one thing that we are sure of is that it kicks the pot off like a start capacitor, and then it changes into a run capacitor and helps to keep down the amp pull on the line. This new type start and run capacitor is easy to trouble shoot. In fact, it is almost a test device in itself. That is, it can be used to test the pot to see if the pot will respond properly.

If you have a PSC pot, use a start capacitor with two

leads bare at the ends. Touch the bare leads to the start and run terminals. If the compressor is good, it should start. If not, the compressor is bad. If the compressor starts, remove the capacitor and it should run. If a new PSC capacitor does not start the compressor then it is time to put the start relay and capacitor on. The compressor has become a hard starter.

NOTE: When using the PSC capacitor, pressure must equalize between the high and low sides. It takes approximately three minutes on the off cycle for the equalization to take place.

Look at the P.S.C. hook up in **Figure 8-3**.

Summing up all that has been said about hermetic pots, we can say that they will fall in one or the other of the following classes:

1. Single phase with plain starting winding, using a relay for the disconnect device.
   a. Used in household refrigerators and some freezers.
2. Single phase with start winding and capacitor start with relay.
   a. Used in household refrigerators, freezers, window units, and a few small commercial condensing units.
3. Single phase, start winding, capacitor start, capacitor run with relay.
   a. Used in window units, central condensing units, and some hermetic package units.
4. Single phase, start winding, Permanent Split Ca-

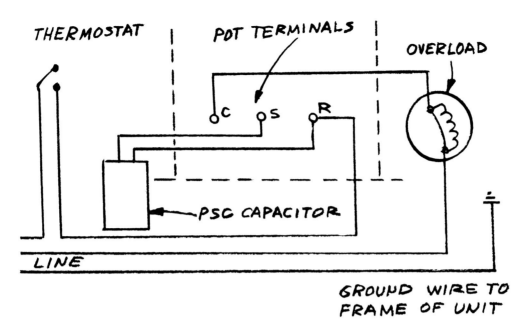

THERMOSTAT | POT TERMINALS | OVERLOAD

C S R

PSC CAPACITOR

LINE

GROUND WIRE TO FRAME OF UNIT

(Fig. 8-3)

pacitor type.

a. Used exclusively in window and central units and in some package units.

5. Three-phase pot. No help of any kind needed.

a. Used in package units, central and commercial condensing units.

You must remember that a run capacitor and a start capacitor are two different things. They may be closely related, but they have a different job to do, and they are built differently. The start capacitor is usually enclosed in a hard bakelite case, and it may be of different sizes or capacity. Its size is usually measured by the MFD and the voltage. If the unit is 240, then you use a 330 V capacitor. The MFD may be whatever the motor is designed to use. You can vary this slightly if you cannot get an exact replacement. However, the voltage must be correct for the unit. Run capacitors are almost always encased in a metal oil-filled can, and the MFD is low. The run capacitor's MFD may be as little as three MFD or as high as twenty MFD or thereabouts. MFD stands for microfarad, and that does not tell us anything much. But we are not studying to be electrical engineers. The permanent-type run-and-start capacitor will look just like the ordinary run capacitor but will generally be larger in size. The MFD will be from two to twenty, and the voltage must be correct. Once in a great while, you will find that a manufacturer has used two 120 volt start capacitors in parallel to take the place of one 240 volt start capacitor. This practice has been discontinued in

most cases. Even though this will work and two 120 volt capacitors are cheaper than one 240 volt capacitor, you would be constantly changing capacitors, both run and start, if you work on many central systems or window units. These starting aids do go bad, and there is no repair for them. They have to be replaced. One of the most common reasons for a pot to fail to start and run is the breakdown of these capacitors. In fact, it is the first thing you check on a capacitor motor after you have established whether there is a voltage available to the unit and that the thermostat is on. One of the simplest tests is to remove the suspected capacitor, whether it be start or run, and spark it out. Sparking out means to put a charge into the capacitor and jump across the contacts to see if it will unload or if it is capable of taking a charge. There is no quicker or simpler test. If the capacitor does not respond to this treatment or seems to be very weak, do away with it and put in a new one.

Here is the spark test: Take the bare ends of a cord plugged into a 120 volt outlet and very quickly flick the wires across the two capacitor posts. This will charge the capacitor. Pull your cord. *Remember: it is hot.* Now take a screwdriver with an insulated handle and short across the posts of the capacitor. If it is good, there will be a hot spark. It will crack like a firecracker. If you are leery of working with the hot ends of 120 volt service, then use a light bulb in the line to reduce the voltage.

See these tests for capacitors in **Figure 8-4**.

What if you were out in the field servicing air con-

117

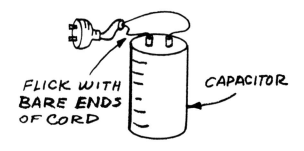

FLICK WITH
BARE ENDS
OF CORD

CAPACITOR

HOT SPARK

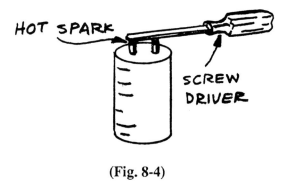

SCREW
DRIVER

(Fig. 8-4)

ditioners and you found a bad start capacitor, say 200 MFD - 330 V. You go to your truck and you look into your assortment of capacitors and can't find anything to match it. But you notice that you do have two 100 MFD capacitors at 330 V. Then you could put these two in parallel and reach 200 MFD. Capacitors in parallel double in MFD. Example: MFD = C1 + C2. MFD = 100 + 100 = 200. See **Figure 8-5** for capacitors in parallel. Now let's say you didn't have the two 100 MFD capacitors, but you did have two 400 MFD capacitors rated at 330 V. You can wire these two capacitors in series and get the 200 MFD you need. See **Figure 8-6** for capacitors in series. The formula for capacitors in series is:

$$MFD = \frac{C1 \times C2}{C1 + C2}$$

$$MFD = \frac{400 \times 400}{400 + 400}$$

$$MFD = \frac{160,000}{800} = 200$$

We have been studying the refrigeration grade motor and its starting devices. We know how to handle the conventional open-type motor, so we can begin now to service hermetic motors since we already understand the devices used for getting these sealed-in motors to rolling. There is one more electrical device which the

refrigeration serviceman should understand if he proposes to work on any equipment larger than one horsepower. This last device which we will look at is the line starter or, as it is sometimes called, the magnetic starter. The line starter is nothing more than a switching device which uses an electric holding coil to do the work of a human being. That is, you would never need a line starter on equipment larger than one horsepower if there were always someone in attendance to turn the unit off and on as it was needed. I agree that the automatic temperature controls will do this automatically, but this won't do where the delicate temperature control would have to take the brunt of the line surge or back pressure of amps which would hit the control when the motor started to run. You may not be aware of this since it is not particularly important to a serviceman, but every motor pulls a terrific amount of amps for about a tenth of a second when it starts. Amps are heat or resistance, and something will give if the back pressure of these amps on the line are high enough. In other words, any motor, no matter what it is connected to, will pull considerable amps when it starts. Most fuses are time delay; that is, they will not blow because of this surge on the line. But they would if it lasted more than a few seconds. As soon as the armature in a motor begins to

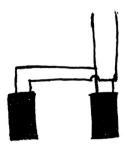

PARALELL

(Fig. 8-5)

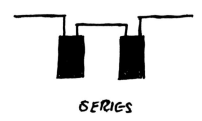

SERIES

(Fig. 8-6)

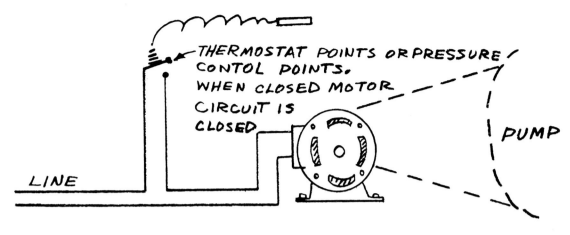

THERMOSTAT POINTS OR PRESSURE / CONTOL POINTS. WHEN CLOSED MOTOR CIRCUIT IS CLOSED

LINE

PUMP

(Fig. 8-7)

roll, this amp load will fall off immediately and will begin the normal amp pull of the running motor. It is the load of starting the armature to rolling that causes this back pressure of amps on the line. Take a look at this thermostat control which starts this motor without a line starter in **Figure 8-7**.

When the contact points on the thermostat close, the whole electrical load is on these points for as long as the unit is running. There is a big load the instant the unit starts, but it falls off directly when the motor begins to turn. Now here you have a small fractional horsepower pump. If the pump pulled three amps when it was running, you could estimate that it might pull as much as 15 amps for a tenth of a second when the motor was first starting. In this case, the points in the thermostat can take a beating of 15 amps without serious injury. But what about a five horse motor starting on a small delicate wall thermostat inside of the home or place of business? In this case, if this were attempted, there would be a flash of fire; and the thermostat would melt off the wall. The thermostat would burn because the five horsepower motor would pull probably as much as 50 amps for a tenth of a second, and this would be long enough to burn up the control. Even if the unit were running, there would not be enough contact surface in the small temperature control points to handle the amp load on a five-horsepower motor. It takes heavy wire and contact surfaces to handle the amp load of a five horsepower motor, especially when it is just starting. But the electrical engineers have designed a switch which will handle this heavy load, both during the start and the run of big motors. They knew that it would be possible to have, say a big knife blade switch and have a man close it every time the load of starting a big motor was put on the line. However, this would require a man

in attendance at all times. So, instead of using an engineer to start this knife blade switch, they decided to use the holding coil. The holding coil is a magnetic coil with a straight ram type armature or rod that will move the minute the coil is energized. They did away with the knife blade type switch and used a flat contact type instead of the blade between the copper contacts.

See the magnetic starting switch **Figure 8-8**.

Now the engineers have a switch that can be thrown with a magnetic coil instead of a human hand. This coil pulls practically no amperage; therefore, the most delicate control can start the biggest motor.

See the hook up of a line starter **Figure 8-9**.

You will find these line starters on practically all equipment larger than one horsepower; and in many cases, they are used to take the brunt of the amp load on the line feeding a motor as small as one-half horsepower. You will find line starters on all central systems of any quality or size, and you will find them on most commercial equipment above one-half horsepower. Window units let the thermostat take the brunt of this load, and you are not likely to find a line starter on any window unit. If the window unit is more than 1 1/2 horsepower, the unit should have a line starter. However, many units of two and 2.5 horsepower still let the thermostat handle the heavy amp starting load. There is nothing mysterious about the function of a line starter. A line starter could be used to open a garage door or a rancher's gate. It is nothing more than an electrical device for doing a job that would otherwise require a manual move by a human being. Do not think of a line starter as having anything to do with the starting devices on a single-phase motor. There is no connection whatsoever between the two devices. You will find line starters in factories on lathes or on an electrically driven

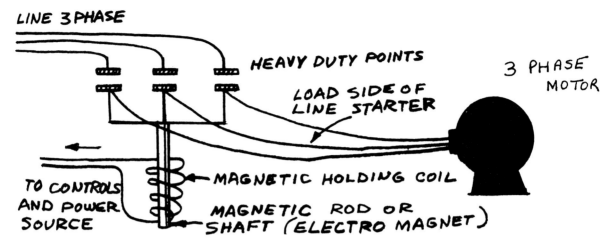

LINE 3 PHASE

HEAVY DUTY POINTS

3 PHASE MOTOR

LOAD SIDE OF LINE STARTER

TO CONTROLS AND POWER SOURCE

MAGNETIC HOLDING COIL

MAGNETIC ROD OR SHAFT (ELECTRO MAGNET)

(Fig. 8-8)

cement mixer. It has one job to do, and it does it as long as there is voltage available to the coil. By using the electrical coil to close or open the switch on a unit, you can see that there would be unlimited opportunities to control the running of the motor from remote places. For instance, you could have one little delicate thermostat in the Empire State Building and when the points of this thermostat closed or called for air conditioning, the whole plant could start up by using line starters. It

would not matter that more than 5,000 horsepower were involved.

In recent years the manufacturers of central systems and many package type air conditioners have not even bothered to use 120 volt coils on their magnetic starters. They are putting a door bell transformer inside of the unit. This transformer gets its power from 120 volts, and then it puts out 24 volts to energize the holding coil in a line starter which may be starting up a

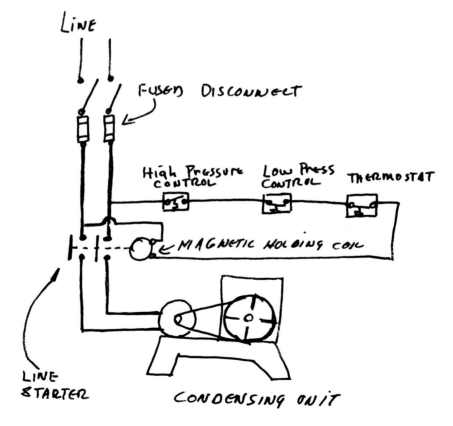

LINE

FUSED DISCONNECT

High Pressure Control   Low Press Control   THERMOSTAT

MAGNETIC HOLDING COIL

LINE STARTER

CONDENSING UNIT

(Fig. 8-9)

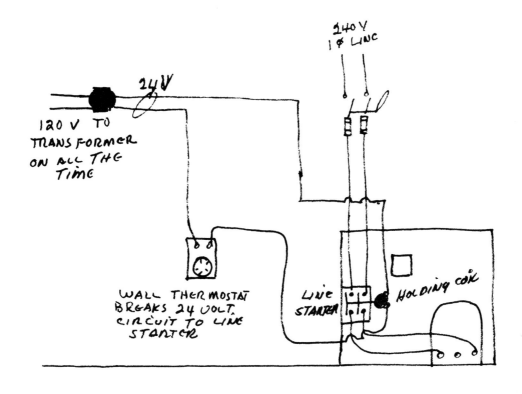

(Fig. 8-10)

ten-ton unit.

The low voltage control system has become very popular with the manufacturers of central air conditioners. One of the advantages is that very small wire can be run throughout the house, and this wire will be required to conduct only 24 volts. The holding coil on the line starter, once energized by this 24 volts will slam contact points together that will close a circuit on a three-horsepower, three-phase pump.

Look at the diagram for a 24 volt control system for a central three-ton unit (**Fig. 8-10**).

See this control system circuit in a large modern home with wire strung along the rafters and up and down the walls between the studs (**Fig. 8-11**).

You will find that the central heating plants used in conjunction with air conditioning use the low voltage transformers and line starters exclusively. There is no electrical advantage to using this 24 volt system. It is

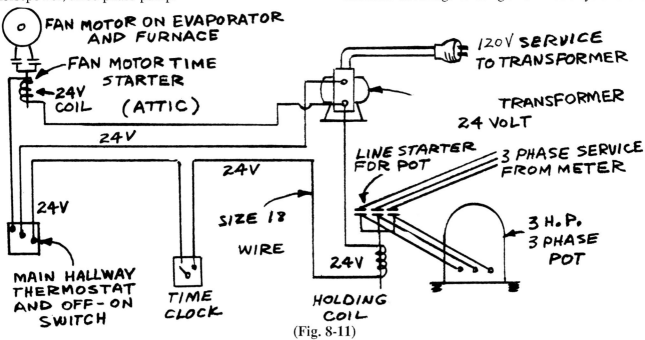

(Fig. 8-11)

121

simply a saving on installation and replacement cost.

There are bound to be other electrical devices which you will encounter from time to time. You could work in this trade for twenty years and still find a new one on a job. Whatever the device is, your job is to first find out for yourself just what the device is supposed to do. Do not take anything for granted. Examine the gadget carefully and trace out its circuit. If the circuit cannot be traced because of a maze of wires or pipe, try to make the control gadget perform for you. In this way you may be able to find out exactly what it is doing tied into the system. You may have some time clock to deal with; and if you do, you are on your own. Most time clocks give complete instructions for setting on the inside of the door. Even the most experienced service engineer will have to read these instructions. In many cases the proprietor can help or explain just what the time clock is supposed to do for the system. Consult with him and ask him if he is familiar with the time schedules on a particular clock. You will find some clocks on commercial equipment, but none on air conditioning equipment, unless they are there to just simply start the system up early in the morning so the building will be cool when it is first opened. One thing you can be sure of when dealing with time clocks is that they will have only one function, that of opening or closing a switch so something will come on or go off. It will be up to you to determine what is to be off and on and when. As you become more familiar with these gadgets, you will see that they are very simple and offer no real problem to the serviceman who is trouble shooting the electrical equipment.

With the knowledge you now have of the motor starting devices and the electrical components on the various systems, you are now ready to trouble shoot equipment both from an electrical as well as refrigerant cycle approach. You can start with the household refrigerator. Suppose you have a household refrigerator in your shop. The box was brought to you for repairs, and you do not even know what the complaint is. What does it matter? You know what a household refrigerator is supposed to do. It is supposed to run, get cold and cycle off at regular intervals when the box is seasoned out. It should make ice cubes in the evaporator, and it should maintain food preserving temperatures in the storage compartment. The box should be reasonably quiet in operation, and it should not leak water on the floor or shock the owner. With this in mind you have nothing more to do than plug it in and see if it will do all of these things. It does not matter that this is a customer's box.

It could be a box that you bought for three dollars out of the back of a furniture store. It could have been a blind trade-in on a box that you sold. It does not matter; it is just a box. Let's go to work. You plug the box in and it is stone dead. Open the door and see if there is a light bulb in the inside light socket. If there is a bulb and it is not burning, unscrew the bulb and test it in another socket. Suppose it burns. You can now say that either the door switch in the refrigerator is bad or the socket is no good. You put the bulb back into the box and jiggle the door switch. The bulb flashes off and on. You can say that there is a defective door switch, which is a small matter. The main thing that you have established without any examination whatsoever is that the cord from the box to the receptacle is good and that there is voltage to the box. You will find out in a minute why it will not run. The fine art of trouble shooting refrigeration equipment is in the ability to eliminate the possible trouble spots. By elimination you shrink the area in which you must work in order to locate the trouble. You have eliminated the cord to the wall. You have found a small sore spot, the door switch, which is not important. If the light had never come on after you jiggled the switch and screwed the bulb in tighter, you would have had to back up to the wall receptacle and the cord. They might have still been good; and something else was the trouble; but, nevertheless, you very quickly established that there was a good cord so that you could get on with your work. Your next move would be to unplug the box and take the screws out that hold the thermostat in the box frame. Tie both wires together that are leading to the thermostat, or put a jumper across the two screws in back of the thermostat. Now plug the cord in again. Suppose the box started. Then you could look to your thermostat for trouble. You know the box is warm; and one thing for sure, the thermostat should have been on. If it was not on, then it is defective. Suppose however, the box still failed to start. You have not wanted to get into the main wiring harness because you see that there are bundles of wires in and around the unit. But now it looks as if you will have to go to the unit to find the trouble. You must go down to the main wiring to the unit and see if there is something wrong there. All of the time you could be thinking to yourself that this customer's pump motor is burnt out, and the motor windings are open (no circuit) or that this box that you bought for three dollars is a bum one. With this thinking in mind, why not go directly to the pot and assure yourself before you waste any more time checking old wire that the pot itself is good. You take the chimney cover off the back of the box. Be careful that you do not kink or break the condenser lines running to the chimney covering the back of the unit. Sometimes the condenser tubing is bonded to the actual chimney. You have now exposed the pot, and you take your flashlight and take a careful look around the pot. You may spot a burnt place on the relay, or the capacitor will look burnt if the unit uses a capacitor. In this case, you will still want to try the pot. So why go down and buy a relay or capacitor and find that, after all the trouble you went to to put it on, the pot still won't run. Suppose you do see a relay that looks mighty suspicious; you will still check the pot. The main thing that you are looking for now is loose wires or broken connections. You are looking for the obvious thing, the simple thing, nothing complicated. You are looking for something that will or might even save you the trouble of checking the pot itself. Suppose you rapped the relay with the handle of your screwdriver, and the unit suddenly kicked in and ran good. There would be no need to check the pot now. It checked itself when you found a defective relay that only needed some rapping to make contact. Even so, there should have been a hum coming from the pot before you kicked the relay. Suppose you rapped the little overload device, and the unit started up and ran. You could say that the overload had become defective. But all of this is wishful thinking in most cases. You will not be lucky enough to establish that the pot is good by just rapping the starting devices.

Suppose nothing happened when you tapped the starting device and the overload. Now you are ready for the acid test. Carefully disconnect the wires leading to the pot terminals, and get your clip cord hooked onto the common and run. You located the common because you knew that the overload was on the common. You located your start connection by following up the wire from the "S" connection on the relay to the terminal which would have to be the start. This leaves only one other terminal; so it would have to be the run. See the follow up in **Figure 8-12**.

Now, this unit does not use a capacitor. With this knowledge, since you did not see a capacitor attached to the pot, you are ready to jump from the run to the start the moment you plug in your clip cord. The unit starts and runs good after you have taken away your jumper. Let it run for quite a spell, thirty minutes to an hour will not hurt. If the pot seems to be running fine electrically, then your other concern is that the pump valves are good and the unit is pumping. After running for a few minutes, you find that there is gas in the system. You find that the evaporator is frosting up nicely, and everything is normal. Now what do you have at this point? You have a good box to sell, as soon as you find the defective electrical device. If it is a customer's box, you know that you are going to have a fair charge and possibly a very nice profit for using your head and not your cutting tools. Never cut into a system until you know whether you are going to have to or not. Had this box run all right electrically and had no refrigeration, then you would have to go one step further and find out whether it was just out of gas or whether the valves were ruptured or not. This would call for cutting into the suction line and putting on a pig tail and your gauges. If, when you cut into the suction line, you found a full charge of gas, you could say right then and there that not only did the box

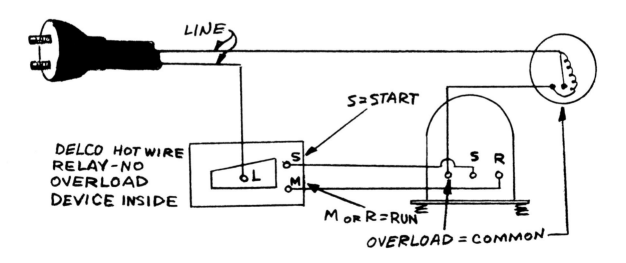

(Fig. 8-12)

123

have a defective electrical device; but it also has defective pump valves. It would then be good only for junk. It does not pay to put new pots in old refrigerators unless they are in warranty. If the box is in warranty, you would be doing the job for the distributor or dealer anyway; and he would make this decision, not you.

Go back now and let's suppose that the box runs nicely with the clip cord. The refrigeration is good and everything is perfect except that the box will not start or run on its own electrical devices. In this case, the most likely thing that will be wrong will be the relay; and the examination and final results that you obtained warrant your trying a new relay without any further examination. Your assuming that the relay is bad is a very fair guess. If it does not remedy the situation and it turns out that the overload is defective, as remote as this might be, you could still say that the relay probably could stand changing out anyway. If this does not cure the trouble, then you, of course, will have to look for the broken wire in the harness or leads running throughout the box from the thermostat and other places. This is not likely to happen in the repair of 100 refrigerators.

You know how to charge household units. If, when you opened up the system because it refused to refrigerate, even though it would run, you found no gas; then you could quickly confirm that the valves were good by noting whether the opening you had made would tend to suck or have a vacuum when the unit was running. If this were the case, you would proceed to evacuate and recharge the unit in addition to replacing the defective starting device.

I believe that it is more important to know what a unit is supposed to do than it is to know what to do if it is not working. You can save countless hours of hard work putting back wiring and welding up holes if you will never cut or take apart anything until you have eliminated the possible trouble spots. You would never tear out the generator or starter on your automobile simply because it failed to start without first finding out whether the battery was up or not. You would never tear out the pot on a refrigerator until you are sure that you have tested every device that could cause it to be dead. In this country the junk yards are full of old refrigerators that have perfectly good units and pumps in them. They are in the junk yards because incompetent servicemen, not being able to find out with any degree of certainty whether the pot was good or bad, pushed the panic button and condemned the pot. Don't you make the pot your spanking boy. In most cases, you can give the pot the benefit of the doubt. It will be the last thing to wear

out on the system. If it will make you more confident, let me tell you that if you were the greatest electrical engineer in the country, you could not proceed to test out the pot on this particular refrigerator with any more degree of certainty than a man with ordinary intelligence who is willing to go step by step and isolate the trouble spot. Whether it be a household box, a window unit, or a thousand-ton commercial unit, the key to being a good trouble shooter is the ability to back off from the job, sit down if necessary, and think before you make a move.

Suppose you and I were standing side by side looking at a household refrigerator that you had just bought from a junk dealer. The box is clean and looks nice. You have a hunch that it can be fixed; and you have the time and know how. You have only three dollars, which is a small gamble, invested in the box. You may even have a ready sale for this box if it proves out. Knowing where you got the box, I would say to you first, "Have you tried to run it?"

"No," you say, "not yet. I am just looking it over."

"Good," I say. "Did you see anything inside that might give us a clue as to why it was in the junk yard?"

"Look for yourself," you say. "It looks like new inside. That's the reason I bought it."

Then I say, "Let's try it out."

You say, "O.K." So you pick up the cord coming out of the box and take a good look at the two-pronged cap. It looks clean

"Here goes nothing," you say and plug it into a 120 volt receptacle in your shop. Nothing happens. We both take a look inside of the box. The light is on, but the box is dead as a door nail. You jiggle the thermostat and nothing happens.

"Thermostat could be shot," I say. "Let's jump across it."

You say, "I'm not even going to worry with the thermostat. It looks too hard to pull out to get to the wires in back. "I'm going straight to the pot and see if it is good and if the whole box is worth wasting another minute of my time."

"You're a sharp serviceman," I say.

You uncover the back of the box and get your cord ready to tie onto the exposed terminals. You look up at me and say "I believe I've found the trouble, but I'm not going to take anything for granted. I'm still going to check the pot." You proceed to tie on your clip cord and get your jump wire ready. I plug the clip cord in for you, and you jump across the right terminals, and the pot starts. You come around to the front of the box and say to me, "So far, so good. Now if I have any refrigeration I can make some money on this deal." You noticed a bum relay when you were down with your flashlight checking the pot terminals. We both listen to the evaporator to see if we can hear the telltale gurgle or hiss of liquid refrigerant beginning to feed into the evaporator from the cap tube. We hear nothing. You believe the valves are gone and that must have been the reason the box hit the junk yard.

I say, "Let's wait. The piston is dry, and the valves are dry because the pump has set up for so long without running that the oil in and around the pump parts has drained down into the crank case of the pot. It may take awhile for the pot to take hold."

You agree since you do not want to cut into the system unless it is absolutely necessary. Sure enough, after about one hour of steady running on the clip cord, the tell-tale hiss-hiss starts to come out of the evaporator. Within another hour the evaporator is all frosted up, and the box is running just like the day it came from the factory. "Well," you say, "we know that there is a full charge of gas; that she pumps O.K.; and, apparently, it should never have been junked." You buy a relay the next day and install it in place of the old one. The box is ready for sale and delivery.

The point of this get together over this household refrigerator is this. You started from scratch. You did not have a customer to tell you that he had punched a hole in the ice cube maker with an ice pick and had heard a hissing sound as the gas blew out of the system. You did not have a customer to tell you that his box had been a hard starter for some time and had finally quit. That is, the customer told you that he heard this box try to start and then click off, hum and click, hum and click, and finally make it after a dozen tries. When a box acts like this you would have a hunch the relay was bad before you ever took the box to your shop. But in the case of the box you bought from the junk yard, you did not have any lead or indication as to what was wrong. You were starting from dead scratch. Yet, you can see it was not difficult to eliminate the trouble spots and get to the main goal. Is the pot good or isn't it? Don't waste time with pots, whether on household freezers, window units or central units. Get one thing established first. Are you wasting your time on a unit that has a burnt-out pot. Once it is established that you have a good pump, then you have no more worries. Many of the so-called servicemen today who have had no formal training in checking out pots will condemn a pot for no other reason than they do not know whether it is good or not,

and they know no way to prove it. You have this formal training, and you know what goes on inside of all single-phase hermetic pots. *Never be in doubt.* Make sure of your pot. As you test more and more pots, you will learn little short cuts that will save you even the trouble of disconnecting the pot terminals. You will get your clip cord on wires and leads that will have the same effect as tying to the pot terminals.

With the knowledge you have now of the household refrigerator, you could begin to take service calls on this particular type of equipment. If you will memorize the following instructions where household refrigerators are concerned, you will never need any further training on this type of equipment. You see if you can repeat the following words whenever you are working on household refrigerators:

"I will take my time and look around before I disturb anything about the box. My first move will be to establish that there is juice available to the unit. My second move will be to make sure that the unit is not dead because of a bum thermostat. My third move will be to locate the starting devices and note whether the unit requires a capacitor or not. I will test any capacitor before I test the pot. If the capacitor tests out all right, I will then test the pot and know for sure whether it will run and, if it will run, whether it will pump or get the evaporator frosted up. I will make every move carefully and with consideration as to what should take place."

If you follow this advice, you will be on your way to being a top serviceman. In most cases, you will be able to ask your customer questions which will help you in your work. Here is an example of good reasoning and what you should think out for yourself. You do not have to be a top serviceman to reason this: Suppose the customer told you that his box was was doing fine until 10:00 o'clock this morning; then, bang!; it quit! It would not take a great deal of reasoning for you to reason that something happened suddenly. Household refrigerators that have been running for years don't just suddenly quit unless something suddenly causes them to quit. Therefore, even though the customer may think the box has just gradually worn out, you should know better. You would look for some trouble that could cause an abrupt stopping of the unit. If your car were running fine and you were cruising along and for some reason it just suddenly went dead, I am sure you would look for a broken ignition wire or even check the ignition switch itself. You would probably look next to the coil wire to the distributor. You live in an automobile world, and you have developed certain reasoning where auto mechanics are concerned. Do the same reasoning where refrigerators are concerned. Take one step at a time and ask yourself if and what this step eliminated. Once you have isolated the trouble, you can fix it even though you may never know what actually took place. Don't make the mistake of guessing that a unit has lost its gas when it has suddenly gone stone dead. Your auto running out of gas would give you some indication by sputtering or dying gradually. Likewise, a refrigeration unit losing its gas would struggle for several days before it died; and, even so, the pot would continue to run. If you are in doubt, back off, sit down and think it out. ∎

TROUBLE SHOOTING GUIDE

SERVICE DIAGNOSIS CHART

SYMPTOMS	CAUSE	REMEDY
A. Compressor does not run.	1. Motor line open.	1. Close start or disconnect switch.
	2. Fuse blown.	2. Replace fuse.
	3. Tripped overload.	3. See electrical section.
	4. Control stuck open.	4. Repair or replace.
	5. Piston stuck.	5. Remove motor-compressor head. Look for broken valve and jammed parts.
	6. Frozen compressor or motor bearings.	6. Repair or replace.
	7. Control off on account of cold location.	7. Use thermostatic control or move control to warmer location.
B. Unit short cycles.	1. Control differential set too closely.	1. Widen differential.
	2. Discharge valve leaking.	2. Correct condition.
	3. Motor-compressor overload cutting out.	3. Check for high head pressure, tight bearings, stuck pistons, clogged air or water-cooled condenser or water shut-off.
	4. Shortage of gas.	4. Repair leak and recharge.
	5. Leaky expansion valve.	5. Replace.
	6. Refrigerant overcharge.	6. Purge.
	7. Cycling on high pressure cut-out.	7. Check water supply.

SERVICE DIAGNOSIS CHART

SYMPTOMS	CAUSE	REMEDY
C. Compressor will not start; hums intermittently (cycling on overload).	1. Improperly wired. 2. Low line voltage. 3. Open starting capacitor. 4. Relay contacts not closing. 5. Open circuit in starting winding. 6. Stator winding grounded. 7. High discharge pressure. 8. Tight compressor.	1. Check wiring against diagram. 2. Check main line voltage; determine location of voltage drop. 3. Replace starting capacitor. 4. Check by operating manually. Replace relay if defective. 5. Check stator leads. If leads, okay replace stator. 6. Check stator leads. If leads okay, replace stator. 7. Eliminate cause of excessive pressure. Make sure discharge shut-off valve is open. 8. Check oil level; correct binding.
D. Compressor starts, motor will not get off starting winding.	1. Low line voltage. 2. Improperly wired. 3. Defective relay. 4. Running capacitor shorted. 5. Starting and running windings shorted. 6. Starting capacitor weak. 7. High discharge pressure. 8. Tight compressor.	1. Bring up voltage. 2. Check wiring against diagram. 3. Check operation manually; replace relay if defective. 4. Check by disconnecting running capacitor. 5. Check resistances. Replace stator if defective. 6. Check capacitance, replace if low. 7. Check discharge shut-off valve. Check pressure. 8. Check oil level. Check binding.

SERVICE DIAGNOSIS CHART

SYMPTOMS	CAUSE	REMEDY
E. Relay burn out.	1. Low line voltage.	1. Increase voltage to not less than 10% under compressor motor rating.
	2. Excessive line voltage.	2. Reduce voltage to maximum of 10% over motor rating.
	3. Incorrect running capacitor.	3. Replace running capacitor with correct mfd. capacitance.
	4. Short cycling.	4. Reduce number starts per hour.
	5. Incorrect mounting.	5. Mount relay in correct position.
	6. Relay vibrating.	6. Mount relay in rigid location.
	7. Incorrect relay.	7. Use relay properly selected for motor characteristics.
F. Starting capacitors burn out.	1. Short cycling.	1. Replace starting capacitor with series arrangement or reduce number of starts per hour to 20 or less.
	2. Prolonged operation on starting winding.	2. Reduce starting load (install suction regulating valve), increase voltage if low.
	3. Relay contacts sticking.	3. Clean contacts or replace relay.
	4. Improper capacitor.	4. Check Parts Catalog for proper capacitor rating, mfd. and voltage.
G. Running capacitors burn out.	1. Excessive line voltage.	1. Reduce line voltage to not over 10% over rating of motor.
	2. High line voltage and light load.	2. Reduce voltage if over 10% excessive. Check voltage imposed on capacitor and select one equivalent to this in voltage rating.

CHAPTER 9

TROUBLE SHOOTING
WINDOW UNITS

You never charge a household refrigerator while the box is in the customer's home. You might change out a relay or a capacitor if that is all the unit needs. The size of the box and unit will dictate your policy. Where some freezers are concerned, it may be more practical to charge them on the premises than to attempt to bring them to the shop. However, if the freezer is small enough to be transported, take it to your shop. Only in case of emergency should you charge a freezer in the home. If you have to charge it right where it is located, come back over a period of two or three days to finish the job. Do not attempt to make major repairs to any household unit, whether freezer or refrigerator, on short notice or on one call.

Take all window units to your shop, if possible. Explain to the customer that you already have a service charge for coming out, and it will cost him no more for you to take it in and check it properly. Tell him that you will call him and let him know exactly what you find and how much it will cost to repair. Remember, if you do a quick job in his home and you overlook some small thing that had nothing to do with the work you were paid for performing, you are still held to make your work good. The customer does not understand that the relay and capacitor are two different things. All he knows is that he has paid you for working on his window unit; and if the job does not hold up, he wants you to make it good. Had you taken the box to your shop in the first place, you would probably have caught the bum relay along with the bad capacitor. It is always tough to have to explain to a customer that you did not look at the capacitors and that all you did was fix the bad relay which was in need of replacing. The main thing that is on his mind is that you did not do your job in the first place. Remember this, charge for your work and give good service and careful workmanship. If you give your work away for nothing and it does not hold up, you will always be remembered for being a poor mechanic. Never will you be remembered as having been charitable. If you charge, and your work holds up, you will always be known as the man who charges but does good work. The very man who may say your charges are high will come back to you, for he wants the best. When you are dealing with the customer in his home, make sure that you both have a distinct understanding as to what is going to be done. Do not take anything for granted. Don't ever be guilty of having to say, "I thought you knew that this was our policy." Make sure that he knows your policy. When dealing with the customer in reference to what is wrong with the unit, always state what you believe to be wrong and tell him that you do not know what caused this failure until you make a more careful examination. Don't give a customer a lecture on the refrigeration cycle. You will only succeed in convincing him that you are trying to find out yourself what is going on. The more you talk, the more you will convince him that you are inexperienced. Be thorough. Be brief. Be sincere. Collect your money when you are finished and be on your way. Don't hang around full of good tidings and joy because you have his money in your pocket. He will become aware of it soon enough after you are gone. If you do your job and do it well, there will be no need to make a high powered sales pitch. The job will hold up or it won't hold up. For this reason, you are again cautioned not to get caught in the trap of doing a fine job on the main trouble and overlooking a small thing that will give the customer the impression that you have never touched the unit simply because it quits again. If you have the unit in your shop, it will offer greater chances for you to find anything that

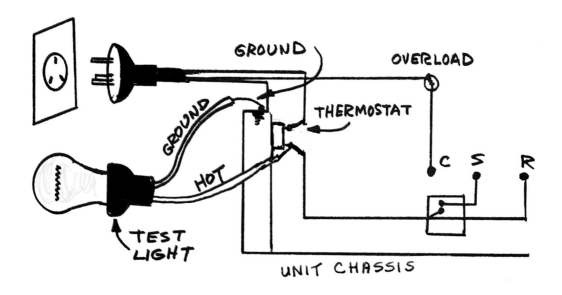

(Fig. 9-1)

might cause a failure after you have declared the job complete.

Let's take a call together on a window unit. We arrive at the home and take in a box of tools. Actually, we really do not need the tools since we do not plan to do any work in the home. But we have to take in the tools since the public expects us to be prepared to work, even if we do not intend to make more than an examination. *Do not set tool boxes down on carpeting.* The bottom of the tool box may be dripping with oil. Watch that wall-to-wall carpeting. Now we are in front of the window unit. We ask the lady of the house a few simple questions which have nothing to do with the technical aspects of air conditioning. A good opening question is, "When did it run last?" If the owner does not seem to know anything whatsoever about the unit, then your job is to find out why she called you. What is the complaint? Suppose the unit won't run. You would immediately check the wall plug to see whether the cord cap is plugged in tightly. The cord being plugged in tightly you jiggle the thermostatic switch, and if the unit attempts to start and dies again, take it into the shop. It probably has a bad thermostat, but that isn't all. It may have run long enough for the thermostat to break down the relay. The unit could be a hard starter because one start capacitor was out and the original unit came with two.

You might also find loose terminals had caused the line from the thermostat to run hot and this had contributed to the thermostat's break-down. But suppose it was simply a defective thermostat, and you change it out

right in the home. If it holds up, fine. If not, you will be called back. You were paid to fix the unit. Now fix it. Back to our unit, suppose it is still dead after jiggling the thermostat. Now you can look to your relay if you have no other way of checking to see if you have any voltage at all. Suppose you do have a pig tail socket and lamp with you, and you check the screw terminals on the thermostat to ground.

See the check for voltage in **Figure 9-1** and see the check points in **Figure 9-2.**

You find that there is juice through the thermostat and on to the relay. There is no use going any further here. You cannot fix this in the home, so just break it off right there. Declare yourself, don't quibble. Say, "I know that there is a defective relay, and I will replace it, but I would like to find out if something caused this relay to go bad or not." Actually, you plan to change the relay, blow out the condenser and evaporator, change the filter, check the pot terminals and the capacitors for leakage. Oil-filled cans can rust out, and the oil will leak out. If this happens, the capacitor is ruined, even though it might start or operate for a few times during the day.

Suppose a customer says that the pump seems to be running, but the fan does not. This will be an observant customer. You will have to check to make sure that he doesn't have it backwards; that is, the pump is out and the fan is on. However, most customers will miss the air flow of the fan very quickly. Now this fan motor is a two-speed capacitor run-and-start type. It may look just like a shaded pole motor; and for all purposes, it doesn't make much difference. This motor is two-speed and

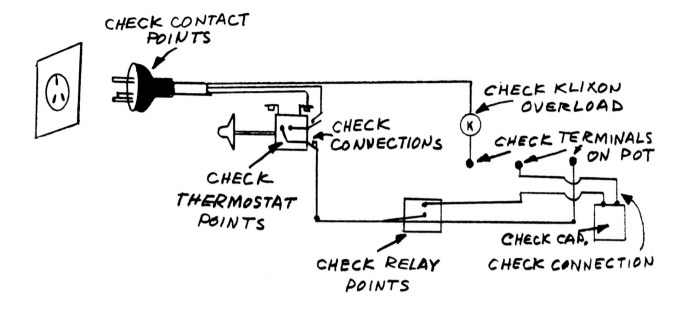

(Fig. 9-2)

looks like this in **Figure 9-3**.

You will note that a two-speed motor has three lead wires. When this is the case, one of the wires will always be on the line. It will be known as the common; that is, common to both the low and high speed windings. See this high-low speed hookup in **Figure 9-4**.

You can always bet that if there are three wires coming in and out of a window unit fan motor, it is two-speed. It will have a two-speed switch, and it will run on either speed. So, if the fan motor changes speed, it simply means that the system operates nicely or correctly on one certain speed and is unbalanced on the reduced speed.

Once in a while you will run into a fan motor that uses a transformer in the circuit in order to reduce the speed of the motor. See this hookup with the transformer acting as a resistor in the line in order to reduce the speed of the motor. **(Fig. 9-5)**

Fan motors are easy to deal with. They either run or don't run, and there is seldom any middle ground. If the motor does not run properly, it will have to be replaced or rewound.

Rewinding window unit fan motors is a gambling proposition. Even if the motor rewind man believes he can do the job and is willing to guarantee his motor rewind, you must consider that it is more work to take a motor in and out than it is to rewind it. It is better to replace burned out fan motors with new motors. Rewinds to these motors are not too satisfactory; and if you have to take a motor out again because it will not hold up, you are risking your reputation with your customer notwithstanding the loss of labor.

You must change a fan motor to the correct voltage, and it would be a good idea to try your very best to get an exact replacement. Do not use substitutes where horsepower and amperage are concerned. Fan blades put just as positive a load on a motor as a piston and crankshaft.

Never take wire colors for granted. There will be a factory wiring diagram on the motor. Take a good look at the diagram to be sure the new motor does not take on

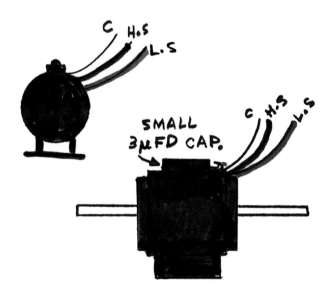

(Fig. 9-3)

131

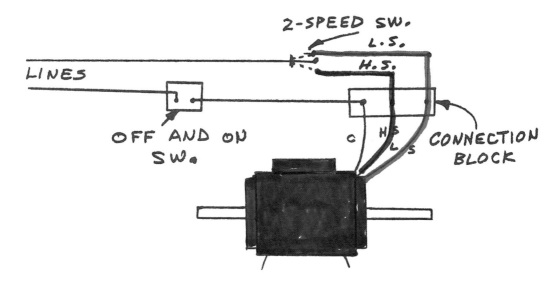

(Fig. 9-4)

a new wire coloring. Do not accidentally hook the line to high and low together. The motor will burn faster than the leads can be removed. See **Figure 9-6**.

Most new two-speed replacement motors will have diagrams with the motor. If a drawing is not packed with the motor then you should check speed to be sure. Run the fan motor on both speeds, and you will know that you have the common located. When the fan connections are made to the switch block or terminal block around somewhere on the unit, the common will always be in the line. The two-speed switch will always be working across the two-speed ends of the motor. **(Fig. 9-7)**

Now let's go back to our original fan motor complaint. The customer has said that the fan wouldn't

work, and you found that he was right. It will not work. You check all of the connections, and you find that the fan motor is apparently burned out or has an open winding. In some cases, it is possible for the fan to work on one winding when the other winding is burned out. Where this is the case, all you can do is to have the whole motor rewound, or replaced with a new one.

You should own a long Allen wrench for taking loose the squirrel cage blower on the evaporator end of double shaft motors. This is a tough screw to take loose, and it will require you to own this particular wrench. See this approach to the Allen screw in the squirrel cage blower in **Figure 9-8**.

Many of the fan blades on window unit fan motors are so designed that they stay in the housing of the condenser or evaporator even though you take out the motor itself. When this is the case, you should be careful not to bend or damage the fan blade when sliding it off the motor shaft. The motor will have to be tilted or turned so you can slide the fan off the motor shaft without putting any undue strain on any part of the fan itself. A bent fan will wobble and cause vibration and in time will loosen on the shaft. Keep fan blades clean, and they will turn better with less load on the motor. Dirt is a drag on a fan. It will cause an overload on the fan motor, and the fan will not deliver enough air if it is dirty. Where double motors are used in a window unit, the motor problem is much simpler; and taking out the motor will be an easier matter. You must always be careful in giving estimates for labor on taking out and replacing fan motors. This work can become very involved. You may find that you will have to charge

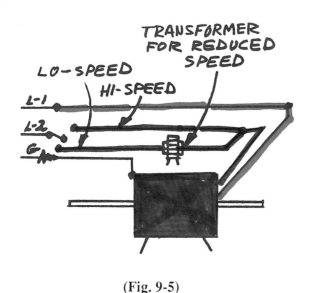

(Fig. 9-5)

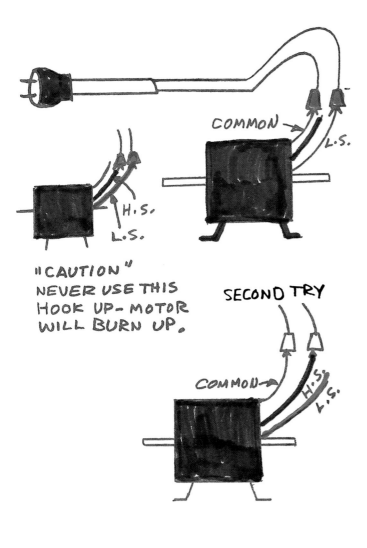

COMMON

L.S.

H.S.

L.S.

"CAUTION"
NEVER USE THIS
HOOK UP- MOTOR
WILL BURN UP.

SECOND TRY

COMMON

H.S.

L.S.

(Fig. 9-6)

twice as much as you estimated in order to break even on your labor.

The same practice applies to the fans used in commercial and package units. Clean the blades occasionally. Keep the fan motor oiled. You should pay special attention to squirrel cage blowers which can accumulate enough dust and dirt to become almost a solid disc of dirt. In many cases, where you cannot get to these blowers, you will have to use high pressure air or even a steam cleaning outfit.

But let's go back to our trouble shooting. We can make a quick check on any fan motor by taking the motor leads loose and using a clip cord.

See this check method to be sure that the fan is all right as shown in **Figure 9-9**.

In most cases, if you make the original call and pick

up the window unit, you will have some idea as to what happened; or at least you will know what the complaint is. For instance, it won't run, no refrigeration, runs noisy, lets water drip on the floor, etc. But suppose you have a helper who picked up a unit and asked no questions. He brings the unit to your shop. You are ready to proceed. First you would take a good look with a flashlight to see if you could locate anything which would tell you why the unit went bad. You would not touch the unit with your hands. While you are having this look around, you would note that the unit is 240 volts and the refrigerant is F-22. If you do not see any metal data tag and you have nothing to go by, you can take a look at the pronged cap that plugs the unit into the wall. If this cap is three pronged or polarized, then you could almost say for sure that the unit is 240. One way or another you can generally determine the voltage. Here is the procedure you can follow to tell you the voltage: See the specification tag, metal data, or name plate. If there is no name plate, look on or about the pot itself for a stenciled letter designating that this pot is F-22. If it is F-22, you can bet it will be 240. Take a look at the cord plug. If it's three pronged, it will be polarized and for use on a 240 volt unit. Another good indicator is to note the horsepower on the pot. If it is one horsepower or better, you can almost bet the unit is 240. Look at the data tag on the fan motor. It may tell you the fan motor voltage, and it will always be the same as the pot. Finally, if you are still in doubt, try running the unit on 120 and see how it responds. Never try 240 volts first. If it is not left on for any great length of time, no harm will be done even if it is a 240 unit. If you have an amp meter, you can always check the amperage against the pot size. If it is not correct for the motor rating, you will have the wrong voltage. This is no problem; forget it. Just note the voltage when you first check a pot. The kind of refrigerant will likewise be found on the name plate of the pot; and if there is absolutely no indicator of the gas, here, too, you can use your own judgment. In most cases, if the pot is one horsepower, it may be F-12. So what? If you put the wrong gas in a unit, it will not run properly, if it will run at all. But no harm will be done. It will take you only a minute to know that you must switch to the other gas. As a general rule, you can say that 99 percent of all window units will use F-22; and 99 percent of all fractional horsepower window units, like 1/2 and 3/4 horsepower, will use F-12.

We noted that this unit was 240 and F-22. You look carefully with your flashlight and find nothing to indicate trouble. Everything looks clean and operative.

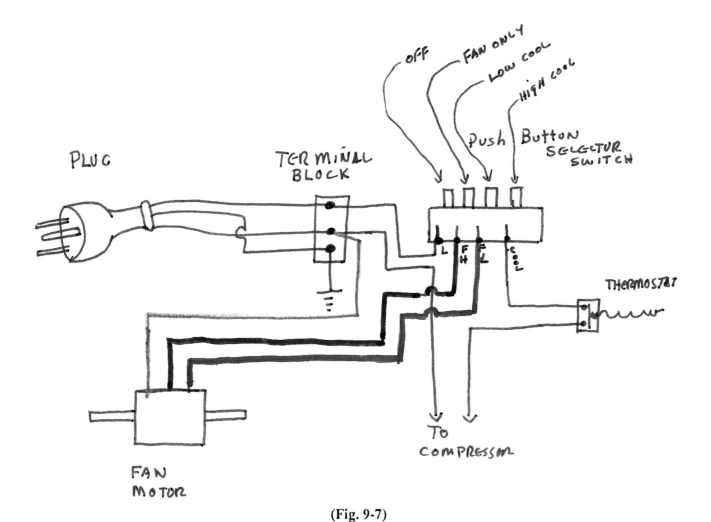

OFF

FAN ONLY

LOW COOL

HIGH COOL

PLUG

TERMINAL BLOCK

Push Button SELECTOR SWITCH

L F H E L COOL

THERMOSTAT

TO COMPRESSOR

FAN MOTOR

(Fig. 9-7)

LONG ALLEN WRENCH STUCK THROUGH THE BLADES OF THE SQUIRREL CAGE BLOWER TO HUB ON THE SHAFT OF THE MOTOR

(Fig. 9-8)

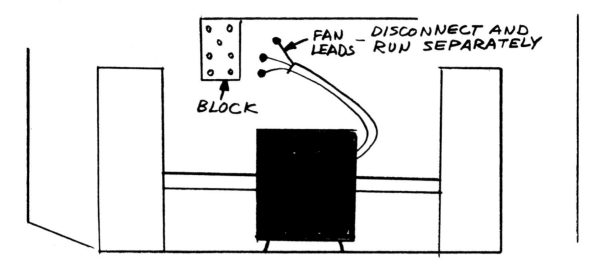

(Fig. 9-9)

Your next move will be to plug it in, try it, and see what happens. With the unit plugged into the correct service, you will flip on the fan switch. If the fan operates properly, you are ready to turn the thermostat up until the pot motor kicks in. When this happens and the pot and fan are running, then you only have to wait and see if there is going to be any refrigeration. Suppose there is no refrigeration. Then you can say that your helper brought in a unit out of gas or with a badly stopped up cap tube. At any rate, you know that the pot runs and the fan runs. Of course, the pump suction and discharge valves could be ruptured. When this happens, the motor runs and the pump turns, but doesn't pump. This is highly unlikely. Generally, if the motor is all right, the pump valves will be all right. Back up again, and let's suppose that there are two fan motors and fans on this particular unit, and the condenser fan failed to start. You can say that if you turned the thermostat up until the pot began to run, the condenser fan would be running at the same time. You never had a pot running at any time without the condenser fan running at the same time. If this should be the case, the pump will run for about one minute. The head pressure will build up because of the dead condenser fan, and the head pressure building up will put a load on the motor. The amps will rise, and the overload will kick the pump motor out. It will begin to short cycle; that is, run awhile, then cut out, run awhile and cut out. Likewise, this same short cycle or kick out can be caused by an extremely dirty condenser. You must remember at all times that more than one factor can cause identical failures. For instance, with no condenser fan running, you have the overload kicking out. In both of these cases, the excessive condenser pressure

was the problem. High head pressure slows the pump down; and the instant the motor slows or is not running full speed, the amperage goes up.

Now, we are trouble shooting a window unit which has been brought into your shop with no information to go on. Right now, without putting a tool on the unit, is where you speed your work and save time and make money. Make a decision one way or another without trying anything. You can say to yourself, "If I will take my time and go over this unit with a fine tooth comb and think while I am examining it, I will have a pretty good idea what is wrong before I ever touch a tool to it." I cannot emphasize too much the importance of taking your time before you cut or open anything up on the unit. Reason out every move you make. It does not matter that you reason out loud as long as you reason.

For instance, there would be no reason for going any further if you found that the cap tube was broken in two and the unit was completely out of gas. You would proceed to get your pig tail on and your gauges installed on the pig tail preparatory to recharging the unit. However, while you were actually putting on the pig tail you could reason to yourself, "Why did this cap tube suddenly break in two?" You will reason that it was probably caused by vibration, and you should make a mental note to tape the tube or tie it tighter to some pipe to stop its vibrating when the unit is running.

Suppose you do not see anything wrong visibly with the unit; but when you run it, it does not refrigerate. You can reason at this point without touching the unit that it either has lost its charge through a slow leak that is not visible, or the cap tube is stopped up, or the valves are gone out of the compressor. Think about this for a

135

minute. Can it be anything else? Not likely. It will be one or the other of the above troubles. Since you are not looking inside of the unit, and you do not know what is going on inside, you have only reasoned that it will have to be one or the other of the problems mentioned here. You are now ready to prove out your reasoning. You must put your gauges on now in order to find which situation exists. Suppose you break or cut off the end of the factory pig tail preparatory to welding on another pig tail. When you make this cut, the gas comes out of the unit under very high pressure indicating that there is lots of liquid refrigerant back of this escaping vapor. Is there any need to check for leaks? No, you can reason here that one possibility, that of no gas, has been eliminated. Suppose you attach the gauges; and with the gas that you found still in the unit, you run or start up the unit. Immediately your compound gauge shows a quick pull down to a vacuum. **(Fig. 9-10)**

You can now reason that only one thing can be wrong; that is, that the unit cap tube is stopped up or restricted so much the proper feeding or metering of gas cannot take place. You know that the valves are good since the pump immediately pulled a vacuum when you started it up. So, in the process of getting your gauges on and starting the unit for just a minute, you have eliminated two possible trouble areas.

1. The possibility of no refrigerant.
2. The possibility of ruptured valves.

These symptoms and remedies apply to all units which use cap tubes and have hermetic pumps. One factor that is present in a hermetic which is no problem in an open-type unit is the fact that you must cut into a hermetic to make a final check; and in the case of the open-type unit, you could put your gauges on twenty times a day with no damage in order to check the internal condition of the unit. Therefore, when you make a decision to go into a hermetic system, you should take your time and think out your problems before you cut. Not that cutting into a hermetic is of so great importance, but it does represent tricky work; and there is no use in making work for yourself. This is especially important where home refrigerators and freezers are concerned.

Let's begin again. The customer's unit is on the work table or dolly in your shop. You are making an examination of the unit. Again, before you touch a tool to the unit, you find out the type of refrigerant and voltage. You look for breaks in lines. You look for oily spots which are a good indicator of a leak. You look for broken wires; you smell and you look carefully. You plug the unit into the proper voltage with all switches in the off position. You move the fan switch, and the fan starts to run. You turn the switch to cool, and all you hear is a strong hum followed by a click, and the unit is dead. You wait a few seconds and the pump tries to start again. Hum, click, hum, click. The unit will not start. Unplug the unit. **Caution!** *Never work on a window unit with only the unit switch or thermostat off.* Unplug the

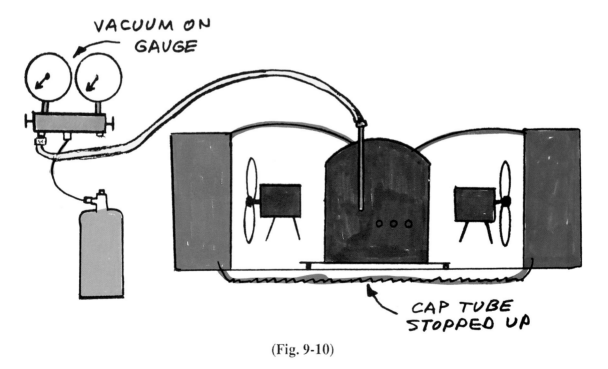

(Fig. 9-10)

unit from the service. Cheaply built window unit controls may break only half the line, and the whole rigging may be hot. You can now reason that something is causing the overload to kick out. You know that the answer here is high amperage. Your next reasoning should concern what is causing the high amperage. You know from previous chapters that if the pot motor does not start to roll, the amperage will skyrocket. So you can say right now that one of several things can cause this condition. You know that the overload is working all right since it will not let the motor sit there and burn up because it can't start. In this order, here is what could cause this pot to be cutting out on the overload.

Bad relay not making contact:

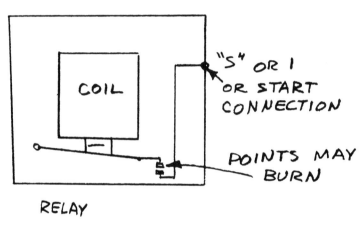

"S" OR 1 OR START CONNECTION

POINTS MAY BURN

RELAY

Blown capacitors:

BLOW OUT HOLE IN START CAP

Loose terminal connections causing low voltage:

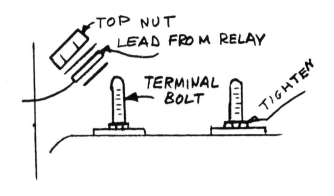

TOP NUT
LEAD FROM RELAY
TERMINAL BOLT
TIGHTEN

Locked rotor mechanical failure:

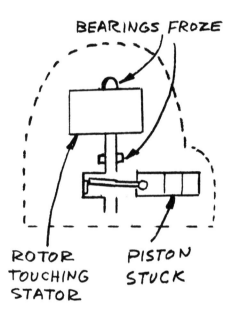

BEARINGS FROZE

ROTOR TOUCHING STATOR

PISTON STUCK

All right, one by one, you eliminate the possible troubles. For one thing, you could take the terminal nuts off, unhook the pot, and use a clip cord to eliminate the whole lot with one test. Right? But suppose we do not have a clip cord today. We have to use our cheap screwdriver to fix this job. The first elimination would be to look the capacitors over carefully; and if they show electrical damage, you could almost give them the benefit of the doubt.

You must remember that a capacitor may look brand new and be burned out or shorted. There is no iron bound rule for knowing whether a capacitor is bad or not. In the case of a combination start and run capacitor, look the oil-filled condenser can over carefully for swells or bulges. Look for rust holes and leaking oil.

Generally these capacitors will look damaged if they are damaged. If all capacitors look all right, move on to the relay. Examine all screws carefully to see that they are tight. If you don't want to bother to take the cover off the relay, smell of it. Generally, if it has suffered any material damage inside, it will have that old electrical burn smell about the case. If you smell no burn, take the cover off carefully and have a look inside at the start points. If the points are burned or rough or show excessive arcing, stop and try the whole system again. Watch the points and the relay in action. You may see poor contact being made in the points; or you may see that the point lever is not moving at all when it should be snapping the points open or closed. If this should be the case, then you have an open coil, even though the relay does not smell burnt. If you find the relay clean and willing to try to start the motor when you plug it in, then you had better move on to the pot. If the leads from the relay to the terminals are in good shape, not broken or loose, then you will have to find a clip cord and make the final decision by running the pot independent of the unit rigging. Then step by step, you eliminate the trouble areas. There is always a possibility (however slight) of mechanical failure inside of the hermetic pot which has nothing to do with the electrical.

Move up to a five-ton package unit, and let's apply some good sound reasoning and make it work. Suppose you are standing in front of a five-ton unit, and the complaint is that it is not doing the job like it used to do. You make sure that the same old load exists; that is, there is no doubling of the building size or increased

load on the unit. Don't forget that a customer might enlarge his place of business twice its original size and call you to find out why his air conditioner will no longer get the job done. You note that the load is the same as always. You might even ask the owner of this five-ton unit if there has been any additional load placed on the unit. If not, you are back to the unit. You are looking at the evaporator, and you notice that it seems to be cooling more near the valve. The coil is dry or not nearly as cool near the suction line side. See **Figure 9-11**.

Your first reasoning is that there seems to be a shortage of refrigerant. You listen at the expansion

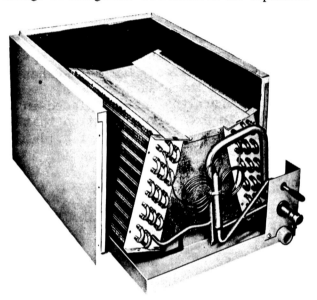

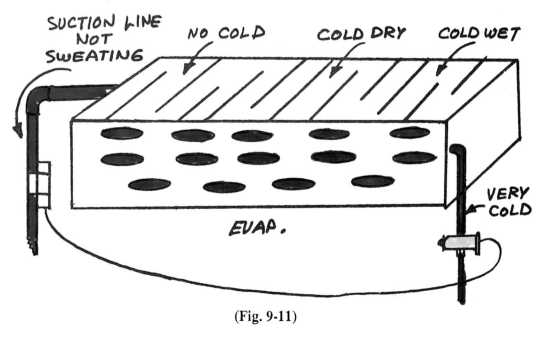

(Fig. 9-11)

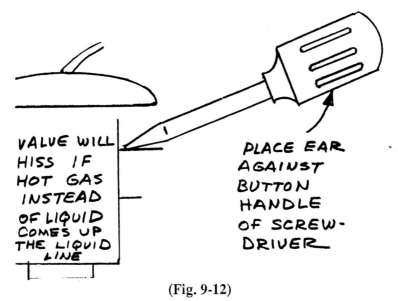

VALUE WILL
HISS IF
HOT GAS
INSTEAD
OF LIQUID
COMES UP
THE LIQUID
LINE

PLACE EAR
AGAINST
BUTTON
HANDLE
OF SCREW-
DRIVER

(Fig. 9-12)

valve with a screwdriver as a stethoscope. See the serviceman's listening device (Fig. 9-12).

You hear an intermittent hiss, and you are almost sure the unit is short of gas. Likewise, you find that the suction line back to the compressor is not cool enough. You find by feeling of the high side line that the heat on the high side is not as hot as it should be since there is a good load on the unit. In fact, you are sure that there is a shortage of refrigerant. You know there is some gas, or there would be no refrigeration whatsoever. But there is not enough to get the job done. So far you have reasoned this from your observations and by feeling of the lines. Now prove your reasoning. Put your gauges on and get a steady reading without adding one drop of refrigerant. Suppose the unit is running steady at 10 psi on the low side. You tie your service cylinder onto the charging line to your manifold and give the unit a long shot of gas. As soon as you close your charging valve note that the unit is running now on a different steady pressure, say 15 psi. What have you proven? You have proved that as soon as you added some refrigerant, there was an immediate response in the operating pressure in the evaporator. Suppose you kept on adding gas until the pressure would no longer change to a higher pressure. Say it leveled off at 40 psi and will not go any higher even though you give it more gas. You can reason right here that this shows that the expansion valve is getting liquid as it is needed and there is even a slight reserve; therefore, there will be no appreciable change in pressure from here on out. The gas that you add now will merely be the reserve that will be in the receiver when the unit is operating normally. Now go back and suppose that you reasoned that the unit was

short of gas and you tried a shot and watched the gauges to see if there would be any change in the operating pressure. There was no change. Could you say at this point that if you continued to put gas in the unit, it will not change anything since there is no change in pressure since the first shot of gas? Try another shot. There is still no change, you will have to look for something else. Suppose you stop right here and say, "I will not put any more gas in this unit. I may have a full charge now, and there can be something else causing the evaporator to starve and the unit not to put out." Suppose you decide that there is a full receiver of gas. Then there must be something stopping the liquid from getting to the valve and on into the evaporator. You follow the liquid line up from the receiver and find out there is a sight glass, strainer and expansion valve. You note that the expansion valve has a strainer on the inlet side. Most of them do.

See the liquid line and possible stop-up points in **Figure 9-13**. Now, suppose you start with the combination dryer and strainer. You know that if there is ever any pressure drop on liquid Freon, there will be an accompanying change in temperature. This means if there is the slightest restriction of the free flow of the liquid, there will be a temperature break. Therefore any suspected pressure restriction will require you to feel for a change in temperature. Here is the way to check a dryer or strainer for stopping up if the unit is running and there is some refrigeration. You know, of course, that if there is absolutely no refrigeration in the evaporator, this test is worthless.

Here is your test.

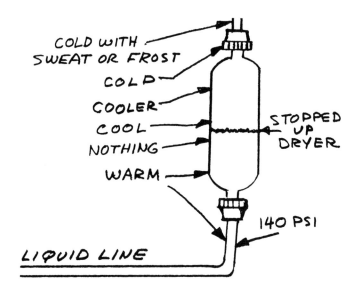

COLD WITH SWEAT OR FROST
COLD
COOLER
COOL
NOTHING
WARM

STOPPED UP DRYER

140 PSI

LIQUID LINE

FEEL IN ALL
OF THESE
PLACES

(Fig. 9-13)

You find this all right, so you move on up to the valve. So far you have used your eyes and sense of feel. Now if you are to disconnect the valve, it will require that you at least make a pump down without letting the liquid in the receiver tank get away when you open up the valve for inspection. Close the receiver to the liquid line valve and let the unit run on the thermostat until your low-side service gauge is just above zero. Pull the switch, close the low-side service valve, and you are ready to go to work on any part of the low side away from the compressor. Should the unit have a low-side pressure control, you will have to make the pump run, even though the LP control is trying to cut it out. Use a jumper around the control. With the unit pumped down, you are now ready to clean out the expansion valve strainer screen.

See this operation in **Figure 9-14**

With the screen clean and replaced, tighten up the valve fittings. You are now ready to purge the system. Since you have the low side service valve shut off, you can bust the nut on the evaporator side of this valve and blow gas through the system from the liquid line valve. When you think all possible air is out, just tighten up the flare nut and put the system back on the line. If this were the trouble, there would be an immediate rise in evaporator pressure and more feeding of the liquid into the evaporator.

Now, if this did happen, and you had one dryer or strainer ahead of the valve, it is strange that so much trash could get up to the valve strainer. When this happens, maybe the lower strainer has a broken screen and is leaking solids through; and they stop at the good screen in the valve inlet.

There is nothing mysterious in the circulation of Freon in the refrigeration cycle. You must treat this circuit of liquid flow just as if it were water. If you were dealing with an air compressor and it gave you trouble, you would soon find the trouble. A refrigeration compressor is identical to an air compressor, and it is compressing a gas just like air. Air can be called a gas. However, air is non-condensable. Air is the enemy of the serviceman. Keep it out of refrigeration systems.

■

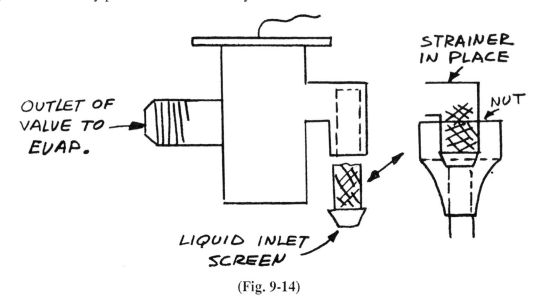

OUTLET OF
VALUE TO
EVAP.

STRAINER
IN PLACE

NUT

LIQUID INLET
SCREEN

(Fig. 9-14)

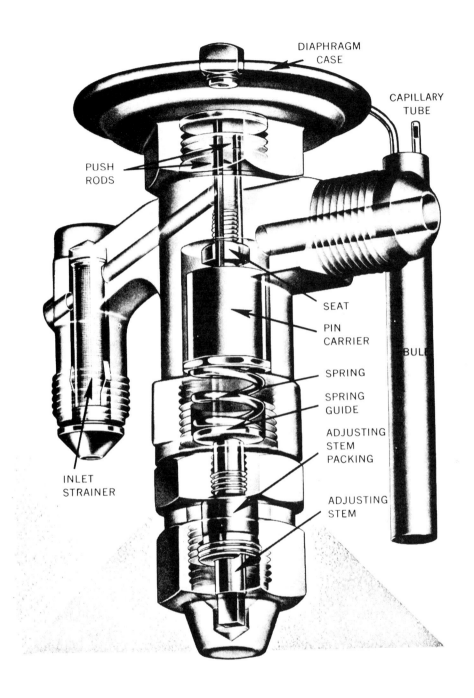

DIAPHRAGM
CASE

CAPILLARY
TUBE

PUSH
RODS

SEAT

PIN
CARRIER

BULB

SPRING

SPRING
GUIDE

INLET
STRAINER

ADJUSTING
STEM
PACKING

ADJUSTING
STEM

SPORLAN

THERMOSTATIC EXPANSION VALVE

141

CHAPTER 10

HOUSEHOLD REFRIGERATORS AND AUTO AIR CONDITIONING

Dry air itself does no particular damage to a refrigeration system other than taking the place of the refrigerant. Big rocks in the gas tank of your automobile would do no particular damage; but if there were enough of them, they would take up the space normally occupied by gas; and you would have an undersize gas tank. Air will fill up the condenser space where the Freon normally condenses, and this will cause a condition that is equal to cutting down the size of the condenser. If you will remember that the manufacturer has already cut the size of the condenser to the minimum to meet competition, then you will be acutely aware of the fact that if any air is let into a refrigeration system, the normal head pressure will be increased in proportion to the amount of air in the system. Said in another way, air in the condenser of a unit is equal to putting rocks in the radiator of your automobile. You will agree that if you put enough rocks in the radiator of your automobile, the temperature of the cooling water may be increased to the danger point. Air displaces the normal condensing refrigerant. The air contained in the charging hoses and gauge manifold could be enough to effect the head pressure of a small system like a household box, home freezer, a window unit if this air is permitted to enter the system. The manufacturer has made the condenser just big enough to get the condensing job done and no more; therefore, if the condenser capacity or efficiency is effected by an alien gas, it would not operate properly if it operates at all.

Never confuse air with water. Air is made up of many gases like oxygen, nitrogen, hydrogen, and so on. Some of these gases contained in air will condense under extremely high pressure, but not at the pressure of the normal refrigerating unit. For this reason, we call air

a non-condensable gas. Water will condense, and it will in turn change into a vapor and from a vapor back into a water state. Water is actually a refrigerant like Freon, but it is of no value as a refrigerant like Freon, simply because it does not evaporate fast enough to do any work in the temperature range that we humans live in.

There is a good chance that you will never let enough air into a system to do any appreciable damage by reason of the air alone; however, air will and does carry large amounts of water. In one cubic foot of air on a humid day, there may be as much as one teaspoon full of plain water. Air soaks up water and becomes saturated. This is something that you already know; however, it would be a wise refrigeration man who remembers that any air that he lets into a system may carry some water in the form of an invisible vapor. This water vapor will condense out of the air when it is put under pressure in the condenser of a unit. Likewise, moisture will condense out of the air if it is cooled to the right dew point in the evaporator. You know this from having taken something cold out of the refrigerator and having water or sweat form on the surface. This sweat was the water condensing out of the air in the kitchen onto the surface of the cold vessel taken out of the refrigerator. If this same air were inside of a unit circulating with the gas and it came into contact with the cold evaporator, the water would condense out of the air, even though the dry air would keep on circulating or end up trapped in the top of the condenser. It is possible to get water into a system by carelessness, and it is possible that a water-cooled condenser could rupture and let water into a system when the gas was all leaked out. Likewise, a serviceman could be using water around a system when he was working and let water in accidentally. However, as a general rule, you can say that when a system is saturated or has moisture in it, the moisture came from a leak. That is, a leak developed and the gas escaped; and when the gas pressure was all gone, the unit continued to run and sucked in air. The moisture in this air is condensed out inside of the system. The unit now has water in it as a result of a gas leak. The colder a system is, the more likely it is to accumulate moisture as a result of a leak; therefore, whenever you open up any system that is still cold or frosty, you must be doubly careful that you do not let any air into the system. In fact, a cold evaporator will condense water just as fast as the air carrying the water enters the coil.

It is important that you know that even though you get the air out of a system, it may have given up its moisture inside the system. This is the case where you let air get into a cold space like a frosted up evaporator. The moisture would condense out of the air very quickly; and when you purge the air out, all you would be getting out would be dry air.

Up to this point we have discussed air and moisture in general terms. Now, it can be nailed down to specific units. Start with a household refrigerator.

These boxes are the worst kind to quit because of moisture. Any water that is in this system will freeze into a solid plug of ice right where the cap tube feeds the refrigerant into the evaporator (ice cube maker). This plug of ice will stop up the flow of liquid refrigerant. The unit will pull down to a vacuum and defrost and will not freeze again until this ice plug is melted and the water pulled out with a vacuum pump. Step by step, let's take a look at a box in a home that is getting a full slug of water.

One of the most common ways that water gets into a household box system is by punching a hole in the evaporator with an ice pick or sharp instrument. Much of the repair work you will do on a household box that has lost its gas is the result of someone trying to defrost the evaporator with an ice pick or trying to pry out a frozen ice cube tray with a sharp instrument.

See this drawing in **Figure 10-1**:

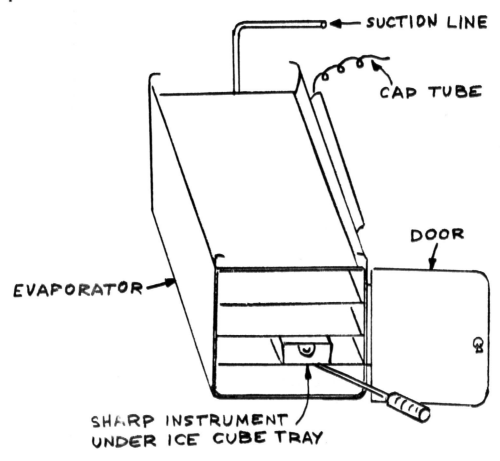

(Fig. 10-1)

Rarely do household refrigerators develop leaks down around the unit. In most cases, the evaporator is pierced by careless defrosting or removing trays. For instance, a man tries to pry out an ice cube tray with a screwdriver or heavy knife. The point of this instrument tears into the evaporator and a small hole appears. Now, no water or air gets in at this instant because the gas pressure inside the evaporator will be leaking out for several hours. The minute the gas starts leaking out, the evaporator starts to partially defrost. When enough gas has leaked out, the whole evaporator will be defrosting. The unit will still be running because it is thermostatically controlled; and the more the evaporator defrosts, the more the thermostat calls for the unit to run. When enough gas has leaked out of the system, the unit will be running with a vacuum inside the evaporator. Now, the hole, instead of leaking gas out, will be letting anything near the hole be sucked into the evaporator. So it stands to reason that if the evaporator is defrosting, there will be considerable water on the outside surface of the metal; and if this water puddles or gets near enough to the hole, it will be sucked in with the air that is getting into the system through the hole. In this case, the system could load up with water if there were a big accumulation of melting ice near the hole. Even if there were no water right around the hole, the air that is being sucked in will give up its water on the cold insides of the evaporator and eventually in the condenser. If a party knew that he had punched a hole in the evaporator, and he immediately shut down the box and let it die a natural death by leaking out the gas, there would be less water or air in the system. However, even so, if repairs are not made immediately, the hole will breathe in air; and moisture will accumulate in the system. The same thing happens to home freezers. Freezers are no different than household boxes where moisture is concerned. Now, if you were to assume that there were no moisture in a unit of this type, and you repaired the leak and started to charge the box, here is what could happen. The moment you got enough gas into the system and a little liquid Freon started to feed into the evaporator, there would be a freezing cold spot start right where the cap tube enters the evaporator.

See the drawing in **Figure 10-2**.

This first cold spot of expanding Freon liquid will have small particles of water suspended in the Freon. These particles of water will freeze into little specks of ice. The moment there is enough of these frozen specks, they will form an ice dam or block; and you might as well have a steel plug driven into the small cap tube

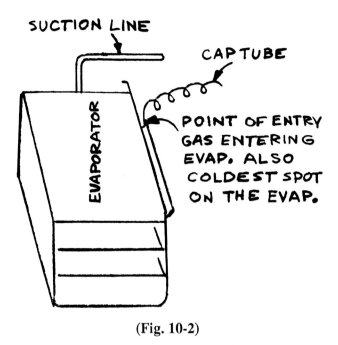

(Fig. 10-2)

feeding the evaporator. This ice block can withstand hundreds of pounds of pressure. No more Freon is going to get through; and if a little liquid Freon did break through, it would only serve to freeze the plug again. Now the box is out of business. The pump will pull the low side (evaporator) down into a deep vacuum with no refrigeration. One sure way to determine if a unit has an ice plug or block right where the cap tube feeds the evaporator is to melt it out with a torch while the unit is running. If it suddenly breaks loose and liquid refrigerant starts to flow, and the evaporator starts to freeze, you know it was a moisture block. Here is the way you would make this simple test: With the unit running (you know there is some gas in the system) you would apply your torch, or even a steaming hot towel, right where this drawing shows in **Figure 10-3**.

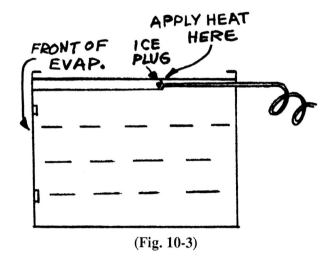

(Fig. 10-3)

Remember that even though you have melted the ice plug, it will be back again in time and block the flow by freezing a plug of ice in the same identical spot. This may take ten minutes or ten days, but the water will be back unless you get it out of the system. It is not uncommon for moisture to take two or three days to work up to the cap tube outlet and into the evaporator in sufficient quantities to cut off the flow of refrigerant. If you will remember that this water will travel with the refrigerant until it gets to a spot where it can freeze solid and close up a small hole, you will always know when it happens. It happens more often when a unit has been worked on, or has not been properly evacuated. Evacuation means to pull everything out of the inside of the system with a vacuum pump before charging.

This is an iron bound rule: *When you charge a household refrigerator or home freezer, you will evacuate the system thoroughly and change the drier.* There will always be some moisture in the system if it ever had a leak. As you now know, if it were a leak around a wet evaporator, there may be as much as a teacup of water in the system; and it takes only a speck of water the size of the head of a straight pin to stop up a cap tube. If the leak were around one of the terminal bolts coming out of the pot, there will be moisture in the system, but not as much as a leak in a wet evaporator. You should remember that even if no actual water got into the system, air can give up water; and it is the same kind of water that is in liquid form.

Another type of equipment that gives trouble because of moisture is commercial equipment. This equipment is generally equipped with a thermostatic expansion valve instead of a cap tube metering device. Moisture freezing into a solid ice plug will stop the flow of refrigerant through the small orifice of an expansion valve just the same as the opening coming out of a cap tube. When this happens on commercial equipment, the unit will pump down and shut off because of the low pressure control; and the ice in the valve will melt on the off cycle. However, there will be poor refrigeration, if any at all; and the moisture must be eliminated. Driers are generally used to eliminate moisture in these systems. A commercial drier is filled with an agent that absorbs water like a sponge or blotter. The drier holds this moisture and will not let it get to the valve. Most commercial installations will have a drier installed in the liquid line, even when they are brand new, just in case there is any chance of moisture. When bad leaks develop from time to time, these driers should be changed out and replaced just like a used oil filter on an automobile. Driers on commercial equipment are always installed in the liquid line feeding the expansion valve.

Window units and other air conditioning equipment are not effected by moisture like the low temperature boxes and commercial equipment. Air conditioning equipment rarely operates with an evaporator cold enough to freeze ice inside of the valve or outlet of the cap tube. However, it can happen and sometimes does. Don't leave moisture in any refrigerating system. Another reason for eliminating moisture is that the moisture not only obstructs the flow of refrigerant by freezing solid in cold places, but also it will contaminate the oil and cause corrosion inside of the system. Many refrigerants that mix with water cause the mixture to be highly corrosive and harmful to the interior of the system. This is much like getting water inside of a fine watch; only in the case of pumps and valves, it will cause acids; and this acid will break down the bearing surfaces as well as the windings of the hermetic motor. Water in a system will also dilute the oil in the pump, and unnecessary wear will result. If you are absolutely straight on the hazards of water in a system, you are now ready to get the water out and do the job properly.

Any person planning to work on hermetic equipment should own a vacuum pump. Your nearest supply house will have vacuum pumps to suit your need. I would suggest one which has a good displacement and can pull a low micron level at a fast pace. If you are only going to work on home refrigerators and freezers and window units, a less expensive vacuum pump could be purchased because not as much displacement would be needed. However in the use of a commercial unit there is more area to evacuate. In this case get the larger vacuum pump. The pumps are going to be an investment, but you will need them. Dollar-wise you could spend as much as $600.00 for a good pump. See **Figure 10-4**. Usually in the case of a commercial system, install an over-sized drier and it will usually catch and hold moisture.

You will always attach the charging line on your gauge manifold to the vacuum pump before you charge a unit as shown in **Figure 10-5**. You will pull as much vacuum as the pump can get; and where a household box or home freezer is concerned, you will heat the evaporator, condenser, pot, and all lines from the pot to the evaporator. If you have a vacuum on the system when you are heating it, the water inside will boil or explode under the heat; and this water vapor will be pumped out of the system by the vacuum pump. Said

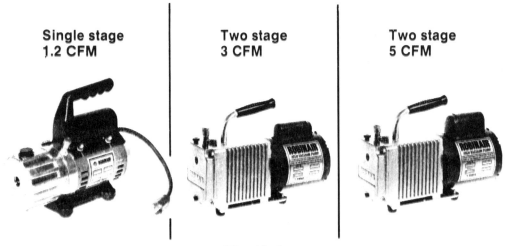

**Single stage
1.2 CFM**

**Two stage
3 CFM**

**Two stage
5 CFM**

(Fig. 10-4)

another way, the water or moisture in a unit will turn to steam very readily under a vacuum and a little heat. Are you aware that water boils or turns into steam at a very low temperature when it is under a vacuum? For instance, if you pull twenty-nine inches of vacuum on a system and heat it up to 115 degrees, the water laying in the evaporator and other parts will explode into steam; and this steam is easily pumped out of the system. Water boils into steam at 212° at atmospheric pressure. But under 29 inches of vacuum, water boils at approximately 115°, just about the right temperature to take a

bath in. Here is a good rule to remember when heating a system. NEVER HEAT ANY PART OF A SYSTEM HOTTER THAN YOU CAN LAY YOUR HAND ON. THIS WILL BE PLENTY WARM WITH A VACUUM TO RID THE SYSTEM OF ALL THE WATER. How long do you evacuate the system? Good judgment will apply here. If you have reason to believe that the system has had a bad leak and sucked in considerable air or water, leave your evacuator on the system several hours and heat it several times until you are satisfied that there is no water or air left in the system. If possible

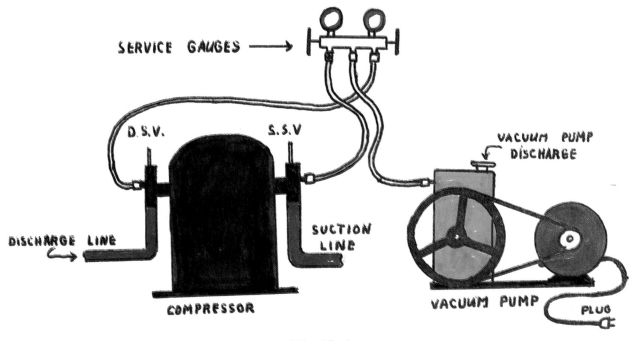

(Fig. 10-5)

leave the vacuum pump on over-night. As you know, big commercial systems and many package units have service valves; and these units can be purged by blowing Freon through the system. Freon will push the air out ahead of the gas. But in the case of the full hermetic, there are no service valves; and the only practical way of getting all air out of the system is to use a vacuum pump. In addition to hermetics, you can evacuate any system with a vacuum pump, no matter how large it is, if it is air tight.

It is important that you remember to evacuate all household units (dry out, too), all window units, and other units that you can get your evacuator hooked to with no trouble. It takes only a few minutes to pull a vacuum on an automobile unit and get rid of all the air and other foreign gases that may be in the system. Where small units are concerned, you do not worry about saving the gas that is left in the unit. Blow it out. You will not be doing the customer any favor by trying to save a few ounces of gas left in any hermetic or automobile air conditioning system.

We can move on to automobile air conditioning now, and you can keep in mind that a vacuum pump is the handiest tool you can have when charging an automobile unit.

The automobile unit is the simplest and easiest of all refrigeration systems to service and trouble shoot. For one thing, you have no electrical problems to cope with. The auto system is driven by the auto engine itself, and you have only the clutch and evaporator fan system to worry about as far as the electrical is concerned. Most of you can figure out why a blower fan will not operate off the auto battery if you will look for fuses and electrical connections. Servicing the evaporator fan motor is no more complicated than finding out why the fan motor on the hot water heater is not working. The pump on an auto unit is driven by belts coming off the engine drive pulley. These belts may also turn the radiator fan and generator. Some units have only one belt, and there are a dozen different ways to drive the pump. Most of these pumps have magnetic clutches. These clutches operate off the same voltage that operates the evaporator fan, that is, the battery. There are so many different kinds of installations that it would confuse you to attempt to describe all of them. The one best way to know whether the clutch is working is to flip the air conditioner switch while the auto engine is running and then take a look at the nut on the end of the clutch mechanism and see if it is turning. This nut is the end of the crankshaft; and in most cases, it will be exposed to view when the unit is pumping or not pumping (idling).

See this drawing in **Figure 10-6**:

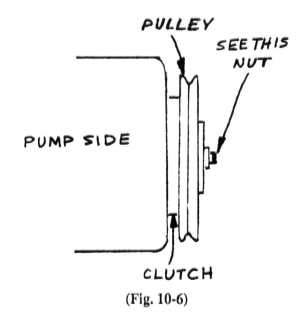

(Fig. 10-6)

An automobile air conditioning system is just like any other system as far as the principle is concerned. There will be an evaporator inside of the car. The pump will be mounted by the engine, and the condenser will be out in front of the radiator. The principle the condenser works on is that it depends upon a stream of air to come through the front grill of the auto while the auto is in motion. This air will cool the hot gas down to a liquid. The liquid leaves the bottom of the condenser, and in most cases, flows into a small receiver much like the receiver on commercial equipment, but much smaller. Instead of being horizontal, it may be an upright tank. The liquid leaves this tank through a liquid line; and in many cases, there will be a sight glass in this liquid line through which you can actually see the liquid flowing to the expansion valve and evaporator.

See the drawing in **Figure 10-7**.

The suction line from the evaporator runs back to the low side of the pump; and generally, you will find a service valve there which can be back-seated. By back-seating this service valve, you can put your gauge hose on with little effort. You will find that these valves have a valve stem cover which must be removed before you can backseat the valve itself. On newer model cars, service valves are no longer used. Instead they have schraeder valves that look like air cores for inner tubes. The gauges can attach to these. You may also find that the suction test port is in an accumulator near the

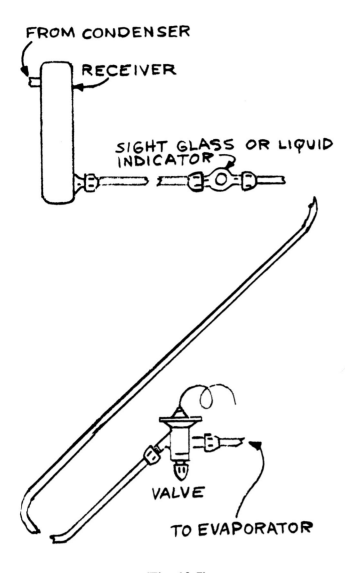

FROM CONDENSER

RECEIVER

SIGHT GLASS OR LIQUID INDICATOR

VALVE

TO EVAPORATOR

(Fig. 10-7)

it fixed. You may find out that the customer has just bought the auto, and the air conditioner has never worked, and he knows nothing about the history of the system. Suppose this were the case; the customer says that he knows absolutely nothing about the unit except that it won't cool. If so, you are ready to take over. First, you would lift the hood and take a good look at the pump and the lines leading to the pump. Look for oily spots or breaks in the lines. Suppose everything looks clean and normal, and you see nothing that looks suspicious like an oily fitting or rubbed place on a line or hose. Many of these systems have flexible neoprene-covered high-pressure hoses for the suction and discharge lines on the compressor. Nothing looks wrong; so you are now ready to look at the evaporator if it is where you can get

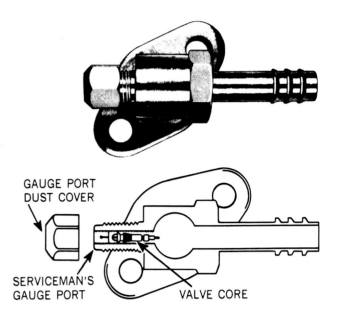

GAUGE PORT DUST COVER

SERVICEMAN'S GAUGE PORT

VALVE CORE

evaporator and that the high side test port is in the liquid line. See **Figure 10-8**.

You do not have to have special training in order to service and repair auto air conditioners. You can follow the instructions given you right here in this chapter and do a good job.

Step by step, here is the way you would handle an automobile air conditioning job.

A customer brings an automobile to you and says that it will not cool. A good question to ask this customer right here is this, "When did it cool last?" What you want to know is, did it just suddenly quit in the middle of the summer, or did the unit run fine last summer but this spring when he tried to run it, it would not cool, so he just waited until he had the money to get

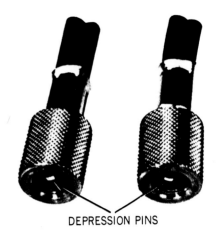

DEPRESSION PINS

(Fig. 10-8)

to it. Usually all evaporators are factory air and built into the dash. Sometime the evaporator is accessible from under the hood and air is directed from inside across the coil and back to the inside of the car. Nothing seems out of place, so now you are ready to check the fan on the evaporator. Without running the auto engine, you should be able to turn on the switch and operate the evaporator blower fan. If this fan is blowing properly and you are satisfied that nothing seems to be out of order, you are ready to proceed with the check on the customer's statement that the unit is not cooling. So far, you have done nothing but look and feel around as if you were looking for something suspicious. This first examination should only take ten minutes of your time. In making this examination you may find broken lines or any of a half dozen troubles that are visible to the eye of a man who has some understanding of the principle of mechanics.

For instance, if you find the belt that is driving the condensing pump is broken or so loose that it cannot turn the clutch pulley on the pump, there is no reason for proceeding further until this is remedied. But suppose you find everything in good order. Then you can make a final check to determine whether there is gas in the system, whether the valves in the compressor are good, or whether the strainer is stopped up where the liquid refrigerant feeds into the expansion valve. None of these troubles may be present. You are just assuming these troubles may be the cause for no cooling. These

last troubles are internal. You cannot see them with the naked eye. Now, have the owner or your assistant get into the auto and start the engine. If no one is there to help you, you start the engine and place a weight on the gas pedal. Try to have the engine turned up at a fast idle. In other words, you are going to make the engine run at a nice easy speed equal to about thirty miles an hour if the auto were moving. This will be just a little faster than a fast idle. With the motor running, you are ready to switch on the air conditioning switch. This switch will turn on the fan which blows through the evaporator and at the same time engage the magnetic clutch that starts the compressor pump to turning.

See the schematic for an auto air condition electrical system in **Figure 10-9**.

Right up to this point, you have not touched a thing. So far, you have just looked and examined the system. Now, with the motor running you will take a look at the nut on the end of the clutch housing and see if it is turning the pump crankshaft at this point. You might switch the unit off and on or have it switched off while you watch the nut to see if the clutch is taking hold or not. Suppose it is doing its job, and the compressor is turning and should be pumping. Now, you will feel of the high side line coming off the pump leading to the condenser. You will note whether there is any heat whatsoever in this line which would indicate that there was a little gas in the system, but not enough to refrigerate the evaporator. You may also feel of the expansion

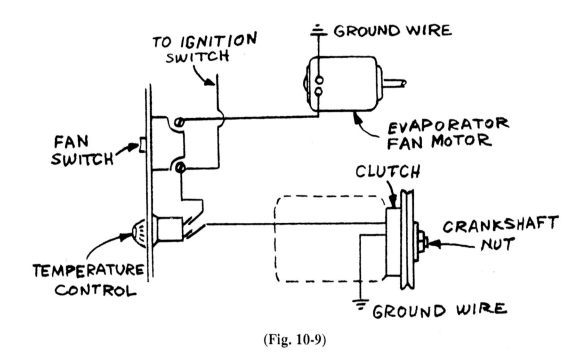

(Fig. 10-9)

149

valve if you can get to it and see if there is a cool spot coming out of this valve. If you make these simple tests, you will at least know whether the leak is a big one which has completely emptied the system or a slow one that has let most of the gas out, but not all of it. You may also take a flashlight and take a look in the sight glass and see if you see a little liquid bubbling or frothing through this glass. If it is so dirty you cannot see anything, clean it off and look. If it still does not show anything, forget it and use your sense of feel on the suction and discharge line of the compressor. You are now running the compressor, and you find that it does not cool. The customer was correct. It is not cooling; but you do not, as yet, know just why it is not working. You can make a fair guess at this point that the system is out of gas. With this in mind, you can proceed to prove this guess. Stop the engine and take the caps off the low-side and high-side service ports. See **Figure 10-10**.

Attach your low pressure gauge to the low-side port and attach your high-side gauge to the high-side port. Now with the gauge manifold valve handles closed, you will read pressure on the high and low side of the system. Now take a look at your gauges. You may have pressure on both gauges, even though it is a weak pressure; or you may have nothing; that is, the gauges may show zero. Take a look at **Figure 10-11** with the gauges attached to the pump. Older auto air conditioners will have service valves on the compressor and also most open type compressors will also.

At this point, you have your gauges attached to the pump, the auto engine is dead, and everything is quiet.

Now purge both gauge hoses. Then you are ready to make a test for bad valves; or at least, you want to see if the pump will pump. Start the engine and switch on the air conditioner. When the clutch engages, note your low-side gauge. If there is no gas in the unit, the low-side will pull down very rapidly into a vacuum; and even if there is a little gas in the unit, it may still pump into a vacuum. If so, there is no longer any need to worry about the pump pumping. It will pump even though there is nothing to pump from the evaporator but a vacuum. Here you can stop the auto engine and get ready to put a test charge into the system. Blow off whatever gas is left in the system and hook your vacuum pump to the charging line on your manifold. Start up the vacuum pump with an extension cord, open the low side valve on your manifold, and let the vacuum pump pull on the whole system until you are satisfied that the system is completely evacuated and you are ready to proceed with charging. You might at this point take notice that you do not want to pull a vacuum on your high-side gauge; so you can again backseat the high-side valve and isolate this gauge. See this hookup of your gauges and vacuum pump in **Figure 10-12**.

Step by step see these drawings of the change-over and the method of vacuum breaking in **Figure 10-13**.

With a deep vacuum pulled on the system, you can shut your low-side valve on your gauge manifold and see if this vacuum holds on the whole system. If it comes up very rapidly, you can say right here there must be a devil of a big leak in the system somewhere; and you should start looking for it. One way to find it would be

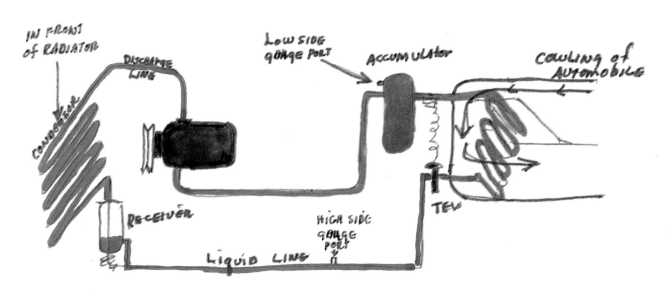

(Fig. 10-10)

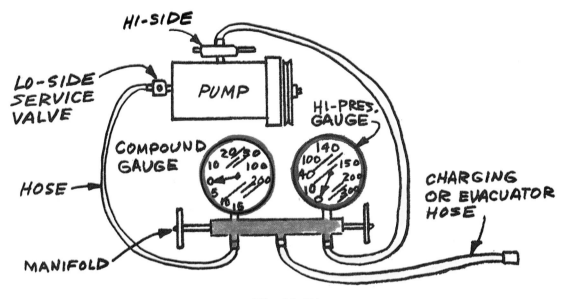

(Fig. 10-11)

to let in a little gas from your Freon bottle until you have some pressure, and then make a leak check all over the system with soapy water or a halide torch. You should forget about any further evacuation because you will have to evacuate all over again when the leak is located and repaired. But suppose the vacuum held nicely and you are satisfied that, if there is a leak, it is a very small one, or it may be that the seal in the compressor has leaked the gas out of the system while the unit was out of operation for a long time. Suppose that the compressor seal leaked all of the gas out while the auto air conditioner was not being used, and it may now hold if the compressor is used regularly.

With all the guess work finished, you are ready to charge up this unit. You would break the vacuum in the system by moving the charging hose over to a Freon drum, purging your hose, and letting gas into the system out of your drum or can.

With pressure from your drum now in the system, you are ready to charge the unit. Remember that you only let Freon vapor into the system from the top of your drum. There is no liquid in the system as yet, only vapor pressure. You will now unbackseat the high-side valve, and you will find that you will have the same pressure on both the low and the high side since everything is equalized. This pressure will be the same as the pressure

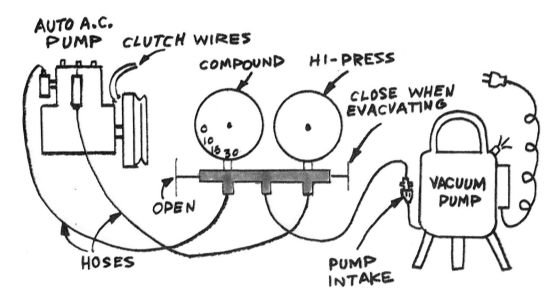

(Fig. 10-12)

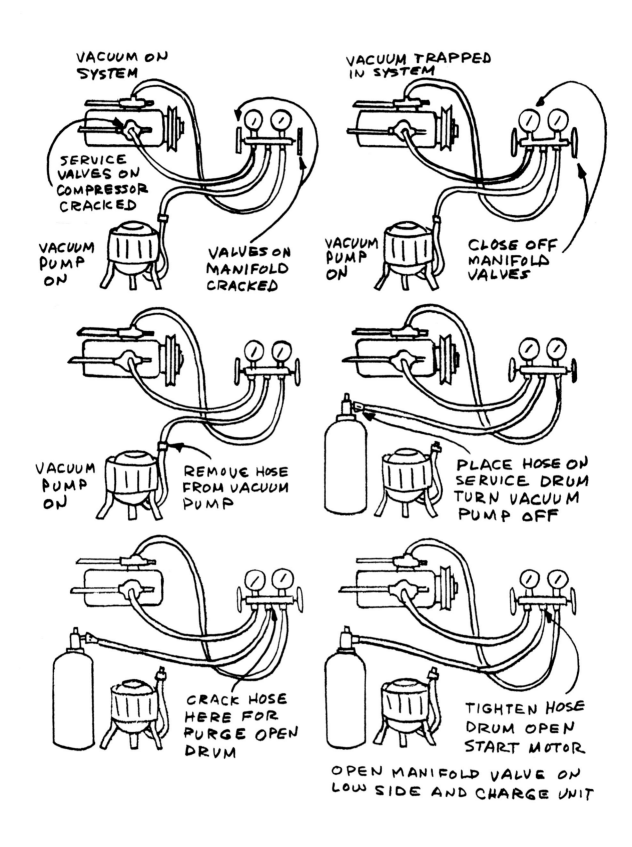

VACUUM ON SYSTEM

SERVICE VALVES ON COMPRESSOR CRACKED

VACUUM PUMP ON

VALVES ON MANIFOLD CRACKED

VACUUM TRAPPED IN SYSTEM

VACUUM PUMP ON

CLOSE OFF MANIFOLD VALVES

VACUUM PUMP ON

REMOVE HOSE FROM VACUUM PUMP

PLACE HOSE ON SERVICE DRUM TURN VACUUM PUMP OFF

CRACK HOSE HERE FOR PURGE OPEN DRUM

TIGHTEN HOSE DRUM OPEN START MOTOR

OPEN MANIFOLD VALVE ON LOW SIDE AND CHARGE UNIT

(Fig. 10-13)

152

in your drum. Now you are ready to charge the unit. Start the engine and turn on the air conditioner switch. The minute you switch on the air conditioning, the unit clutch will engage, and the pump will start to pump. You will note that the low-side gauge will start to come down and the high-side will start to go up. All right, now give it a big slug of gas by opening your low side valve on your manifold; the drum valve is already open. As soon as you give the unit enough gas to form some liquid in the condenser, this liquid will start to feed through the liquid line to the expansion valve; and you will notice some heat in the high side line; and the sight glass may start to show some activity. By shutting off your charging manifold once in awhile, you will note a steady rise in the low-side operating pressure. Now you are ready to keep a careful watch on the suction line coming back to the compressor. The moment this line flashes cold or begins to cool off, you can say that there is a bare minimum charge in the system. At least you know that there is expansion (refrigeration) taking place in the evaporator, or it would never have a cool suction line. See **Figure 10-14**.

With a cool suction line and a hot discharge line out of the compressor, you know you have refrigeration going on. At this point you can now make a decision as to how much gas you will put in, over and beyond this minimum charge. First set a pedestal fan in front of the radiator. The auto fan blade does not move enough air when the car in not moving. If you are in doubt as to whether you have too much or not, then keep a sharp eye on your high-side gauge and see if it suddenly goes very high. Also watch your sight glass very closely. For

instance, if it were running along with a head pressure of about 150 psi, and you gave the unit a shot of gas, and the head pressure jumped to 200, or even 300, then you know you have added too much gas. There is no need to panic, but shut down immediately or something may blow out. Now, bleed off that last shot you gave it; and you know you are fully charged when the gauge pressure on the high-side comes back to the steady normal operating pressure that you had with a minimum charge. Say about 150 psi, and the sight glass is clear. You are ready now to backseat and disconnect your gauges, put the cap and covers back on the valves, and you are finished. (Some older cars still have service valves.) You may have to tighten the belt on the unit. Remember these belts run fairly tight.

Here is some extra service you can give the customer. If the seal were leaking, then it undoubtedly leaked out some of the compressor oil. Add some oil. Use 300 viscosity refrigeration compressor oil or what is recommended by the factory. Do not use motor oil. Be generous with the oil. Most automobile air conditioner pumps are reciprocating, and they will take oil very nicely. Here is the procedure for adding oil to an auto compressor: Close the suction line service valve on the pump by front-seating until nothing can get into the pump from the evaporator. Run the pump and pull a vacuum on your low-side manifold. Now place the charging line in a can of oil. Now crack the manifold gauge valve and let the vacuum in the pump suck in its oil. You may want to purchase a hand-type pump from your nearest supply house. It will pump the oil in even under pressure. About a pint of oil is as much as should be added to the average auto unit.

See the steps in **Figure 10-15**.

Now one word of caution. You have been taught that you can use oil generously in a reciprocating pump. Now where cylindrical type compressors are used on General Motors autos (not all GM cars), you will not add oil unless it is absolutely necessary, and it would be advisable to let the pump be a little short on oil rather than add oil. This GM type compressor that I am referring to looks very much like an automobile generator. It is round and built like a cylinder, not like a piston-type pump.

Charging automobile units is easy work and simple if you take your time, and do not hurry. You can do a good job by just following the instructions in this lesson. There is one factor that must be considered when you are charging an auto unit. You must keep in mind that the design of the unit called for the auto to be

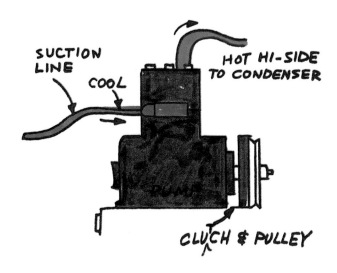

(Fig. 10-14)

moving if the condenser is to be properly cooled when the unit is operating. If you are charging a unit and the engine gets hot, it will cause the condenser to run hotter than it should. The fan on an automobile engine is never large enough to take care of both the radiator and the condenser. Many servicemen use a garden hose to run a little water stream on part of the condenser while they are charging it in order to keep down the head pressure.

Where the addition of gas does not effect the gauge pressure on the low side, that is, it still goes into a vacuum, then you may have a stopped up dryer or defective valve. Don't keep adding gas if it does not feed through the evaporator.

Take your time and try one, and the next one will go twice as fast. The only tools you will need to do an auto charge job are: your charging manifold, crescent wrench, valve wrench and a drum of gas. YOU SHOULD HAVE A VACUUM PUMP. ∎

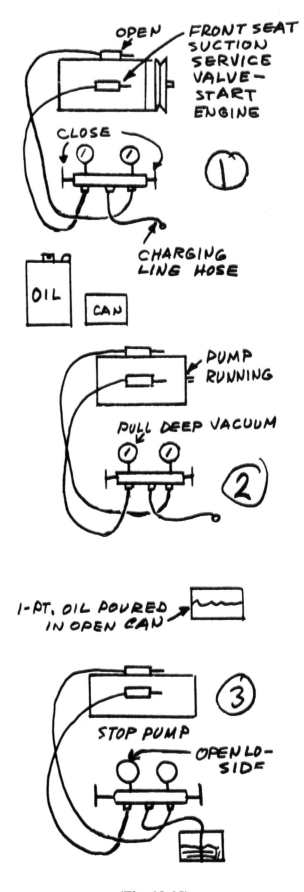

(Fig. 10-15)

CHAPTER 11

CENTRAL SYSTEMS

Many years ago the iceman stopped at every other house on his route. In some neighborhoods he made every house in the block. The iceman was known by his first name and mechanical refrigeration was practically unknown. In fact, it was almost a novelty. Today an iceman would starve to death trying to sell ice to homes for a living.

Electric refrigeration is as much a part of the household as the electric lights. Many years ago a few dentists and physicians had electric console model air conditioners and a few fancy places of business had remote air conditioning systems. Today you cannot do business with the public unless your business is fully air conditioned. Air conditioning is as much a part of a modern store as the lighting or heating. In some areas of our country, there may be less need for air conditioning; but, on the whole, it is everywhere. Years ago window units were a novelty. They were thought of as a fad and the manufacturers of big console model air conditioners said that these little pip-squeak air conditioners would never catch on. How wrong they were! Today there are many window units in homes around America and sometimes two or more units.

Many years ago the first practical central systems were being installed. Today any builder of new homes installs central heating and cooling. It is as much a part of the home as the kitchen would be. Imagine a modern home in the city today sold without an air conditioning system of some kind.

Of course, there are rural homes that still do not have inside plumbing. But for every one of these homes, there is a farm home with complete refrigeration. A modern farm will have household refrigerators, home freezers and in most instances central heating and air conditioning.

Let your mind wander a bit and just imagine how much ice would be needed to take care of a modern supermarket in order to take the place of the commercial refrigeration fixtures. Commercial refrigeration is necessary to operate any business that stores food or perishables. Many advances have been made in the field of commercial refrigeration in recent years. However, they are not significant when one thinks of the advances made in air conditioning. Many years ago, commercial refrigeration was the mainstay of the refrigeration mechanic. Today there are a thousand times more horsepower pulling air conditioning units than commercial units. Many of the commercial men have moved into the air conditioning field because of the increase in its use and better pay. Basically the two systems are alike inasmuch as they are both conditioning air. The commercial unit is conditioning air in order to preserve food with cool air, and the air conditioner is conditioning air for human comfort. The greatest advances in the air conditioning field in recent years has been in the central systems. If the home has electricity there will be air conditioning. Air can be conditioned in winter as well as summer. The only difference is that in the winter the air is being heated for comfort cooling, and in the summer the heat is being removed for human comfort.

The modern central system is a combination of both heater and cooler. Both systems may be installed independently or individually and/or combined. The more common installation is the combination heating plant and cooling plant. Basically, the cooling end or air conditioning end of a central system is just simply an air conditioning unit installed in conjunction with a heating plant. The AC unit will consist of a condensing unit and evaporator or cooling coil. The condensing unit will set outside of the home and the cooling coil will be installed in the duct work in the attic or in the duct coming out of the furnace in a closet.

Take a look at this outline of a central system as shown in **Figure 11-1**.

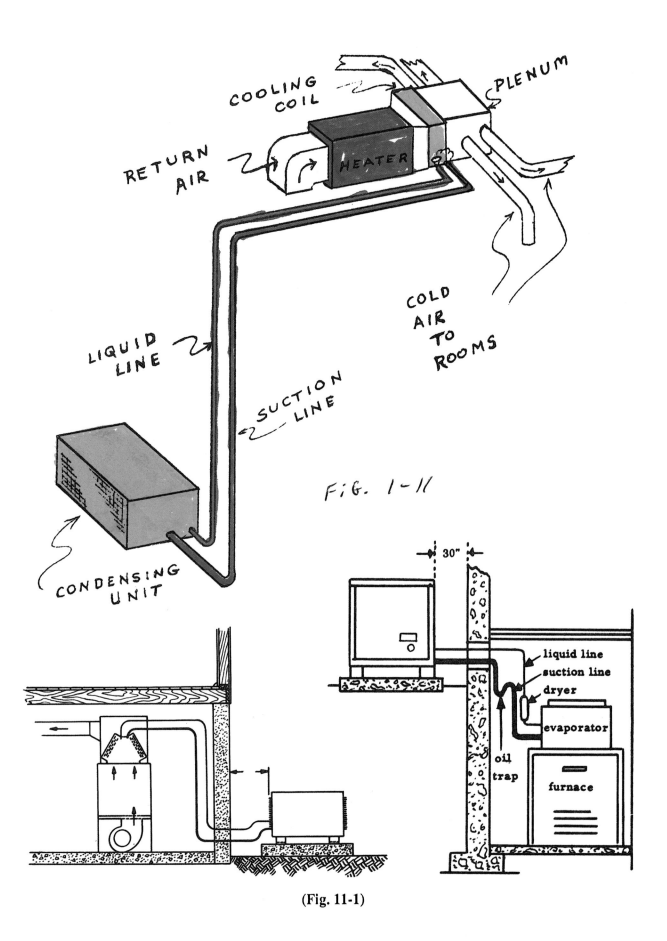

COOLING COIL

PLENUM

RETURN AIR

HEATER

COLD AIR TO ROOMS

LIQUID LINE

SUCTION LINE

Fig. 1-11

CONDENSING UNIT

30"

liquid line

suction line

dryer

evaporator

oil trap

furnace

(Fig. 11-1)

156

The central system is much like a commercial unit. In fact, a central system could be converted into a commercial system if it were powerful enough and the house were insulated like a refrigerator. In the central system the only part of the heating plant that is of any importance to the air conditioning is the furnace blower. The blower that circulates air over the evaporator is the same blower that circulates warm air in the home in the winter time. The same ducts are used for both winter and summer air circulation. The only other parts of the system that are combined are the thermostat and the low voltage circuits.

A serviceman can treat a central system just like he would a window unit. The condensing unit is generally air cooled and it will look like **Fig. 11-2**.

The condensing unit may be water cooled; and if so, it will have a water tower, evaporator condenser or a waste water system. However, the water cooled unit has just about lost its popularity altogether. Most central system condensing units will be equipped with schraeder valves and it is a simple job to put on the gauges. Most central condensing units will be hermetic up to five hp in size and semi-hermetic if over five hp. The largest percent of all home condensing units are in the three horsepower class and are hermetic. You will find that they are almost always 240 volts and in some cases, three phase.

See the three-ton hermetic condensing unit as shown in **Figure 11-3**. The schraeder valves we talk of are located just outside of the unit, where the suction and liquid line couple on. Remove the 1/4" caps and attach your gauges. Usually these units have no service valves at all and cannot be pumped down. However there are still some of the older units around that have service

(Fig. 11-2)

valves.

Do not try to read something complicated into a central system that does not exist. The unit is the same old condensing unit that is used in a window unit, only now it has been jumbo sized and is bigger in horsepower and pumping capacity. The serviceman should approach these units just like he would a window unit. That is, he should take his flashlight and look around

Easy to service, because entire cabinet and discharge grille is removable.

Refrigerant shut-off valves aid in easy servicing.

All controls are factory wired and color coded for easy installation.

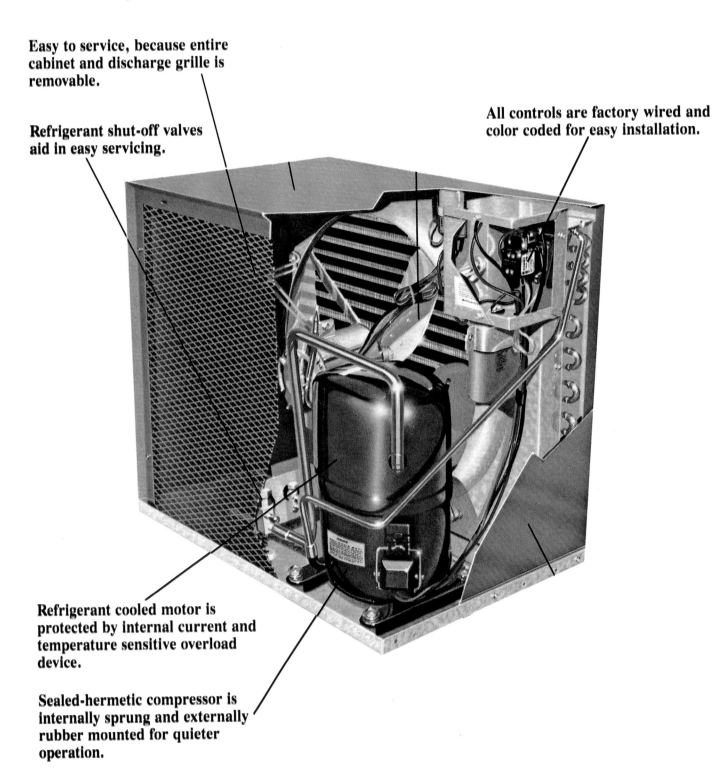

Refrigerant cooled motor is protected by internal current and temperature sensitive overload device.

Sealed-hermetic compressor is internally sprung and externally rubber mounted for quieter operation.

(Fig. 11-3)

and make a thorough inspection of the whole system before he goes to work. In almost every case, he will find the condensing unit out beside the house; and he will see the liquid line and suction line running down to the unit from the attic. There will be electrical service running to the condensing unit, and the control wires from the thermostat inside of the house will be running out to the unit.

Step by step, you would approach a central system just like you would a window unit. First, you would ask a few questions if there is someone who can give you the answers. As soon as you have established what the complaint is, you are ready to go to work. Your first move would be to take the cover off the condensing unit and take a close look at the pot and the lines and wiring going into the unit. Now you will not touch a thing as yet. Get your flashlight and make a trip to the attic and find the evaporator; and have a look at the expansion valve to see how it is located and what condition the coil is in. You might, while you are in the house, find the air intake to the furnace fan and examine the filter to see if it needs attention. Also, while you are in the house, see if the switch is on summer cooling and that the thermostat is set properly. If so, the furnace fan should be running. Go back out to the unit, and you are ready to make some sure-fire checks to find the trouble. Suppose the complaint is that the unit is running all the time and the house is warming up; that is, not cooling down. You can say to yourself at this point that one of several things can be wrong: either the unit is O.K. and the furnace blower is not blowing; or the evaporator coil is iced up, or the unit is out of gas and running with no expansion taking place in the evaporator, or that the unit has gas, but it is not getting to the cooling coil.

The best way to prove this is to put your low side gauge on and see if the normal low side pressure is present. Suppose the unit showed a vacuum on the low side; then you could say that the unit must be out of gas, or the liquid line is stopped up. One thing for sure, if it pulled down into a vacuum, you would know that the pump was O.K. and ready and willing to pump if it had some gas to pump. A pump with bad valves would not pull down into a vacuum on the low side. It would show a high pressure on the low side if it were fully charged. If you pulled the switch and let the unit set for awhile and the low side came up to a pressure of a few pounds, you could almost say for sure that the gas was gone; and there was just enough to overcome the vacuum, but not enough to feed the evaporator properly. If this were the case, you would shut the unit down, add enough gas to

bring the pressure on the system up to about forty or fifty pounds and then start looking for the leak. A good place to look would be around the expansion valve. The nuts on the lines hooked to this valve will loosen up in time and bleed off the refrigerant charge. When the expansion valve nuts leak, there will be oil around the fittings; likewise, if the fittings or welds leak, there will be the telltale oil present. You could use soap, water or anything that would bubble to establish a leak. A halide leak detector would sniff a leak and show the presence of Freon on the reactor plate flame. Find the leak and fix it. You are ready then to blow the whole system down and evacuate and recharge.

To charge a central system, you would follow the same rules for other units. Watch the sight glass and fill until the sight glass is completely full of liquid; or if there is no sight glass, put in just enough gas to stop the expansion valve from hissing and for the suction line to come back cool to the compressor.

Charging central systems is no longer guess work. We can no longer sniff and feel our way around or use rule of thumb tactics. We now have to charge by the "super heat" method to reach a proper charge. Understand this and that is if the evaporator has an expansion valve and a sight glass, charge it as you have been instructed. If the unit is capillary tube, or uses a restrictor device, the superheat method is recommended. Refer to the charging chart on the following page. To the left you see "outdoor ambient." At the top you see "Suction Pressure at Compressor—PSIG," reading from 42 psi to 80 psi. Below that you see "Suction Line Temperature at Compressor -°F." To properly charge a unit take the outdoor temperature, and strap a temperature sensor to the suction line near the evaporator and get a temperature reading. Example: Let's say the outside ambient is 95°F, the suction pressure at the compressor is 68 psi and the suction line temperature is 53°F. You would have a perfectly charged system. Another example: The outside temperature is 95°F, the suction pressure is 58 psi and the suction line temperature is 58°F. You would have an undercharged unit.

In many cases you will find the unit dead when you arrive, and the first check you would make is the thermostat. At least, you would want to establish that the unit was not cut off by a defective thermostat or a thermostat that a child had tampered with or someone had got out of adjustment. These combination heating and cooling thermostats are tricky; but an experienced serviceman will not do more than give them a quick look; and if it seems normal enough, he will go out to the

OUTDOOR AMBIENT °F		SUCTION PRESSURE AT COMPRESSOR — PSIG																
		48	50	52	54	56	58	60	62	64	66	68	70	72	74	76	78	80
		SUCTION LINE TEMPERATURE AT COMPRESSOR - °F																
C	105+										41	43	44	45	47	48	49	51
O	105									41	42	44	45	46	48	49	50	52
O	100									45	46	48	49	50	52	53	54	56
L	95							48	50	51	53	54	55	57	58	59		
	90						52	53	55	56	58	59	60	62	63			
M	85				54	56	57	59	60	62	63	64	66					
O	80					59	61	62	64	65	67	68	69					
D	75				63	64	66	67	69	70	72	73						
E	70				65	67	68	70	71	73	74	76						
	65			69	70	72	73	75	76	78	79							

unit and find the wires running from the thermostat back to the unit and cross them or jump them. By doing this, he can isolate the thermostat and establish immediately if the trouble is in the thermostat. See this simple test for a bad thermostat in **Figure 11-4**.

If nothing happens when the thermostat is jumped, then the next check will be to see if the heaters are kicked out on the line starter. Push the reset button. Note: Most modern day units do not have line-starters, but have contactors with no O.L. heaters installed. If nothing happens, then lift the line starter manually with a screwdriver or stick. You will be lifting the part that the line-starter holding coil normally would lift. If the

unit started to run, then you may have a bad line-starter holding coil. Check and see if voltage is to the coil. However, if the unit remained dead, you now must check the fuses somewhere back up the line.

Many good quality central conditioning units are equipped with pressure controls, both high pressure and low pressure. **(Fig. 11-5)**

If the manufacturer has cut cost to the bone, he may have eliminated the low-pressure control or the high-pressure control; and in extreme cases, he may have left off both controls. This is poor practice and will lead to many, many service calls. You are not concerned about this; however, if you were dealing in units, it might be

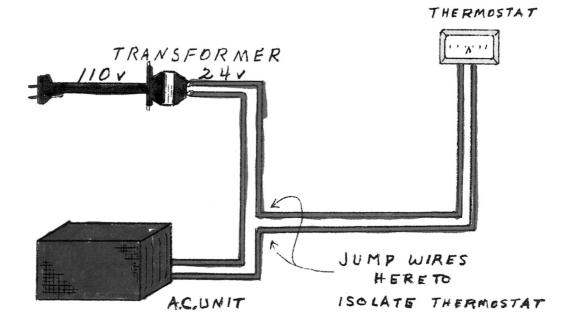

(Fig. 11-4)

a good idea to make sure the line you carry is fully equipped with both controls. Suppose the low-pressure control was left off the unit. Here is what could happen as a result: The filter could get very dirty and the coil freeze up, and there would be no cool air going to the ducts because of this wall of ice on the evaporator. Would the thermostat down in the hall cut the unit off? No, in fact, it would make the unit run more because the house would be warming up. Suppose the high pressure control were left off the unit. Here the unit is running; and if the condenser got dirty, the head pressure would climb; and the compressor overload did not kick it off, the pump motor would burn up. Even with an electrical overload, it would be hard on the motor windings to rely on the amperage load to cut the unit off instead of the pressure safety control. In the case of a water cooled unit without a high-pressure cut-out, the unit could possibly build up pressure on the high side with a water failure until the condenser or receiver were in danger of rupturing a relief valve. Where both controls are left off a unit, you would know that the only safety device left is the overload and the line fuses; and this is poor policy. It will be well at this point to remember that pressure controls on air conditioners are safety devices and not temperature controls. You must be perfectly clear on this point. The pressure controls on an air conditioner are not for controlling the temperature of the work being done. These controls are used for one reason and one reason only, and that is **SAFETY**. The low-pressure control is a lower limit control which will cut the unit off if the gas charge is lost or if the evaporator freezes up for some freak reason. A dirty filter will permit an evaporator to freeze over with ice; and when this happens where a low-pressure control is installed, the unit will pump down, cut off, and stay off until the ice is melted off the evaporator.

When the ice is off the evaporator, the unit pressure will rise again on the low side; and the control will permit the unit to go back into operation. This may not happen once in several months; yet it can happen; and when it does, the low-pressure control will take over and kill the unit until it is properly defrosted. This being the case, the low-pressure control on an air conditioner will always be set at a cut-in pressure slightly lower than freezing. Look on your temperature-pressure card; and you will note that if the unit is using F-22, the control will have to be set to cut off at a pressure of about 55 psi. A good setting for the low-pressure control on almost all F-22 air conditioners would be a cut-in point about 90 psi. For units using F-12 the pressure would be lower, but you would use the same chart according to a pressure just below freezing, say at 30°F.

The high-pressure control on an air conditioner is strictly a safety device, and it acts as a safety device on all refrigeration equipment regardless of whether it is commercial or air conditioning. This control is no different than the pressure control on an air compressor tank in a garage. It simply cuts off at a certain setting and comes back on at another setting. You know that when air is used out of the air tank in a filling station, the compressor comes back on when the pressure comes down to a certain setting. The same applies to the high-pressure control on air conditioning. The control cuts the unit off if the pressure ever comes up to the cut-off point; and it will stay off until the pressure comes down to the cut-in point. Now here is the reason the high-pressure control is installed on most big units and most central condensing units. If there is any failure of the condenser, the pressure would naturally go up in the unit; and the control will keep it from going so high that it could blow out the receiver or condenser relief valve or fusible plug. Where units are water cooled, they must have a high-pressure control as a safety device. If the water were suddenly cut off from the condenser, there would be an immediate increase in the head pressure; and something would have to give; usually it will be the

Low Pressure Control

(Fig. 11-5)

High Pressure Control

relief valve or fusible plug. In the case of the hermetic unit, the excessive head pressure resulting from a condenser failure could cause the unit to labor and slow down until it either burned up; or the electrical devices let go. Most units will have electrical overloads on the power supply. The overload switch will kill the unit if it starts to labor under an excessive head.

Once in awhile you will run into a high-pressure control with a reset mechanism; and once the control kills the unit, the reset button will have to be pushed manually in order to close the circuit again. Generally though, the high-pressure control will be an automatic resetting device. You will make many calls to units that are cutting out on the high-pressure control, and the customer may think he has a bad thermostat and will tell you that the thermostat is cutting up. You should know that a thermostat does not generally cut up. It is either on or off; and if there is something wrong with the system, the thermostat will be on and calling for refrigeration when the unit cannot refrigerate. In the case of the pressure control, the control will be governed by the internal pressure in the system; and it does not care whether the house is cold or hot. It functions according to the pressure in the system. High pressure controls are generally set by the man installing the unit. He will set the high-pressure control to cut out about 50 to 75 pounds above the normal head pressure that the unit is running under. Suppose you hooked up the unit and you were on the line and the unit were running smoothly and you wanted to set both controls. You would set the low pressure control at approximately 28 psi cut-out and approximately 60 psi cut-in for F-12. For F-22 you would set the low-pressure control to cut out at about 55 psi and to cut-in above 80 psi. You would determine the high-pressure control setting by the way the unit ran. Suppose the F-12 unit ran with a seemingly normal head pressure of about 140 psi with an air cooled condenser holding this pressure steady. You would choose a setting of about 40 to 75 psi higher than this normal running pressure. Now if the condenser fan motor burns out or the condenser gets dirty with dust and lint, the pressure will climb to your setting; and the unit will cut off before any damage is done. If the control is an automatic resetting control, you would choose a differential of about 50 pounds and let it go. Now the unit will cut off at about 200 psi head and will come back on if the pressure comes down to 140 or thereabouts.

Never confuse the operation of the pressure controls on air conditioners with the pressure controls on commercial refrigeration units. The high-pressure control on both installations are for the same purpose, but the low-pressure control on the commercial refrigeration unit can be a temperature controller, and its job is to maintain a steady temperature in the refrigerator by controlling the evaporator pressure. You will remember that practically every air conditioner in the world is controlled by a thermostat as far as the temperature of the air conditioned space is concerned. This is not so in the case of commercial installations. The temperature in the refrigerated space is controlled by the low-pressure control, and this control permits a constant off-and-on cycle that holds the ice formation on the evaporator to a minimum. This control of the defrost cycle is the principle of most blower-type evaporator coils used in commercial refrigeration.

The electrical devices you will encounter on central condensing units are in a sense the same as those on large well equipped window units. There may be more than one start capacitor, and there may be more than one run capacitor, or there may be any of a half-dozen combinations of electrical starting devices. But one thing is for sure, you will still have your old common start and run terminals like the single-phase 120 and 240 volt pots. You will have a relay and capacitor or PSC capacitor but the one thing that is sure to be on these units that is not used on the small window units is the contactor. The contactor is the electrically operated switch that is built to withstand the brunt of the electrical load when a large unit starts up. Three-phase central condensing units do not have any electrical starting devices like capacitors or relays. These units are started by the three-phase line starter with heaters built into the line starter. These are the most simple of all electrical pumps. Note the difference in these two systems. One is a single-phase 240 volts hookup, and the other is a three-phase pump. See **Figure 11-6**.

It is a good policy, no matter how experienced a serviceman you are, to make a drawing if you are going to disconnect any of the wires in and around a central system. Draw a picture that you will understand. Use tags on the wires if necessary. Use colors and numbers on the relay. You can chase out the circuits on a unit in time, but the smart serviceman makes a quick sketch of all the wire he is going to disconnect. Relays wear out, and starting capacitors and running capacitors wear out or blow out. It is a good policy to change these devices if there is the slightest suspicion that they may not be functioning right. Never try to save an old relay if a new one is available, and the customer can afford a replacement. This is the same as replacing the points on an

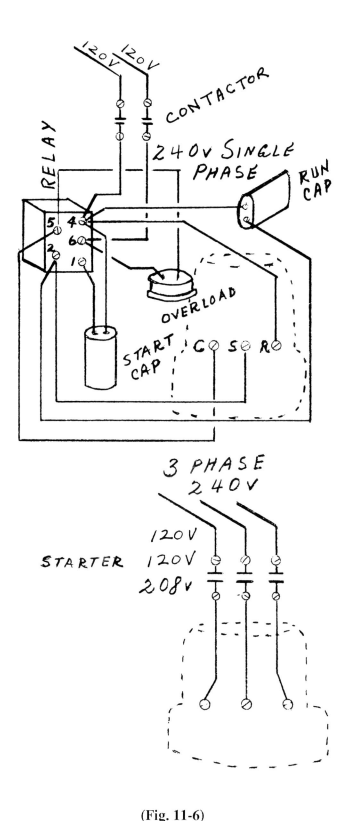

RELAY

120V 120V

CONTACTOR

240v SINGLE PHASE

RUN CAP

5 4
2 6
3 1

OVERLOAD

START CAP

C S R

3 PHASE 240V

STARTER 120V
120V
208v

(Fig. 11-6)

automobile timer. You would not think of tuning up an auto and leaving or filing old ignition points. If the relay has not been in use for very long, then you can use your own judgment; however, if the unit has been giving trouble, and the capacitors are blowing out, and it is a hard starting unit, then get rid of a weak and worn out relay. You should be able to handle the electrical repairs on a central unit just like a window unit. Where the contractor is used, as in the case of the central unit, you will have to remember that this gadget will have to kick in on the holding coil in order to have service to the pot. If the unit is dead and will not even try to start, you will have to look to the contactor for your trouble if the fuses are all good.

The line starter is generally operated on low voltages from the transformer, and the circuit is made through the thermostat in the house. See the simple drawing shown in **Figure 11-7** of the principle of the line starter in conjunction with a low-voltage system.

The low-voltage method of controlling a central system is tied in with the heating plant. Basically, the low-voltage is nothing more than a 24 volt transformer circuit. Once this transformer is tied in on 120 volts, it will set there for years and be relatively trouble free. See the transformer hookup to the power source as shown in **Figure 11-8**.

Notice the two wires coming out of this transformer carrying the 24 volts or low voltage. This 24 volts is used throughout the central plant to energize and operate the relays and holding coils. The reason for the use of low-voltage is not that it imparts any quality to a circuit that could not be obtained with 120 or 240 volts, but because it permits the manufacturer to use small relays, holding coils, and wire sizes. In fact, low-voltage wires can be strung throughout a house like the door bell wires; and this is a considerable saving over lines which would be big enough to carry 120 or 240 volts.

Take a look at the drawings shown in **Figure 11-9** and you can see the possibility of using low-voltage to do almost any electrical starting job.

Where low-voltage is used to operate a central system, the voltage will actuate the starting devices on the air conditioning unit; and it will be necessary to understand the furnace tie-in with this low-voltage system. The furnace controls are interlocked with the air conditioning controls. The word "interlocked" is not a scare word. It simply means that the furnace controls are in the same low-voltage lines that service the air conditioning end of the central system. Therefore, you

163

must understand the furnace safety devices and the controls that operate the furnace in order to service the system as a whole. The interlocking is simply the precaution the designer has taken so that the furnace cannot run if the air conditioner is on and vice versa.

In summarizing the servicing of the central air conditioning unit, you will attach your gauges if there are service valves and charge the unit just like you would charge any commercial unit. The amount of gas the unit holds will be determined by the size of the

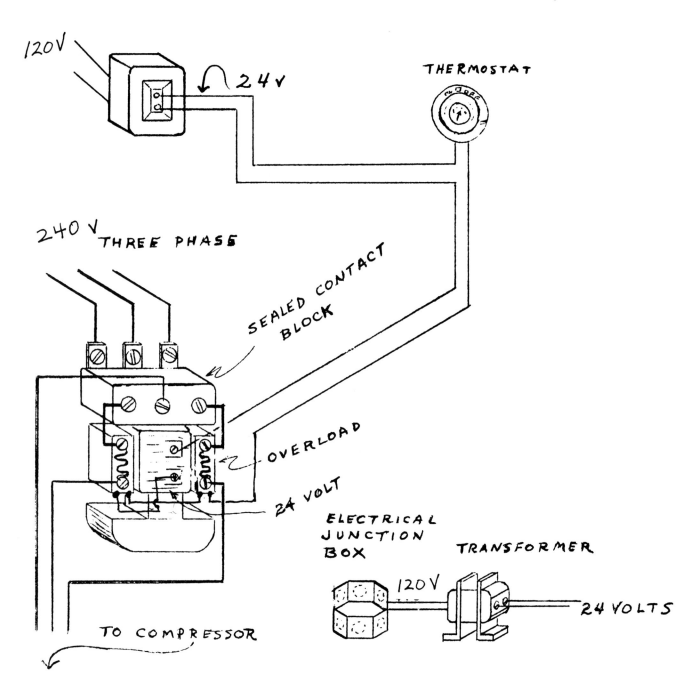

(Fig. 11-7)

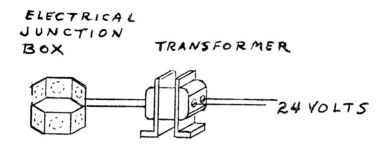

ELECTRICAL
JUNCTION
BOX

TRANSFORMER

24 VOLTS

(Fig. 11-8)

receiver and the distance from the evaporator to the condensing unit. If there is considerable distance, like fifty feet or more, you should make sure that there is ample gas in the unit to keep the liquid line full all the way from the receiver to the expansion valve. These units are difficult to purge if major repairs have been made. If there is a service valve on the receiver where the liquid line comes out, you can trap the Freon in the receiver while you are working on leaks and making repairs. When this is the case, assuming that there is a charge of gas in the system, you can purge through the whole system by letting the Freon go to the valve and evaporator and blow out at the suction line nut on the pot. If the suction line is welded to the pot where the low-side valve is located, then you will have to use a vacuum pump. Most central systems hold from four to ten pounds of Freon. ∎

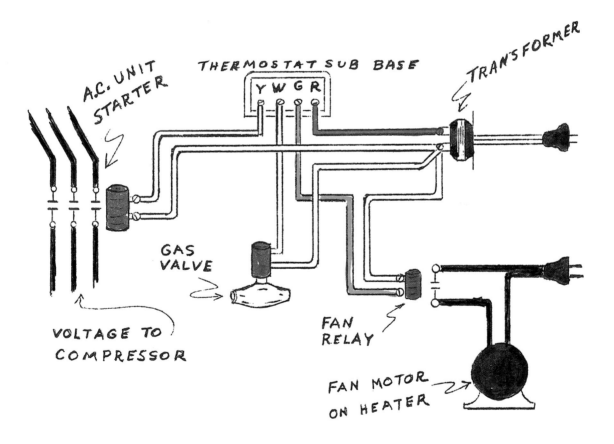

THERMOSTAT SUB BASE

A.C. UNIT STARTER

Y W G R

TRANSFORMER

VOLTAGE TO COMPRESSOR

GAS VALVE

FAN RELAY

FAN MOTOR ON HEATER

(Fig. 11-9)

CHAPTER 12

GAS AND ELECTRIC HEATING

Heating is the process by which a specific area is supplied with heat to bring that area up to a desired temperature and maintain that temperature automatically. The most common heating system found in residential and some commercial application is the gas fired, natural or butane, electric operated, forced-air heating system. A heating unit is centrally located with a series of ducts to each room. A squirrel cage blower in the heating unit forces air through this duct system and into each room. It is a very common practice to install a cooling coil in this duct work, locate a condensing unit outside the house, and have a combination heating and cooling unit.

Furnaces are rated by "input" and "output." The input rating is the amount of heat the burners are capable of putting into the furnace heat exchanger and the output rating is the amount of heat you can actually get out of the heat exchanger. For instance a furnace rated at 100,000 BTUs input would probably have an output rating of approximately 80,000 BTUs. You can see right off that 20% of the input heat is lost. This is due to the loss of some of the heat out of the vent stack. Today, furnaces are manufactured to save more gas. You can still purchase the 80% furnace, but now furnaces have efficiencies that can reach 97%. This efficiency is rated as A.F.U.E. (annual fuel utilization efficiency). High efficiency furnaces will be covered later.

The most practical approach to heating service is a study of the various controls used on central heating systems. If the operation of each control is understood, and the part it plays in the automatic control of the heating units, you are on your way to becoming an excellent serviceman. No matter how simple each control may be, it must be set right for that unit to function as it was designed to, whether it is a wall furnace, floor furnace, or a central heating system.

Central heating systems are automatic. They start when the temperature is down and stop when the desired temperature is reached. For that reason, we need a pilot light as shown in **Figure 12-1**, which is a small flame located in the heating unit for the purpose of lighting the main burner when heat is desired. The pilot burner is located in such a place that it may be easily seen and safely lighted. It must also be in such a location as to light the main burner as soon as heat is called for.

All pilot burners are simple in construction and are easily serviced. It is very important that the burner have a blue flame. If the air intake hole on the pilot burner becomes stopped up with dust and lint, the pilot light will burn yellow and will not be hot enough. Be sure to clean the pilot burner whenever you make a light-up on a heating system.

Why is it necessary to have a hot pilot light? A pilot burner has another important function. It is used in conjunction with a thermocouple as a safety feature on heating units.

What is a thermocouple? It is two unlike metals, iron and nickel, which are fused together at one end. Whenever heat is applied to the junction of these two metals, a small amount of voltage is generated. It would be necessary to have a special meter to read this voltage. A millivolt meter would be used. In checking the thermocouple with a meter, if the thermocouple were good, it would generate a little above 25 millivolt. If the reading dropped below 17 millivolt, the thermocouple would be too weak and need replacing. A poor flame on the pilot burner could also be a reason for the thermocouple not to generate the full voltage. Again, be sure the pilot burner is clean and burning properly.

By using a number of simple thermocouples in one unit and having all thermocouples connected in series, it is possible to make a unit that would generate a greater voltage than one single thermocouple. This would be known as a pilot generator (**Fig. 12-2**) since the pilot burner and thermocouples are in one unit. The flame is in the center of this unit and heats all simple

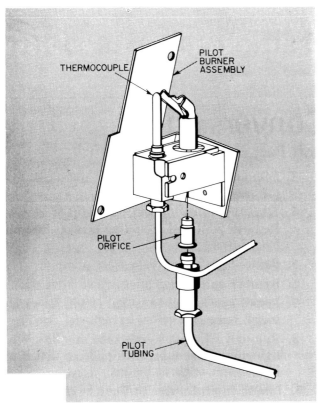

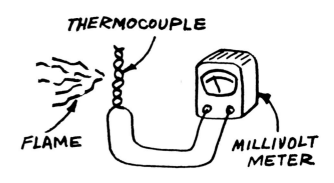

(Fig. 12-2)

thermocouples at one time. See **Figure 12-3**. The voltage generated by this unit would be above 250 millivolts and would be used on wall furnaces and floor furnaces. By using a pilot generator it would not be necessary to use another voltage supply for the operation of the control circuit. A pilot generator would generate enough voltage to operate a small solenoid valve.

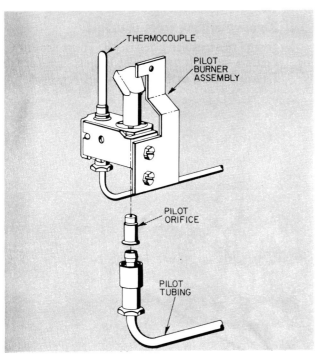

Styles of Mounting Brackets Used

(Fig. 12-1)

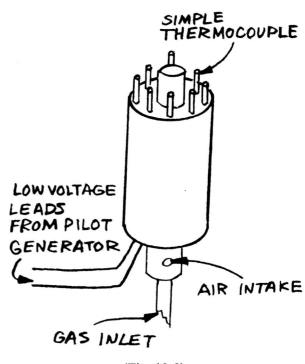

(Fig. 12-3)

Each heating unit has a gas valve installed in the gas line to the main burner. The purpose of the gas valve is to control the flow of gas supply into the main burner. See **Figure 12-4**. Natural gas leaves an oily film on the valve plunger and seat. During the summer season when the valve is not in operation, this film will become sticky and cause the valve plunger to stick; and the valve will not open. Tap the valve slightly to see if it will open while voltage is applied. This only indicates that the valve was stuck and will not remove the sticky substance. Never disassemble the valve. Replace it. Burned magnetic coils will also cause the valves not to operate. If this should happen in a diaphragm type valve, it would be necessary to change out the complete valve.

Whenever any service is done on the gas valve, be sure you check the operation of the valve and are sure the valve will open and close properly.

What controls the opening and closing of the gas valve? The thermostat which is located in the heated area and set on a desired temperature controls this valve. Whenever the temperature drops, the thermostat makes contact and starts the heating unit. When the temperature rises to the desired setting, the thermostat opens, or breaks the control circuit, and cuts off the fire.

Most thermostats are sealed mercury bulb thermostats and operate at a very close differential between cut-in and cut-out. They give better control than the old snap action stat. See **Figure 12-5**. If cool air passes over the metal spiral, the metal will contract and cause the mercury bulb to tilt forward and make contact. If warm air passes over the metal spiral it will expand and tilt the mercury bulb back and break contact. It is necessary to keep the thermostat free from dust and dirt. Small particles of dirt can cause the operating parts to malfunction. The mercury bulb is sealed and cannot be cleaned. If the thermostat is not opening and closing properly, replace it. Just the natural circulation of the warm air rising and circulating through the thermostat will deposit enough dust and lint in the thermostat to cause it not to function properly.

The close differential of the thermostat is controlled by the heating anticipator located in the thermostat. It should be set at the same amperage of the gas valve. This is what gives the close differential. The anticipator is shown in **Figure 12-6**. The anticipator is adjustable. If the amp rating of the gas valve is .4 amps, then set the anticipator accordingly. If the amp draw on the gas valve is unknown, then take your amprobe and wrap ten turns of thermostat wire around the jaws of the meter. Remove one wire going to the gas valve and hook it as shown in **Figure 12-7**.

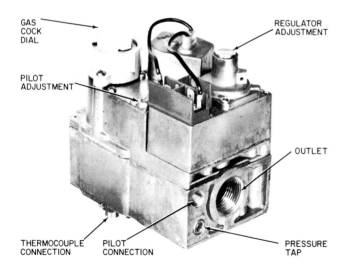

(Fig. 12-4)

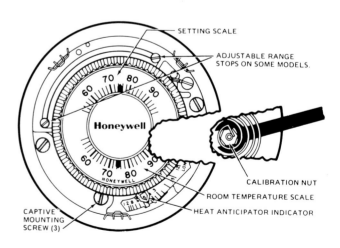

(Fig. 12-5)

(Fig. 12-6)

Let's say the reading on the amp meter was 6 amps. Divide by ten (ten wraps of wire) and you would get .6. Set the anticipator at .6 amps. The heating side of the thermostat has an adjustable heat anticipator because of various valve ratings. When replacing gas valves be sure the new gas valve does not have a different rating. It could cause the thermostat to cut in and out too soon. Cooling anticipators are fixed and do not need setting. Thermostats on floor furnaces and wall furnaces can be snap-action contacts or mercury bulb. Generally these thermostats do not require a thermostat with an anticipator.

By combining a part of the controls covered here it is possible to have the simple control circuit used on wall furnaces floor furnaces and some closet-type furnaces. (Fig. 12-8) Remember a voltage supply, a valve and a switch in the circuit are needed. This switch would be the thermostat. So by combining a pilot generator, a diaphragm gas valve and a thermostat, a single control circuit can be made.

First it is necessary to turn on the gas at the manual gas valve. This is just a line valve and is required to be installed on all gas equipment. Next in line was added

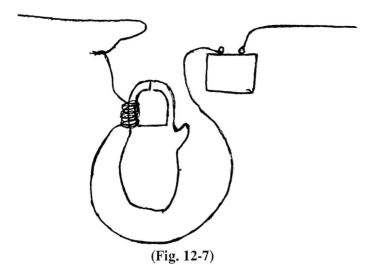

(Fig. 12-7)

a regulator valve. This valve is required on all heating equipment also. It comes from the factory set; and it is very seldom if ever necessary to change the setting of the regulator valve. In case of a ruptured diaphragm or for some reason the regulator valve will not let gas pass it will be necessary to replace it.

With the gas valve on we have gas to the solenoid valve and Pilot Generator (P. G.). Now light the P. G. and in a few moments it will generate enough voltage to operate the control circuit. Now close the thermostat and on comes the main burner. When the desired temperature is reached the thermostat opens and out goes the main burner. The P. G. remains burning.

Check the heater closely now for trouble points. Dirty air ports on the P. G. would cause it to burn yellow and not heat enough to generate the proper voltage to operate the unit. A dirty thermostat might keep the system from coming on. Clean both thermostat and P. G. Also check for any bad connections.

Another operation of the P. G. can be discussed here. If for some reason the P. G. went out and someone turned the thermostat on nothing would happen since the voltage supply to the control circuit would be dead. Just think for a moment what would happen if the P. G. went out and it were still possible to open the solenoid valve. The heater would fill with gas; in fact the whole house could fill with gas. This could be very dangerous. But as it is, if the P. G. went out it would be necessary to relight it and put it back in service before the system could be turned on.

Going now to larger closet furnaces and attic furnaces we find a 24 volt control circuit. Here it is necessary to cut the normal 120 volts supplied to a heater down to 24 volts by the use of a low-voltage transformer.

These transformers are simple in construction and will last only a couple of minutes if a short circuit occurs in the low-voltage side.

In using a 24 volt control circuit the P. G. is eliminated. It is replaced with a pilot burner and a thermocouple; and it will be necessary to add another valve in the gas line, a safety pilot valve. This valve will operate off the thermocouple. The safety pilot valve or pilotstat has a small power unit in it that the thermocouple will operate. After the pilot burner is lit the thermocouple heats up but it is necessary to set the pilotstat manually. The power unit in the pilotstat will hold the valve open and the unit can be operated. If the pilot burner goes out the pilotstat valve will automatically close; and even though it is possible to operate the

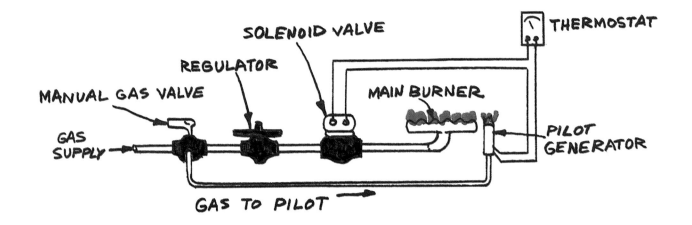

(Fig. 12-8)

solenoid valve, no gas can pass into the main burner.

Check back over all the controls now and remember the main service points covered. The thermostat must be free of all dirt and dust and must have a proper set anticipator. The solenoid valve must be free of any sludge that will cause it to stick. This is important because a sticky valve could cause real trouble. The pilot burner must be free of any dust or dirt and have a hot blue flame. The safety pilot or pilotstat must hold open properly to allow the gas to come through to the main burner when the thermostat calls for heat. Check each valve for proper operation. Remember a clean pilot burner and clean thermostat means a whole lot to proper operation of the heater.

Checking back on the heater remember the fan chamber. The fan is to circulate the hot air throughout the house and it must operate automatically also. What starts the fan? A fan switch mounted on the heater near the outlet end of the heater.

The fan control operates as a pressure control. Having an element that is filled with gas and by applying heat to the element pressure builds up and causes the points to close. This will start the fan motor and warm air is then blown from the heater into the space. When the thermostat opens and closes the gas valve the fan element cools, the pressure drops; and the points open and stop the fan motor.

Now let's follow the complete heating cycle. The thermostat calls for heat. The points close, the solenoid valve opens and on comes the main burner. In about two minutes the temperature in the heater reaches 140 degrees. On comes the fan and warm air is circulated around the house. The thermostat is satisfied, the points open and the solenoid valve closes and cuts off

the main burner. The fan will continue to run until the temperature in the heater falls to 100 degrees; the fan switch then opens and the fan motor stops.

Remember the settings now. The fan comes on at 140 degrees and cuts off at *100* degrees. It may be necessary to spread the setting. That is, if the fan, after running through a normal cycle, cuts on and off several times. The thing to do then would be set the cutoff at 95 degrees. This would cool the heater five more degrees and would not have as much heat present in the heater to build up to the point at which the fan cuts on. Whenever a setting is made on a fan switch, be sure to check the operation of the heater before you leave. If a fan control sticks and the fan does not stop, change out the control. Don't try to repair it, replace it.

There is one more control to cover, the limit switch. This is another safety control and should be checked whenever a service call is made on a heater. The limit switch is wired in such a way as to cut the fire off if something should happen in the fan chamber, say a broken belt or burned-out motor. The fire will still come on; but when the temperature closes the fan switch, nothing happens. The temperature still continues to climb and at 180° the limit switch will open contact, thus cutting off the fire. As the temperature falls to 150°, the control closes contacts and on comes the fire again. Even though the fire continues to cut off and on, the temperature will not get high enough to cause any damage.

As you can see by the sketch in **Figure 12-9**, the control circuit is very simple. When the limit switch opens, it ends the high voltage supply to the transformer; and it would be impossible to operate the solenoid valve. The operation of the limit switch can be

170

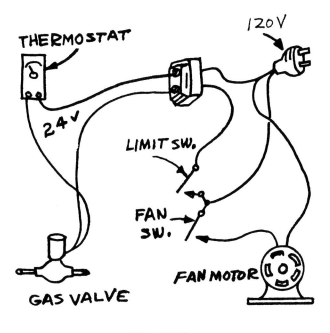

THERMOSTAT

120 V

24 V

LIMIT SW.

FAN SW.

GAS VALVE

FAN MOTOR

(Fig. 12-9)

multi-speed fans, usually four or five speeds. Since one furnace may be able to handle from two tons of cooling to three tons of cooling, the air movement may have to be changed by selecting different speed wires. Nevertheless, we have to use a fan relay to operate in conjunction with the fan control. With this set up we are able to use the low speed during heating (less air required for heating) and high speed for cooling. See **Figure 12-10**.

The thermostat should always be set in the "auto" position at the stat. If so, during heating there will be no power to the fan relay coil from the stat. The fan relay contacts would be positioned to supply voltage through the fan control and on to the low speed of the fan motor. In the cooling position, terminal "G" or "F" will be made hot when the outside condensing unit comes on. This will magnetize the magnetic coil in the relay and cause the other set of contacts to close and complete a circuit to the high speed. The fan relay is to switch for high and low speed and is also used for continuous or automatic fan operation. The fan relay is also to be sure the high and low speeds separate to prevent a motor burn-out should the speeds get together.

Be sure to check any heating unit over very thoroughly. Each control on that heater serves a purpose, and it's up to the serviceman to set the controls so they will operate properly and do the job they were designed to do. This by no means makes you an expert heating serviceman, but it will give you a real good start.

checked by taking the belt off the fan or disconnecting the common wire on direct drive fans, and operating the burner. When the temperature reaches 180° the limit switch opens and cuts the control circuit off. Do not force the limit switch on high efficiency furnaces. Replace the belt or put the common wire back on and put the furnace back into service. If, for any reason, the limit switch fails to operate, don't attempt to repair it; replace it.

Central heating and cooling systems of today have

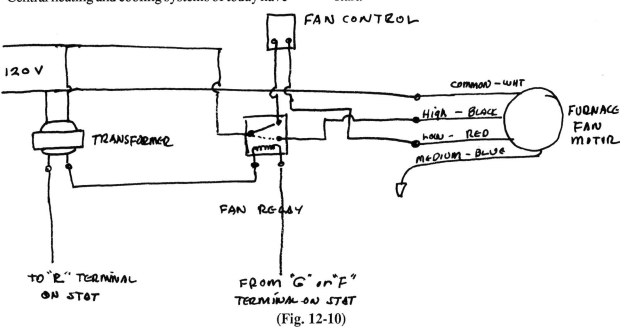

FAN CONTROL

120 V

TRANSFORMER

FAN RELAY

common — WHT

High — BLACK

Low — RED

MEDIUM — BLUE

FURNACE FAN MOTOR

TO "R" TERMINAL ON STAT

FROM "G" or "F" TERMINAL ON STAT

(Fig. 12-10)

See **Figure 12-11** for a basic schematic for today's conventional furnace.

Servicing heating plants is a profitable business of the air conditioning serviceman, especially in the winter months when air conditioning is at a standstill. Refer to the heating trouble shooting chart which follows immediately. We will then move into high efficiency furnaces. ∎

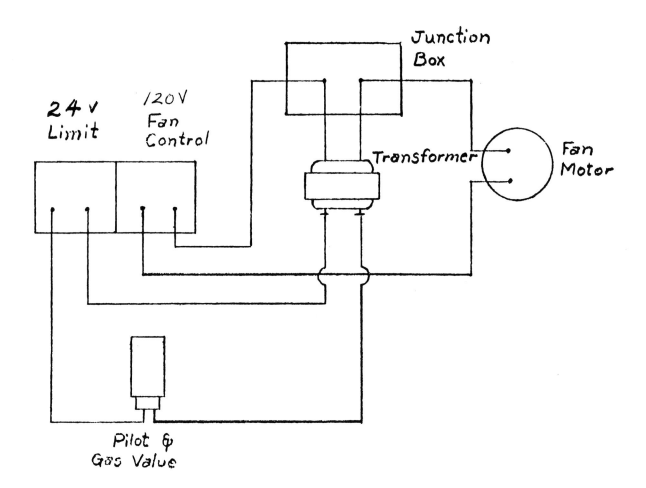

(Fig. 12-11)

CENTRAL HEATING SERVICE CHECK LIST FOR FALL LIGHT-UPS

1. Check air filter. If the filter is dirty, be sure to clean or replace.
2. Inspect fan and blower motor. Oil both if not self-lubricated.
3. Inspect belt. If the belt is frayed or cracked, replace it.
4. Clean pilot burner and light. Be sure to have a clear blue flame.
5. Check safety pilot and be sure it is operating properly.
6. Check the operation of the solenoid valve and be sure it will open and close properly.
7. Check the differential and make changes on the heat anticipator if needed.
8. Check out the operation of the fan control. Be sure that the fan will come on and go off properly.
9. Check and be sure the limit switch will operate. This can be done by removing the belt or removing the common wire on direct drive motors and turning on the main burner.
10. Let the heater operate through several heating cycles. You can do this by changing the setting on the thermostat manually.

SERVICE POINTERS FOR CENTRAL HEATING

1. Heater will not come on.
   A. Check pilot light. If it is off, clean and light.
   B. When pilot light all right, check thermocouple. It may be bad and need to be replaced.
   C. Check voltage supply to transformer; if all right, check transformer to be sure of voltage to the control circuit.
   D. Check and see if solenoid valve will open.
   E. Check thermostat and be sure the points are making contact.
2. Heater cycles off-and-on too often.
   A. Check heat anticipator settings.
   B. Check the setting of the limit control. If set too low, it would cause the fire to cycle often.
3. Heater takes too long to heat the house.
   A. Check for dirty filter which would cause a restriction to the air flow.
   B. Check for loose fan belt, fan not running at full speed.
4. Heater will not cut off, house overheating.
   A. Check thermostat to see that the points are opening.
   B. Check and make sure the solenoid valve is not stuck open. If it is, be sure to replace it.
   C. Check the control circuit and be sure there is no short circuit.
5. Fan runs constantly.
   A. Fan control stuck closed; replace it.

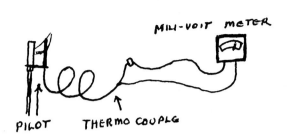

Checking thermocouple with milivolt meter. A reading of around 30 milivolts indicates good thermocouple.

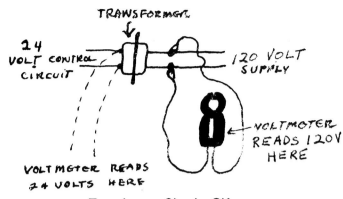

Transformer Checks OK

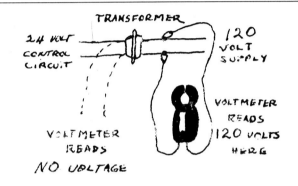

Transformer Checks Bad

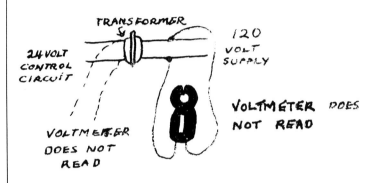

No Power To Transformer

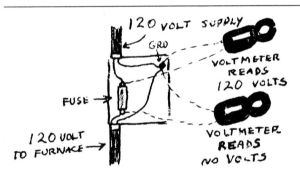

Fuse Shows Blown if top meter reads 120 volts and bottom meter reads no voltage fuse is blown

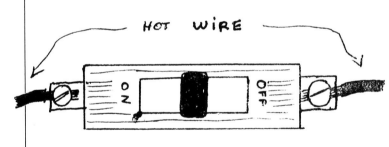

When breaker is in half way position, breaker is tripped. Turn to off position then back on, to reset breaker.

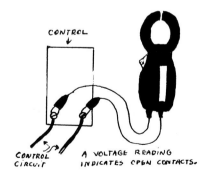

A reading of no voltage indicates contacts are closed.

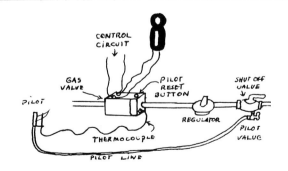

If voltmeter reads 24 volts with pilot reset good gas valve is bad

174

HEATING TROUBLE SHOOTING GUIDE

COMPLAINT	CAUSE	REMEDY
A. No heat.	1. Pilot light off.	1. Check and see if pilot is on 100° shut off.
	2. Thermocouple bad.	2. Replace thermocouple.
	3. Milivolt coil in pilot reset bad.	3. If milivolt coil is built into valve replace valve. If pilot safety is remote, replace pilot safety only.
	4. Pilot orifice dirty not letting enough flame burn across the thermocouple.	4. Clean pilot orifice.
	5. Draft in furnace burner, causing thermocouple to fail. Pilot blows away from the thermocouple in this case.	5. Check for cracked heat exchanger first. If OK, eliminate unusual draft.
	6. No power to furnace.	6. Replace fuse or reset breaker, and find reason for failure of either.
	7. Bad transformer.	7. Check size and voltage of transformer. Check for shorted 24 volt circuit.
B. Burner will come on but fan will not run.	1. Fan control contacts not closing.	1. Replace fan control
	2. Setting of fan control too high.	2. Set fan to cut in at approximately 140 degrees.
	3. Fan motor defective.	3. Check power at fan motor terminals, and replace if defective.
	4. Fan motor will come on but fan will not run.	4. Check for broken fan belt.
C. Burner cycles on thermostat normally but fan will not cut off.	1. Fan control contacts stuck.	1. Remove one wire of fan control. If fan cuts off, replace stuck control.
	2. Fan control cut out set too low.	2. Set fan control to cut out at approximately 110 degrees.
	3. Fan set in manual position.	3. Check to see if fan has manual position. Turn to automatic.
	4. Thermostat set on continuous fan.	4. Set thermostat to auto position.
	5. Fan relay stuck.	5. Replace fan relay.

HEATING TROUBLE SHOOTING GUIDE cont.

COMPLAINT	CAUSE	REMEDY
D. Burner cuts off but comes back on too soon.	1. Heat anticipator in thermostat out of adjustment. 2. Burner cycling on high limit control. 3. Defective thermostat	1. Set anticipator amperage to match gas valve amperage. 2. Check for broken fan belt, bad fan motor, dirty filter etc. Check setting of limit control. Set control around 190-200 degrees cut out. 3. Replace thermostat
E. Burner comes on and controls furnace, but furnace does not heat enough.	1. Gas pressure to burners not high enough. 2. Undersized furnace. 3. Not enough air through ducts. 4. Main gas meter valve not open enough.	1. Increase gas pressure at regulator. 2. Check BTU rating of furnace according to square footage of building. Approximately 30,000 BTU per 500 square foot of area. Needs to be approximately 40,000 in northern climates. 3. Speed up fan motor 4. Open valve fully open.
F. Burner stays off too long before coming back on.	1. Heat anticipator out of adjustment. 2. Thermostat defective. 3. Heat from television or table lamp keeping thermostat off.	1. Set anticipator at thermostat according to amperage rating of gas valve. 2. Replace thermostat 3. Relocate thermostat or device causing problem.
G. Pilot will light when pressing reset button, but will not stay on after releasing reset.	1. Thermocouple out.	1. Check milivolt output, and replace thermocouple if output is low.
H. Flame pushes outside and up outside of furnace wall when burner comes on. (FAN OFF)	1. Soot build up inside of heat exchanger. 2. Flue stopped up.	1. Clean soot out of heat exchanger. 2. Disconnect flue and clean.

HEATING TROUBLE SHOOTING GUIDE cont.

COMPLAINT	CAUSE	REMEDY
I. Flame burns normally when burner comes on but burns outside of furnace wall when fan comes on.	1. Crack in heat exchanger. 2. Heat exchanger not sealed properly when installed, and air leaking back into combustion chamber.	1. Replace heat exchanger or install new furnace whichever is cheaper 2. Check for defective workmanship and repair and seal any leak found.
J. Pilot light keeps going out.	1. Weak thermocouple. 2. Low gas pressure. 3. Dirty pilot orifice. 4. Pilot flame not burning against enough of thermocouple.	1. Check milivolt output of the thermocouple and replace if MV's are not up to 30MV's. 2. Increase gas pressure to pilot. Call city if main gas pressure is too low. 3. Clean orifice. 4. Clean orifice to pilot light, or increase gas pressure to pilot light.
K. Furnace burner cycles on high limit control.	1. Fan motor not running. 2. Fan control contacts open. 3. Belt broken on fan. 4. Belt slipping. 5. Filter dirty. 6. Not enough air moving through furnace. 7. Gas pressure too high. 8. Dirty evaporator. 9. Too many room outlets closed.	1. Check voltage supply at motor, replace if defective. 2. Replace fan control. 3. Replace belt. 4. Replace belt or tighten. 5. Clean or replace. 6. Increase speed of fan motor. 7. Adjust pressure regulator to give approximately 40° to 70°F temperature rise across furnace. 8. Clean evaporator. 9. Open for proper air flow.
L. Windows sweat and water runs down on window inside of building.	1. Not enough running time on fan after burner cuts off during extremely cold weather. 2. Heavy drapes drawn across windows preventing air circulation across windows.	1. Lower fan control setting. 2. Instruct customer to open drapes so air movement can get to windows (especially in colder weather).

HEATING TROUBLE SHOOTING GUIDE cont.

COMPLAINT	CAUSE	REMEDY
M. Burners make roaring sound when gas valve opens.	1. Air adjustment too far open on burners. 2. Too much gas pressure to burners.	1. Adjust air to burners until roaring ceases. 2. Decrease gas pressure to burners until roaring ceases. Set gas for proper temperature rise across furnace.
N. Burner flame burns yellow.	1. Air adjustment closed off too much. 2. Flue stopped up. 3. Heat exchanger stopped up. 4. Not enough combustion air.	1. Adjust burners for more air. 2. Clean flue 3. Clean heat exchanger 4. Make opening in furnace room so burner can receive ample combustion air. (One square inch opening per 1,000 BTU of furnace.)
O. Flame lights with a boom when burner ignites.	1. Dirty burners. 2. Pilot light not close enough to burner. 3. One pilot light out on two pilot burner.	1. Remove burners and clean. 2. Move pilot closer to burner. 3. Light pilot and establish reason for its failure.
P. Furnace is noisy.	1. Belt cracked. 2. Motor pulley and fan pulley out of line. 3. Bearings in fan section dry or worn.	1. Replace belt. 2. Line pulleys. 3. Oil bearings, or, replace if worn.
Q. Air is noisy coming out of room outlets.	1. Fan running too fast. 2. Room outlet too small.	1. Slow fan down. 2. Install larger outlet

178

HEATING TROUBLE SHOOTING GUIDE cont.

COMPLAINT	CAUSE	REMEDY
R. Combustion gas can be smelled in home.	1. Flue stopped up 2. Too many turns in flue, causing improper venting. 3. Not enough combustion air. 4. Heat exchanger cracked.	1. Disassemble flue and clean. 2. Reroute flue in straighter run. 3. Make opening in furnace room for combustion air. 4. Check heat exchanger.
S. Transformer continues to burn out.	1. Voltage too high to transformer. 2. Transformer too small. 3. Short in 24 volt wiring	1. Select transformer with correct voltage rating. 2. Install correct size transformer (20 VA for heat only, 40 VA for heat-cool combination). 3. Check for incorrect wiring of circuit or short in wiring.
T. Burner will not shut off in any position of controls.	1. Thermostat contacts stuck. 2. Gas valve stuck in open position. 3. Valve in manual position.	1. Replace thermostat, cleaning of contacts not recommended. 2. Remove control wire from valve terminal and wait five minutes. If valve does not close, replace stuck valve. (Some valves have a delay action in opening and closing.) 3. Check and see if valve has manual device. If in manual, turn to automatic.
U. One part of the building is warm and the other cool.	1. Too much air coming to one side of building and not enough to the other.	1. Adjust volume dampers to even air distribution. (Every 500 sq. ft. of area needs approx. 300 CFM of air for heat; 400 CFM for cool; 450 CFM for heat pumps.

HIGH EFFICIENCY FURNACES

Never before has anyone been able to get 100% efficiency from anything involving heating homes. Over one hundred years ago when settlers constructed their homes and built fire places out of clay, they could not get even 30% efficiency because most of the heat went up the chimney. The wood stove was somewhat better. It could be set in the middle of the room where wood was burnt inside and vented outside the house. There was an improvement of efficiency but still most of the heat was vented away. After the fire place and the wood burning stove, it was found that fossil fuels could be used for heat. Along came butane, coal, and propane heaters and the old kerosene stove. Efficiencies were somewhat improved but even higher efficiency was needed. Then came the gas fired, automatic furnace and natural gas became popular. Furnaces were set in hallway closets or put in the basement. Duct work was attached and carried heat to certain areas of the house either by gravity of the heat or by use of a furnace blower. Efficiency and cost of gas was really not that big of an issue then. Furnaces were finally being manufactured that would get as much as 70 to 80% efficiency. What a miraculous achievement. Only 20 to 30% of the total heat was being wasted. Everyone was satisfied. Today it is different. Manufacturers are striving to do the best they can to reach 100% efficiency. A 100% efficient furnace has not been developed yet but they are coming very close, right up to 97%.

Input and output heat — The more modern heaters used today are rated by the input and output ratings. If you take a look at the nomenclature tag of a furnace, along with the model number and serial number you will find the ratings. For example you may find that you have a 100,000 Btu input and a 80,000 Btu output. What this means is that the gas burners, by burning gas in the heat exchanger are able to generate 100,000 Btu of heat (input), but you only get out of the heat exchanger 80,000 Btu (output). By simple arithmetic and percentages you can see that 20% of the heat was lost. More realistically you could say that as much as 26% was lost. This 26% loss went up the flue pipe See **Figure 12-12**.

Notice that the flame burns up through the inside of the heat exchanger. Air from the atmosphere mixes with the gas and when burned creates heat. The heat is confined to the inside of the heat exchanger cells and continues up due to venting through the flue pipe. Heat will travel through the walls of the heat exchanger into the air stream making the air warm. In the process of exchanging gas to heat and heat to air we lose some of that heat out of the flue pipe, anywhere from 20 to 30%. Manufacturers are trying to do better and bring the efficiency up to 100% if possible. This brings us up to the efficiency rating of the high efficiency furnace and all other furnaces as far as that goes. AFUE (annual fuel utilization efficiency). Furnaces today have to have an efficiency rating. A tag has to be glued on to the unit showing that efficiency. The efficiency is derived at by bringing the output rating as close to the input rating as possible. If you could build yourself a furnace and construct it so that you could have an input heat of 100,000 Btu and get an actual output of 97,000 Btu, you would have a 97% efficiency. That is great for that particular moment, but that rating must not be taken for a moment. A true time test would be for a full season. Then you would have an AFUE of 97%. Just remember, whatever the AFUE rating is, that is basically the amount of heat you are getting out of the fuel you are putting into the furnace for heat. If your AFUE is 97% as compared to another furnace at 67%, your savings would be 30%. Over a long winter season of several months the dollars saved could easily be several hundred.

How the efficiency is improved — Since 20% of the heat from a furnace is going out of the flue pipe, wouldn't it make sense to reclaim some of that heat and reuse it. Over the years, many devices have been tried and at the time were successful in what they were supposed to do. One of these devices was a control damper in the flue pipe. It would close when the burners cut off. There was a 24 volt motor on the damper that would allow the damper to close and hold heat in the heat exchanger. The fan continued to operate and cool the heat exchanger down before it was cut off. Most all of the heat would be blown into the area to be cooled and not vented out the flue pipe. See **Figure 12-13**.

With this device the efficiency of the furnace could be improved by a few percentage points. Another energy saving tactic was the S shaped heat exchanger. (Keep in mind before we go any further that these furnaces are very modern and used extensively). This type of heat exchanger was brought about

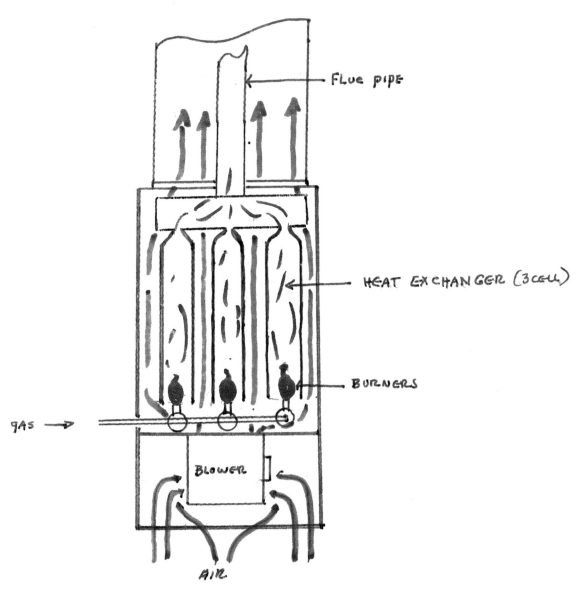

FLUE PIPE

HEAT EXCHANGER (3 CELL)

BURNERS

GAS →

BLOWER

AIR

(Fig. 12-12)

because someone had the idea of rather than let the heat go straight out of the flue pipe, it could be slowed down by moving the hot air back and forth through an extended heat exchanger. See **Figure 12-14** of an S curved heat exchanger.

Notice the hot gases now have to make all of the curves, therefore utilizing more of the heat exchanger surface. At the same time the hot gases are slowed enough to get even more heat out of the heat exchanger. This heat exchanger is still efficient but usually has an AFUE of around 82%. This is still far from the efficiencies that can be arrived at today.

The more the heat exchangers are restricted, revised and added on to however, causes problems. One

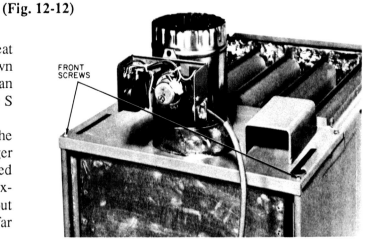

FRONT SCREWS

(Fig. 12-13)

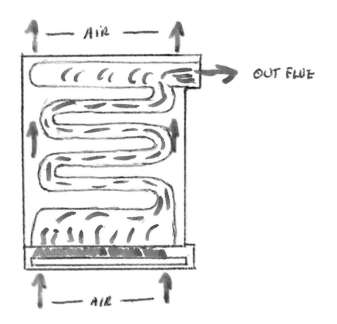

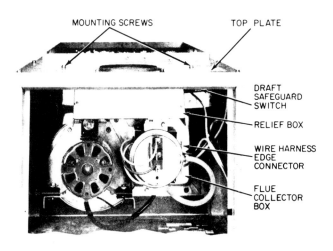

(Fig. 12-14)

is that on a natural draft furnace, the more efficiency reached for, the more the gases will be slowed down to a point that the heat exchanger overheats. Then it will be necessary to put a draft fan on the furnace to help vent the hot fumes. Manufacturers have taken care of that problem. See **Figure 12-15**.

Combustion Air — As you have noticed on conventional furnaces, they always vent the gas fumes to the outside. The air taken into the burners comes from outside. If the furnace is in a closet, combustion air will be pulled in from the attic through a screen. A hole is usually cut through the sheet rock. Some local codes may have you bring a metal pipe from the attic down

(Fig. 12-15)

even with the burners. They may even have you make an upper and lower combustion air intake. The way you size combustion air openings is to have a one (1) square inch opening for each 1,000 Btu rating (input). If you had a 100,000 Btu input furnace, then you would have a 100 square inch opening or a 10" x 10" opening.

Reclaiming vented heat — Instead of wasting the vented heat, why not use it? See **Figure 12-16**.

Here you can see that fresh air is being pulled into the heat exchanger and upon entering the heat exchanger mixes with the burning fuel at the burner. The air is heated as it travels through the heat exchanger. By the time it gets to the bottom of the heat exchanger the flue gas temperature is very hot. Normally these hot gases would be vented out the flue pipe at this point, but not so here. The warm gases are pulled through the secondary coil. The furnace fan blows across the secondary coil, then across the primary coil. The air is then blown into the area to be heated.

When the secondary coils are cooled, there is some moisture given up by the air (latent heat). This is due to condensation. There must be a small drain line to carry the moisture away (down a drain). The combustion fan continues to draw the air through the coils and discharges the combustible gases outside. This type of furnace is called a "condensing furnace." So as you can see, the 20% flue heat that would be wasted in a conventional furnace is reclaimed to a large extent by using the condensing coil (secondary coil).

The Pulse Furnace — The Lennox Pulse Furnace creates a lot of interest about the trade, mainly because it is different than the normal run of furnaces. It is a very efficient furnace and has proven to be the highest efficiency furnace yet developed **(Fig. 12-17)**.

Notice that gas and air mix by coming through two flapper valves, one for air and one for gas and then enter the combustion chamber. To begin a cycle a spark from the plug is used to ignite the gas and air mixture (this is one pulse). When combustion takes place a positive pressure takes place and closes the flapper valves and also forces exhaust gases down the tail pipe. When the exhaust gas starts leaving the chamber, a negative pressure is created causing the flapper valves to open again and let air and gas mix. At the same time part of the pulse reflects back causing the new gas and air mixture to ignite. No spark is needed. This is another pulse. There are approximately 60 to 70 pulses per second forming consecutive pulses of 1/4 to 1/2 Btu each. A small blower is used to purge the combustion chamber before and after each heating

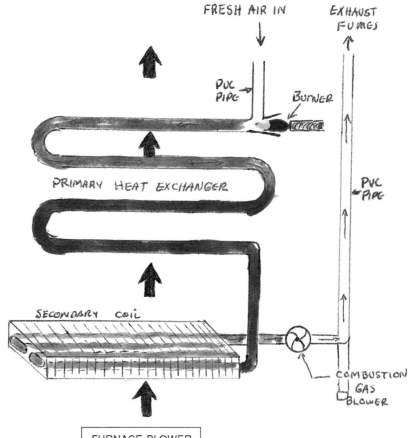

FRESH AIR IN

EXHAUST FUMES

PVC PIPE

BURNER

PRIMARY HEAT EXCHANGER

PVC PIPE

SECONDARY COIL

COMBUSTION GAS BLOWER

FURNACE BLOWER

(Fig. 12-16)

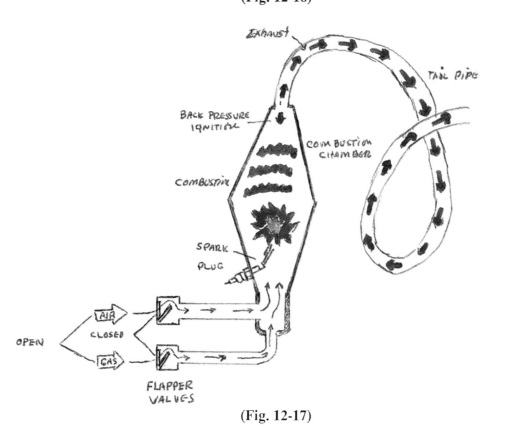

EXHAUST

TAIL PIPE

BACK PRESSURE IGNITION

COMBUSTION CHAMBER

COMBUSTION

SPARK PLUG

AIR

CLOSED

OPEN

GAS

FLAPPER VALUES

(Fig. 12-17)

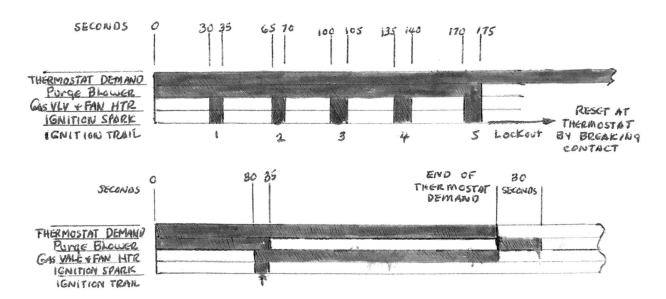

cycle. Here is a normal ignition sequence for the pulse furnace.

Take a look at "thermostat demand." This means the room thermostat closes and brings the purge blower on. In 30 seconds the gas valve and fan heater are energized along with the ignition spark. The ignition spark lasts for 3 seconds. When the ignition occurs a flame sensor proves it and the spark and purge blower are de-energized. At the end of the heating cycle, the gas valve and fan heater are de-energized, purge blower

is started and continues to purge for 30 seconds after the heating cycle is stopped. Now let's suppose the ignition fails. Here is what happens. The thermostat calls for heat and purge blower is energized. At 30 seconds the gas valve, fan heater and ignition spark are energized for 5 seconds. If there is no ignition the purge blower continues to run. After another 30 seconds the ignition tries for another 5 seconds. If no ignition the purge blower continues to run another 30 seconds. This is repeated for 5 tries. Without ignition

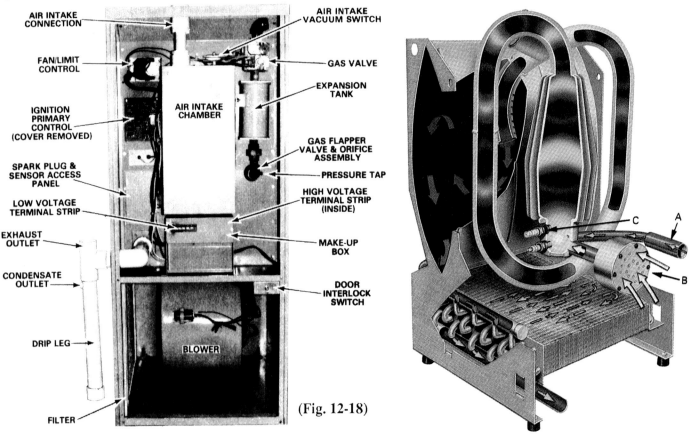

(Fig. 12-18)

184

the control locks out and cannot be started until the thermostat is turned off and then back on. This resets the control. Now take a look at the combustion chamber, the heat exchange assembly and the condensing coil (**Fig. 12-18**). Here again the combustion gases are not vented out and wasted, but pushed across the coil to give up heat.

Using a wiring diagram of the pulse furnace, let's review the sequence of operation.

1. Line voltage feeds through the door interlock switch. Blower access panel must be in place to energize unit.
2. Transformer provides 24 volt control circuit power.
3. A heating demand closes the thermostat heating bulb contacts.
4. The control circuit feeds fan "W" leg through the exhaust outlet pressure switch (C.G.A. units only), the air intake switch (A.G.A. and C.G.A. units) and the limit control to energize the primary control.

5. Through the primary control the purge blower is energized for approximately a 30 second purge.
6. At the end of the pre-purge, blower continues to run and the gas valve, fan control heater and spark plug are energized for approximately 8 seconds.
7. The sensor determines ignition by flame rectification and de-energizes the spark plug and purge blower. Combustion continues.
8. After approximately 30 to 45 seconds the fan control contacts close and energize the indoor blower motor on low speed.
9. When heating demand is satisfied the thermostat heating bulb contacts open. The primary control is de-energized removing power from the gas valve and fan control heater. At this time the purge blower motor remains on.
10. When the air temperature reaches 90°F the fan control contacts open - shutting off the indoor blower.

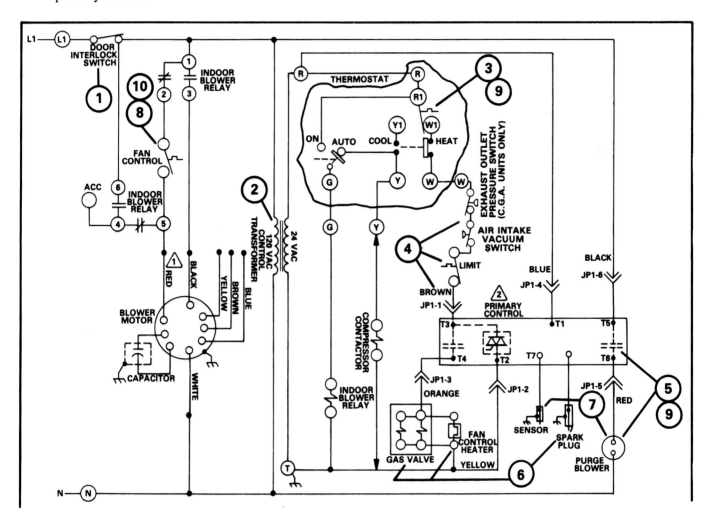

The induced combustion gas-fired furnaces — The induced combustion gas fired furnace shown here in **Figure 12-19** is a high efficiency furnace but is not a condensing furnace. Its efficiency is not as high as the condensing furnace, but is higher than the conventional furnace. Look at the S curved heat exchanger in the picture. You will see an exhaust fan moving combusted gases out the metal vent pipe and to the outside. Now look at the burner section. You can see the gas valve to the pilot. This pilot is totally automatic. It burns only when the furnace is calling for heat. On a call for heat, power is sent to the gas valve where gas

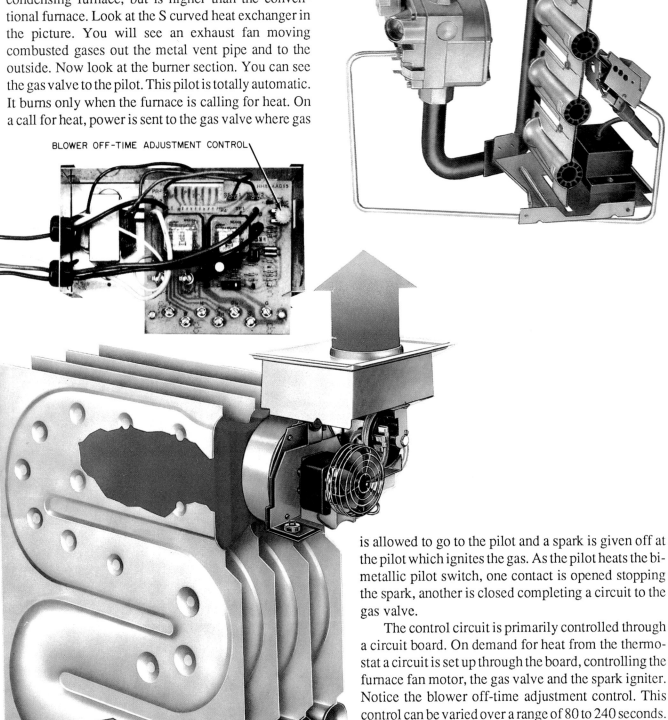

BLOWER OFF-TIME ADJUSTMENT CONTROL

is allowed to go to the pilot and a spark is given off at the pilot which ignites the gas. As the pilot heats the bi-metallic pilot switch, one contact is opened stopping the spark, another is closed completing a circuit to the gas valve.

The control circuit is primarily controlled through a circuit board. On demand for heat from the thermostat a circuit is set up through the board, controlling the furnace fan motor, the gas valve and the spark igniter. Notice the blower off-time adjustment control. This control can be varied over a range of 80 to 240 seconds. The factory setting is 240 seconds. The on-time of the fan is non-adjustable (60 seconds). For this particular furnace let's use the wiring diagram shown here and study the sequence of operation. Follow **Wiring Diagram 12-1** to identify the controls and the note reference. The control circuit of this furnace results in the following sequence of operation for the heating cycle.

1. When the blower door is in place, door switch 9G will be closed. Transformer 1A is energized. There is also power to one side of relay 2E and 2D but it is not flowing through with the relay contacts open.

2. The wall thermostat is identified by terminals R, Gh, W, Y, G and C. The heating circuit is made from R through the stat back to W. This supplies power through 7H3 and 7H2. Stop here and back up to the TH1 switch. In that circuit notice that if the fuse 11B, 11C and TH1 are open there will not be any power to the R terminal or to one side of the coil 2E.

3. Continuing through TH2 and TH3 the inducer-motor relay coil 2D is energized through switch 7V. When the 2D coil is energized there are two 2D contacts that close. One starts the 3A motor inducer and the other locks in the circuit when 7V opens. This position will be maintained until the thermostat opens.

4. When inducer motor 3A comes up to speed, flow sensing switch 7V goes to another position and makes pilot solenoid coil of the gas valve 5F (this would be the hold coil). At the same time the time delay relay in spark generator 6F is energized. Notice at the same time relay 6H has closed a circuit to the 5F gas valve the "pick" coil.

5. When the pilot solenoid coil of gas valve 5F is energized, gas blows to the pilot. The internal pressure switch within gas valve 5F senses the pilot gas pressure and closes, completing the hold circuit. Normally open time-delay circuit closes after a 10 second purge delay, energizing spark generator 6F. The pilot gas is ignited by a spark produced by spark generator 6F.

6. After a short time delay, during which the pilot flame heats flame sensing contact 6H, the normally closed contacts open. The pilot solenoid coil remains energized through the hold circuit (contact 2D). Spark generator 6F is de-energized a few seconds after flame is sensed at spark electrode. The normally open flame sensing contacts 6H close 5 to 20 seconds later,

energizing the MGV solenoid coil of gas valve 5F. The MGV solenoid opens 6 to 15 seconds later, allowing gas flow to the main burners, where it is ignited by the pilot flame.

7. Simultaneously the time-delay circuit in the solid state time delay relay is energized. Approximately 50 seconds after the gas valve 5F opens, heating relay coil 2E is de-energized, closing the 120 VAC contacts of heating relay 2E and starting blower motor 3D on heating speed (low to medium low). The electronic air cleaner (EAC) terminals are energized with 120 VAC whenever the blower is operating on either heating or cooling speed. The electronic air cleaner is an optional piece of equipment.

8. When the thermostat is satisfied, the R and W circuit is broken, de-energizing gas valve 5F, inducer motor relay 2D and the solid state time delay circuit on the main printed board. Gas flow to the pilot and main burners stops immediately. After approximately 80 to 240 seconds, heating relay 2E is energized and blower motor 3D stops.

There are some things you need to know now and that is, if you get a call that the furnace fan runs continuously, here is what you look for.

1. **A burned out transformer.** Check at terminal PR1 and PR2 with your volt meter. If you have voltage there then check SEC1 and SEC2. If there is not 24 volts, transformer is bad. Put a new transformer on and snap your amp meter around one leg of the 24 volt wires coming out. If the amps exceed 1 amp, there is a short. Do not leave the circuit on because the transformer will burn out again. Check for shorts in the wiring.

2. **11C Limit Fuse blown.** With this fuse link blown and power on, the relay 2E will be de-energized and stay de-energized letting the 2E contacts continue to supply power to the 3D motor.

3. **Limit Switch 7H1 open.** You would have the same symptoms as the 11C Limit Fuse. See **Figure 12-20** for location of limit switches.

4. **Limit Switch 7H2 open.** You would have the same symptoms as 2 and 3 above. This limit has a reset switch and can be reset by pushing in. On Limit Fuse 11C the fuse link would have to be replaced.

5. **Bad Circuit Board.** If there is power to SEC 1

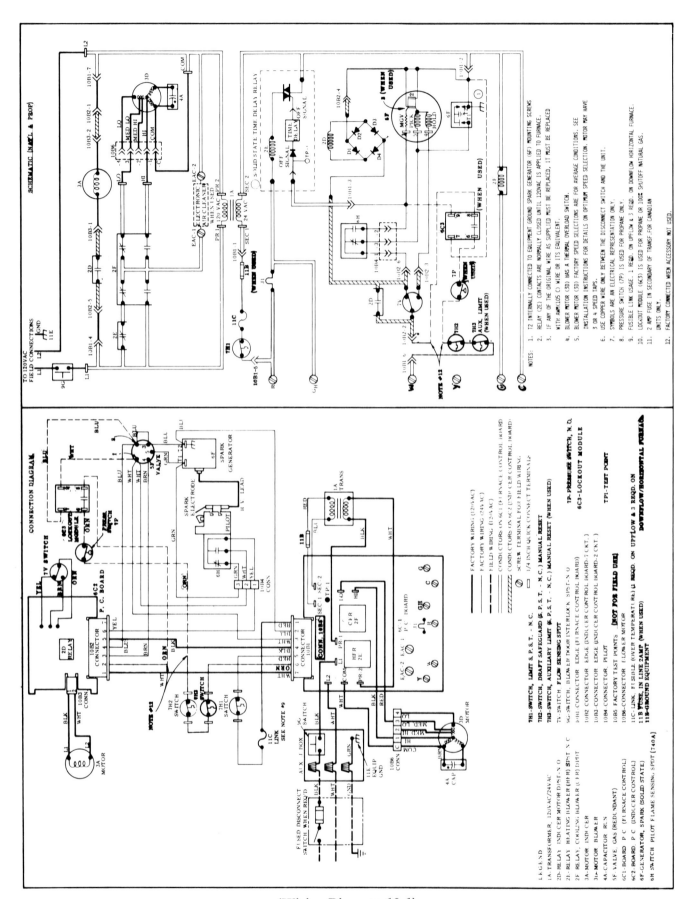

(Wiring Diagram 12-1)

188

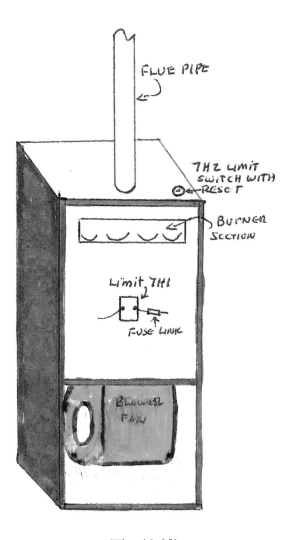

FLUE PIPE

7H2 LIMIT SWITCH WITH RESET

BURNER SECTION

Limit 7H1

FUSE LINK

BLOWER FAN

(Fig. 12-20)

and SEC 2 and power to "C" and "W", there is power to the board. If the 3D fan continues to operate when the "W" terminal is killed then you need to replace the 6C1 circuit board.

How to adjust the pilot flame — Disconnect terminal 1 on the gas valve. Avoid letting the loose wire touch any metal for it will blow out the anticipator in the thermostat. The heating circuit would be dead and a new thermostat would have to be installed. With wire 1 off, the main burner cannot come on. This will give you ample time to adjust the pilot flame. The adjustment is found on the gas valve near the pilot opening. If an adjustment does not make the pilot burn high enough, check the orifice at the pilot. It may be restricted. Use very small strand wire to open up the orifice ports.

Stopping the furnace for winter — If you plan on having the furnace off during winter months, then the condensing coil may freeze because of the moisture. Pour a 50-50 mixture of anti-freeze down the

drain tube from bottom of inducer outlet box. Using a funnel, pour the anti-freeze into the furnace until you see it start running out the condensate drain.

Further understanding the diagram — You have been shown two diagrams on one sheet. The one to the left is a schematic and the other is a ladder type diagram. Most servicemen find the ladder diagram easier to read, but you will have to understand the terminal lettering and numbering. On the ladder diagram find connection 10B1-6 and 10B2-2. On the other diagram you see that 10B1 and 10B2 are connectors. Then 10B1-6 would mean that this connection is found on 10B1 connector and it is terminal No. 6. Then 10B2-2 would mean that we have connector 10B2 and terminal No. 2. Note the symbol for a connection. Notice you have 10B1 connector, 10B2 connector, 10B3, 10B4, 10B5 and 10B6. Find them and know what they mean.

Always refer to your footnotes. For example, look to the diagram on the left and find where it says "11C Link see note #9". Now go to the right and see what note 9 will say. Read it and it tells you of fusible link usage. Ninety percent of your ability to correct control circuit problems will come from how well you can read the diagram. You should have a diagram glued to one of the panels on the furnace. If you have no diagram, you would be in the same boat as an experienced serviceman, unless he had worked on the furnace many, many times. Even then he may not be able to repair it without a diagram. Don't be bashful about finding a diagram if one is missing.

You may have noticed that we have only covered two types of high efficiency furnaces, one being the Lennox pulse furnace and two, the Carrier high efficiency furnace. The Lennox pulse furnace is one of a kind, however the Carrier furnace gives you a general idea how most all condensing furnaces operate both mechanically and electrically. Manufacturers vary their way of controlling different devices. Time delays may be different and switches may be different. It is next to impossible to put into this chapter exactly how every furnace manufactured today operates. What you have read so far is general. When you run into other manufacturers equipment, you will need the wiring diagram for that particular furnace. In years past, a heating serviceman worked on all makes of furnaces because they were so similar and the control circuitry wasn't nearly so complicated as the circuitry for high efficiency furnaces. As years pass we can probably look forward to heating dealers specializing more. The dealer will know the brand of furnace he sells but not be so familiar with another brand. The answer to the problem is to keep yourself well informed. The distributor of certain brands of equipment usually have short seminars on this equipment as it comes out. Try and make yourself available for these training sessions and stay up with the trade. Another way is to go to the distributors of this equipment and ask for service pamphlets and wiring diagrams. In your spare time you can study and be ready for most problems.

ELECTRIC HEAT

Electric heat is still a very common type of heat. In many of the larger cities it is more practical cost-wise for builders and utility companies to use electricity for heating since it doesn't require a lot of digging to lay pipe for gas. Many homes in cities and rural areas are all electric. The only fault of electric heat, if there is any, is that it costs more to heat with it than it does with natural gas. In areas where electricity is the only choice, heat pumps are becoming popular. In areas where electric furnaces exist, heat pumps are taking their place. Even though you are still heating by use of electricity, the heat pump will operate as much as 30% less in cost. However, since electric heat is still around you need to know something about it. You need to know how it works and what it takes to service it.

The Electric Furnace — The electric furnace today is adaptable for summer cooling. It is like any other furnace whether it be gas or whatever and that is it must be sized to put out enough heat. It also has to be able to move enough air for the size of the air conditioner that is attached to it. Here is an electric furnace shown with a built-in coil. Take a look at the parts and where the controls, etc. are located (**Fig. 12-21**).

Electric Resistance Heat (elements) — The part that creates heat is the element. When voltage is applied to the element the resistance of the element wire creates a lot of heat. See element in **Figure 12-22**.

Elements are most commonly 5KW each in rating (5000 WATTS) and 240 volts. The heating capacity of a furnace may have several elements in order to reach a certain capacity. For example, a 20kW furnace would have 4, 5kW elements. If you need to know the Btu output, multiply the watt rating by 3.414, since there are 3.414 Btu per watt. Convert 20kW to watts. There are 1,000 watts per kW. There are 20,000 watts. Btu = Watts X 3.414. Btu = 20,000 X 3.414 = 68,200. A 20kW furnace would have a Btu capacity of 68,200 Btu. Not all elements are 5kW, but could be 2.5kW or 7.5kW. If you are not sure of the kW and cannot find the rating on the furnace, check the amperage draw of the element. Take the amp draw and multiply it by the voltage. Watts = Amps X volts. Suppose the amp draw was 21 amps. Then watts = 21 X 240 = 5,040. In kW ratings 5,040 = 5kW approximately. Suppose you needed to know the amp draw of a furnace. Maybe you have just installed a 20kW furnace and the electrician

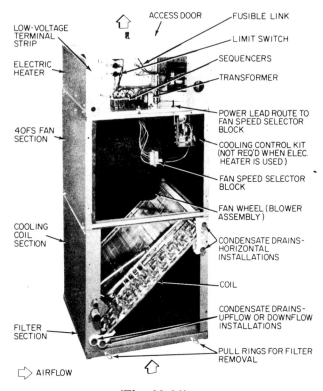

LOW-VOLTAGE TERMINAL STRIP
ELECTRIC HEATER
40FS FAN SECTION
COOLING COIL SECTION
FILTER SECTION
AIRFLOW

ACCESS DOOR
FUSIBLE LINK
LIMIT SWITCH
SEQUENCERS
TRANSFORMER
POWER LEAD ROUTE TO FAN SPEED SELECTOR BLOCK
COOLING CONTROL KIT (NOT REQ'D WHEN ELEC. HEATER IS USED)
FAN SPEED SELECTOR BLOCK
FAN WHEEL (BLOWER ASSEMBLY)
CONDENSATE DRAINS-HORIZONTAL INSTALLATIONS
COIL
CONDENSATE DRAINS-UPFLOW OR DOWNFLOW INSTALLATIONS
PULL RINGS FOR FILTER REMOVAL

(Fig. 12-21)

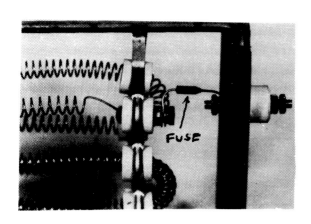

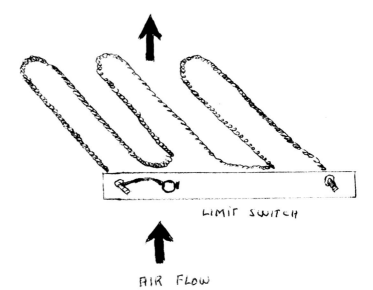

LIMIT SWITCH

AIR FLOW

(Fig. 12-22)

is running a circuit to it. He asks you what the amperage draw of your furnace is. Quickly you divide the watts by the voltage. First, 5kW is 5,000 watts.

$$Amps = \frac{Watts}{Volts}$$

$$Amps = \frac{5,000}{240} = 21 \ amps$$

This is for one 5kW element. Four 5kW elements would draw 84 amps.

The Thermostat — Thermostats for electric heaters are basically the same as any other thermostat (**Fig. 12-23**). A thermostat that controls a cooling and heating system for gas heat could work as well for a cooling system with electric heat. Later on you will see that certain thermostats are required for different situations. There are electric heaters that do not have a sequencer to bring on the fan motor. If this is the case, you must get a thermostat specifically for electric heat. This type of thermostat will bring the fan on with the heating elements.

The Sequencer — The sequencer is a time delay element used to bring on the heating elements in sequence (**Fig. 12-24**). The sequencer is also used to bring the fan motor on. Notice "A" in the illustration. This is a resistance type heater fed by the thermostat on demand for heating. When the thermostat contacts close between R & W a 24 volt circuit is completed to coil "A." A resistance causes the coil to start heating. There is a delay until finally enough heat has been created to cause contact "B" to close and shortly contact "C" will close.

There can be more than one sequencer on an electric heater and the sequencer above could use contact "B" to close a circuit to the fan and the other one or two heating elements.

The Limit Switch — The limit switch is a control that will open a circuit to the heating element if there is an over-heating situation, such as fan motor not running or dirty filter. The switch will open and automatically close after cooling down. See picture of

(Fig. 12-23)

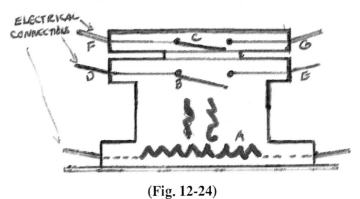

ELECTRICAL CONNECTIONS

(Fig. 12-24)

191

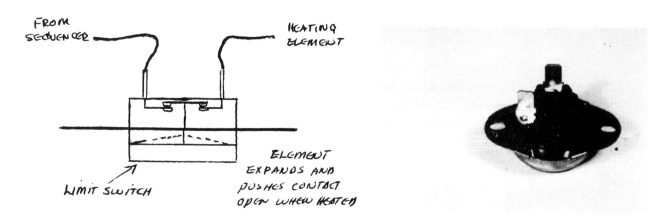

(Fig. 12-25)

(Fig. 12-26)

limit switch in **Figure 12-25**.

The Transformer — Like most air heating systems the control circuit is usually 24 volts. To get the 24 volts a 240V to 24 volt step down transformer is used (**Fig. 12-26**).

The Fan Relay — The fan relay is used primarily to run the fan on continuous or automatic operation (**Fig. 12-27**). Upon turning the fan "auto - on" switch to the "on" position the fan will work continuously. When on "auto" the fan will cycle either on heat or cooling. The illustration to the right shows the contact arrangement. One contact is normally open and the other is normally closed. This is the normal position of the contacts when there is no power to the magnetic coil. The N.O. (normally open) contact is for the high speed of the fan for cooling and the N.C. (normally closed) contact is for the low speed of the fan.

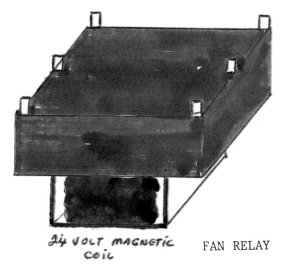

FAN RELAY

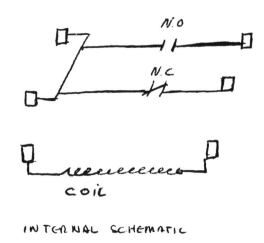

(Fig. 12-27)

The Wiring Diagram — Take a look at the **Wiring Diagram 12-2**. Notice in the bottom right hand corner the legend. This identifies the controls. Now take a good look at the diagram. Lets go through the sequence of operation. On call for heat, power from the transformer feeds to R on the thermostat, through the thermostat contact to W, then on to sequence 1. A 24 volt circuit is completed to the common side of the transformer. Now find the sequencer contacts. After a delay, one of the sequencer contacts will close and complete a circuit to the heater and at the same time feed through the normally closed FR (fan relay) contacts and the fan is started. A few seconds after the first contact closes the second sequencer 1 contact closes and brings the other heater on. There is full heating capacity. Notice the fan motor. It has 4 speeds. See how the high and medium speed wires are hooked up. Yellow is common and has power to it as long as L2 is hot and the fuse is good. When the fan relay is de-energized the contact between 5 and 6 is closed for medium speed (heating). When 24 volts is put to the fan relay, the contacts close between 2 and 4 bringing in the high speed. When the contact closes between 2 and 4, the contact between 5 and 6 automatically open. Medium speed is used for heating and high speed for cooling. When the thermostat opens "W" is killed and the sequencer is turned off. The sequencer starts cooling off. In a few seconds Seq. 1 contact between terminals 7 and 8 open. Then in a few more seconds the other sequencer contacts open killing the other heating element and the fan motor. Be sure you know what all controls are and what they do. For instance, "What is the switch marked LS2?" It is the limit switch on the 2nd element. Use your legend. Also note the component arrangement. Pay close attention to your wiring diagrams. They will show you things that will keep you from going to extra trouble. ■

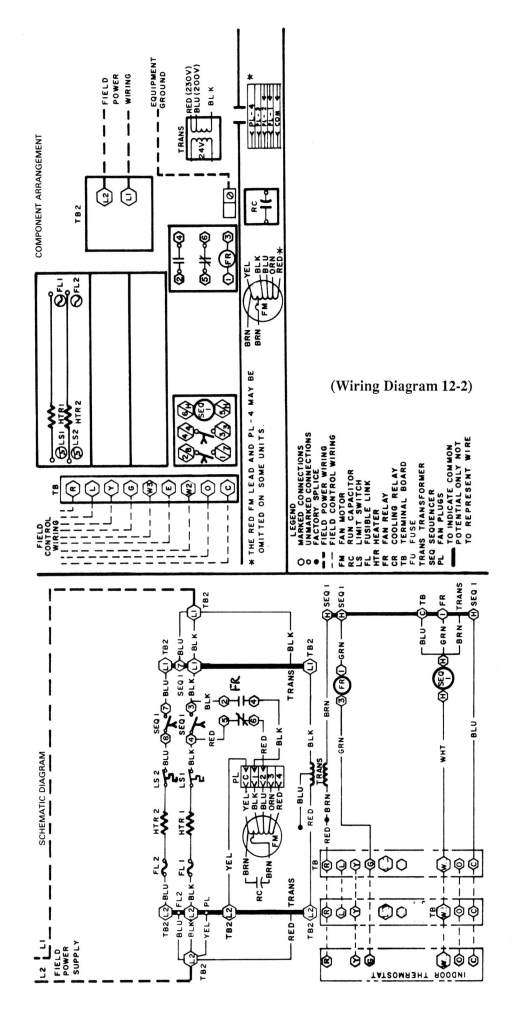

(Wiring Diagram 12-2)

194

CHAPTER 13

HEAT PUMPS

Heat pumps are an intricate piece of machinery, requiring more knowledge than what the average serviceman has. Refrigerant charges have to be accurate and the heat pump is more complex than the common Central Heating and Cooling systems.

Heat pumps have been around for many years especially making a big show around the late 1950s and early 1960s. At this time they were more popular in warmer winter climates such as Florida, Southern California and deep southern states.

Today, heat pumps are making big advances in all climates. High fossil fuel costs (butane, propane and heating oil) have turned people's interests in its direction. Where only electricity is available heat pumps are common, since the electrical cost to operate a heat pump is as much as 30% less as compared to heating with electrical resistance heat.

Heat pumps can be complicated when servicing them. If you do not have a complete understanding of how the heat pump operates and you can't readily answer questions by the customer, they will quickly diagnose you as inexperienced. This section will give you the savy to answer questions, make quicker observations and more knowledgeable in making the repair.

HOW THE HEAT PUMP WORKS

Cooling — During cooling season, the heat pump works like any other summer air conditioner. It uses an indoor coil, a compressor and an outdoor coil to remove heat from inside the home or office to outside. Fans move air across the coils and circulate air in the conditioned space. A thermostat turns the fans and compressor on and off as cooling is needed. Hotter weather means more cooling is required, so the unit will run longer. When the temperature is highest, the unit may run continuously for several hours.

Heating — In the heating season, the use of the coils are reversed. The outdoor coil picks up heat from outside air (remember there is heat in air down to absolute zero -460°F) and the indoor coil releases this heat to warm the conditioned area. Colder weather increases heat needed and the unit runs longer. In most areas the temperature will sometimes drop low enough that the heat pump will run continuously. This outdoor temperature, at which the heat needed is equal to the heat pump's capacity, is known as the system's "Balance Point." This temperature will vary with each installation, depending on the heat loss of the area and the size of the heat pump selected. Below the "Balance Point," the heat pump will run continuously and the auxiliary resistance heat will be cycled on and off by the thermostat. Heat pumps can continue to operate effectively at outdoor temperatures below 0°F.

Always remember this, and be sure the customer knows that discharge air temperatures in the conditioned area, using heat pumps, will only operate 15° to 30° warmer than room temperature. The air will feel drafty if blown directly on a person. People who have used gas heat or some other type of heat will have difficulty in understanding this. However, the temperature of the conditioned area should have no problem maintaining its temperature set point at the thermostat. At this point you will find it better to refer to the condenser as the *outdoor coil* and the evaporator as

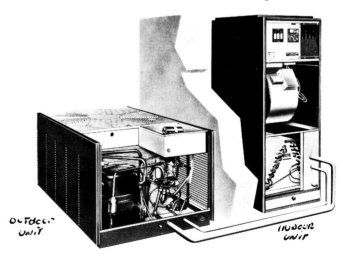

outdoor unit   indoor unit

the *indoor coil*. The outside unit should be referred to as the *outdoor unit* and the inside fan and coil the *indoor unit*. This terminology will make heat pump conversations easier to understand.

Defrost Cycle — When the outdoor temperature drops below 45°F, frost may occur on the outdoor coil. Frost build up will be heaviest on damp days with the temperature at 35° to 40°F. The heat pump has an automatic control which will reverse the system and stop the outdoor fan to defrost the coil when needed. When the reversal takes place, we are actually going back to an air conditioning cycle whereas the outdoor coil becomes the condenser and melts the ice away. Some units operate on a timer at 45 to 90 minute intervals. Others have an electronic control, which senses coil and air temperatures to determine when a defrost cycle is needed. They may go as long as 6 hours between defrost. The coil may be almost completely covered with frost at times. Don't worry unless it continues to build up a thick layer with areas of hard clear ice. If excessive ice build-up occurs, then service is needed.

When the heat pump is defrosting, a cloud of steam may rise from the outdoor unit for a short time. This is normal and harmless, but may be diagnosed as smoke by the customer, more especially if they have never experienced this before. The water, which runs from the defrosting coil, must be drained away from the unit. In areas where snow is common, the unit must have legs long enough to be kept above the snow level.

IMPORTANT THINGS TO REMEMBER ABOUT THE HEAT PUMP

1. Water must drain away from the outdoor coil.
2. The outdoor unit must be above snow level.
3. Install the unit so roof dripping will not build up on the outdoor coil.
4. Do not install the unit where snow most commonly drifts.
5. Select the sunny side of a building, away from the north wind to make use of the sun's radiant heat.

THE THERMOSTAT

The heat pump thermostat is much different than the conventional heating and cooling stat. It will have switches to select some or all of the following functions (Fig. 13-1).

Cool — Turns cooling on when temperature rises above the set point. Contact is made through a sealed mercury bulb in the stat.

Heat — Turns heat pump heating on where temperature drops below set point. If room temperature drops another 2°, turns on auxiliary resistance heat with heat pump continuing to run.

Auto — Turns on cooling or heating as required to maintain set points. Most thermostats have at least 4° separation between heating and cooling settings.

Off — Turns heating and cooling modes off (fan may still run in Fan/On).

Fan/On — Turns fan on for continuous operation (leave in auto for best electrical conservation and humidity control).

Fan/Auto — Fan cycles on and off with cooling or heating operation.

Emergency Heat — Turns heat pump compressor and outdoor fan off and provides heat from electric resistance heat only. Use this switch to manually turn the heat pump off and change to auxiliary heat in case of heat pump problems.

A lever is used to set the temperature desired. Some stats have two levers, one for heating and one for cooling.

Lights may be used to indicate that the auxiliary electric heat is operative. The lights may be different colors on different types of stats. Typical lights will be blue or green for normal auxiliary heat and red for emergency heat.

To operate most economically, do not change the temperature setting up more than 2°. Above 2° will bring on the auxiliary heat with the heat pump operating. If a higher temperature is required, be patient and go 2° at a time.

Night Setback Thermostats — Night setback thermostats are available to automatically turn the temperature down at night and back up in the morning.

(Fig. 13-1)

Only setback thermostats with gradual, incremental or intelligent recovery should be used with heat pumps. Setback stats without gradual recovery will use the auxiliary heat strips to warm air in the morning, and may use more electricity than was saved at night. Ask your parts supplier for recommendations before installing a setback or "Energy Saving" stat.

Heat Pump Monitor — A control called a heat pump monitor may be installed with the heat pump system. If so it will check the performance of the heat pump and turn it off if a problem occurs. It will switch to auxiliary heat and turn on the emergency heat light on the thermostat to tell you that the system requires attention.

Operating Economically — To prevent nuisance calls from your customer and to help the customer economically, here are some tips to improve system efficiency.

1. Keep all grilles and registers open and clean of obstructions. The system is designed to deliver 450 CFM of air per ton. Restricting that air could damage the unit or cut off on a safety control.
2. Keep doors and windows closed. Fresh air is not needed.
3. Be sure all air ducts are well insulated (2 inches minimum) and all joints taped tight.
4. Advise the customer to let sun through windows in winter and keep out in summer.
5. Be sure clothes driers are vented to outside away from outdoor unit.
6. Fireplaces are pleasant, but fireplaces infiltrate more outside air for combustion and flue draft than they heat. The heat pump may actually run more. Even though the thermostat setting may be keeping the area warm, the excessive draft may make the occupant feel cold. This is even more common with a heat pump since the registry air discharge is cooler than most heating systems.
7. Use kitchen and bathroom exhaust fans only when necessary.
8. Add insulation, storm windows and insulate outside doors. Seal cracks to prevent air leaking. This is advise you would give to the customer.
9. Keep filters clean. Reduced air flow reduces efficiency.
10. Operate the indoor fan on "AUTO." It will cost less and will provide better humidity control in the summer.
11. Keep lamps, TV and other heat sources away from the thermostat.

EFFICIENCY

EER — Most people are very conscious of the EER or energy efficiency ratio, and rightly so. The higher the EER the more savings received from operation. The EER is used when the compressor is used and is nothing more than how many Btu you are getting out of every watt of power put in. It is determined by dividing the total capacity (in Btu/hr) of the system by the total electric power consumed in watts/hour. For example, if a three ton heat pump operating on a 40°F outdoor temperature has a heating capacity of 39,000 Btu/hr and is using 4,380 watts per hour, then the EER would be 39,000 Btu/hr over 4,380 watts/hr= 8.9 Btu/watt or EER = 8.9.

$$EER = \frac{39,000}{4,380} = 8.9$$

It would be easy to understand that an EER of 10.0 is more desirable than an EER of 7.0, because the EER of 10.0 means that you are getting 10 BTU's of cooling for each watt consumed.

Heat pumps are more economical than straight electric heat. We can measure this economy using either of the following rating methods.

One method is the energy efficiency ratios or EER, which is used to rate any air conditioning system. Another method, which has been used for years when discussing the efficiency of a heat pump, is the coefficient of performance or COP. No matter which method is used, a knowledge of how they are determined is important.

COP — This is the coefficient of performance for a heat pump. To determine COP of a heat pump, the following formula can be used. COP = Btu/hr capacity over unit wattage X 3.413 Btu/watt. If we use the same three ton unit as in the last example, the COP would be 39,000 Btu/hr over 4,380 watt X 3.413 Btu/watt = 2.6.

$$COP = \frac{Btu\ hr}{Watts\ x\ 3.413} \quad or$$

$$COP = \frac{39,000}{4,380\ x\ 3.413} = 2.6$$

This indicates just how efficient the unit is when compared to electric resistance heat. When electric heat is used it generates 3.413 Btu of heat for each watt. Therefore, the COP of straight electric heat is (1.0) and can never be any higher. When comparing the COP of the heat pump with that of an electric heater, we find that the heat pump can deliver more heat per watt of power used. The heat pump is therefore more efficient.

Under the operating conditions in our example, the heat pump will deliver 2.6 times as much heat as our electric resistance heater using the same wattage.

THE REFRIGERATION CIRCUIT

The first step in understanding the heat pump is to know how the refrigeration system works. Since the basic concept used for changing the unit from cooling to heating cycle is to reverse the refrigerant flow, each coil will act as a condenser or as an evaporator depending on the direction the refrigerant is flowing, so the coils are referred to by their location in the air stream rather than their function (**Fig. 13-2**). The indoor coil is located in the indoor air stream and the outdoor coil is in the outdoor air stream.

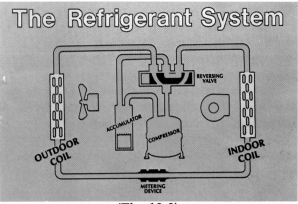

(Fig. 13-2)

Let's trace the flow of refrigerant through a typical heat pump during the cooling cycle when the unit is removing heat from the conditioned space and rejecting it outdoors (**Fig. 13-3**). Starting at the compressor discharge, the hot gas is discharged to the reversing valve, which directs it to the outdoor coil. The outdoor coil acts as a condenser. Outdoor air circulated over the coil removes heat from the refrigerant and causes it to condense to a hot liquid.

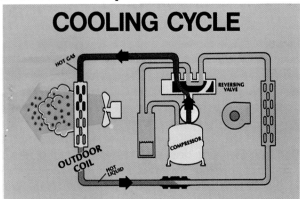

(Fig. 13-3)

The liquid refrigerant now passes through a metering device (**Fig. 13-4**), where the pressure is reduced causing some of the liquid to flash off and cool the remaining liquid to a lower temperature. This cool liquid passes to the indoor coil which acts as an evaporator. The liquid boils and absorbs heat from the indoor air passing over the coil causing the refrigerant to evaporate into a cool vapor.

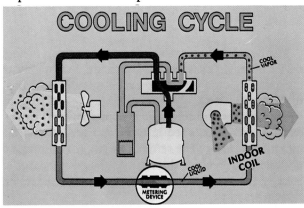

(Fig. 13-4)

It is important at this time to know that heat from the air crossing the evaporator is enough to make a refrigerant boil since the boiling points of refrigerants are lower than air temperatures and therefore absorb heat.

In **Figure 13-5** the cool vapor now passes through the reversing valve which directs it to the accumulator, where any liquid refrigerant will be trapped and then back to the compressor where the gas is compressed to a high pressure, high temperature gas and the cycle is repeated.

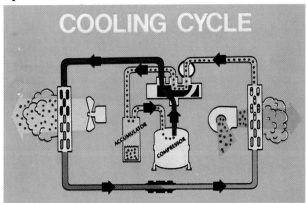

(Fig. 13-5)

During the heating cycle, (**Fig. 13-6**) when heat is removed from the outside air and rejected indoors, the hot gas discharge from the compressor is directed by the reversing valve to the indoor coil. The indoor coil now acts as a condenser. The indoor air picks up heat

198

from the refrigerant as it passes over the coil and causes the refrigerant to condense to a hot liquid.

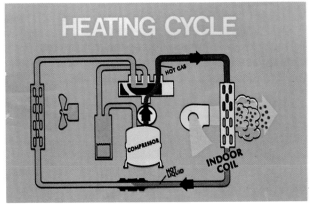

(Fig. 13-6)

In **Figure 13-7** the liquid refrigerant now passes through a metering device where it flashes to a low pressure, low temperature mixture of liquid and vapor. The mixture now passes to the outdoor coil.

The outdoor coil acts as an evaporator. Heat from the outdoor air is absorbed into the refrigerant as it boils into a cool vapor. The refrigerant gas goes through the reversing valve which directs it to the accumulator and back to the compressor where it is compressed and the cycle is repeated.

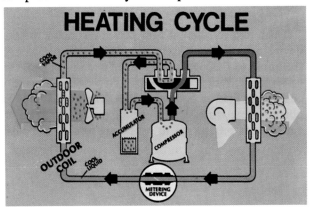

(Fig. 13-7)

From the cycle diagrams, you can see that heat is in fact moved by the heat pump. The total amount of heat rejected by the coil, (indoor coil) which is acting as a condenser, is equal to the total heat absorbed into the system plus the heat of compression.

When a heat pump is operating in the higher temperature range, the heat of compression is only a fraction of the total capacity of the system.

As the outdoor temperature drops, so does the heat pump's ability to absorb heat. This results in a decrease in capacity of the unit. For example: A three ton heat pump may be capable of putting out 39,000 BTU's of

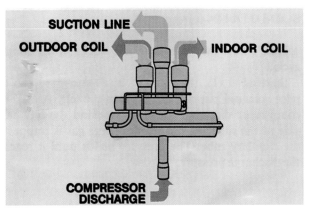

(Fig. 13-8)

heat with the outdoor temperature being at 45°, but on a 20° day, it may only put out 22,000 Btu of heat. At some temperature, the heat pump will not be able to supply enough heat to maintain comfort conditions, so some form of supplemental heat must be provided, generally electric resistance heat. This will be discussed in detail later.

In order to make the process of transferring heat work for both heating and cooling cycles, the direction of refrigerant flow must be controlled. The device that controls the direction of refrigerant flow in a heat pump is called the reversing valve (**Fig. 13-8**).

There are four piping connections on the reversing valve. There are two directions of flow that are never changed and they are the hot gas discharged from the compressor and the suction line back to the compressor.

The remaining two ports connect to the outdoor and indoor coils. The direction of flow depends on the cycle.

The valve is solenoid controlled. A typical reversing valve directs the refrigerant flow by using a free floating slide inside a cylinder (**Fig. 13-9**). The slide changes position by refrigerant pressure. As this slide

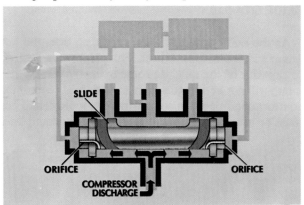

(Fig. 13-9)

shifts, it directs the refrigerant flow to and from the coils.

The compressor discharge is connected to the single port on the cylinder. There is always discharge pressure bleeding between the two ends of the slide, represented by arrows. At each end of the slide there is a small orifice, which allows the discharge gas to bleed behind the slide.

At both ends of the cylinder there are capillary lines that connect to a pilot solenoid chamber (Fig. 13-10). Also connected to the pilot solenoid chamber is another capillary that connects to a compressor suction line. Here it's the center capillary.

The pilot pin carrier changes position when the solenoid coil is energized or de-energized.

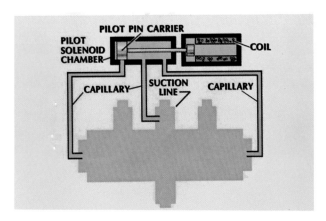

(Fig. 13-10)

Figure 13-11a shows how the reversing valve works in the cooling cycle. Starting at (1), the solenoid is de-energized, so the pilot solenoid pin carrier moves to the left. At (2), the compressor discharge gas bleeds behind the left hand side of the slide. At (3), the discharged gas is trapped in the capillary tube. The

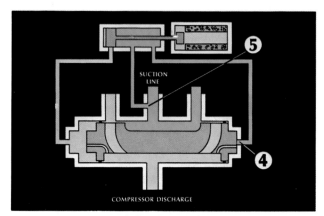

(Fig. 13-11b)

pressure builds until it reaches the discharge pressure at (4) (Fig. 13-11b). The pressure behind the right side of the slide equals the compressor suction pressure, which passes through the pilot solenoid chamber and down through the center capillary to (5).

Since the pressure behind the right side of the slide is less than the pressure behind the left side, the slide moved right.

The flow of refrigerant is directed to the outdoor coil and the heat pump is operating in the cooling cycle (Fig. 13-12).

When energizing the solenoid valve, at point (1), the pilot pin carrier moves to the right. Compressor suction pressure now passes through the center capillary and the pilot chamber down to the left side of the slide at (2). At (3), compressor discharge bleeds through the orifice behind the right side of the slide where it is trapped. The pressure behind the left side of the slide is now less than the pressure behind the right side and the slide moves to the left. The hot gas is now directed to the indoor coil and the unit is operating in the heating cycle.

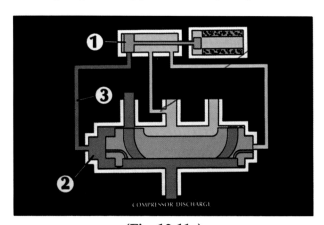

(Fig. 13-11a)

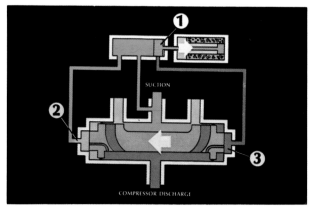

(Fig. 13-12)

METERING DEVICES

Refrigerant is metered to the coil which is to absorb heat. During the heating cycle the refrigerant must be metered to the outdoor coil and during cooling to the indoor coil. Several types of metering devices are used to control refrigerant flow. Some systems have only one, located in the liquid line allowing flow in either direction. The most common way of metering refrigerant is to use two metering devices, one at each coil. Expansion valves, capillary tubes or a combination of the two could be used (**Fig. 13-13**).

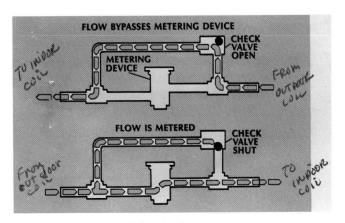

(**Fig. 13-14**)

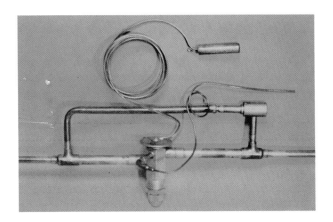

(**Fig. 13-13**)

Figure 13-14 shows where check valves are used to control the flow of refrigerant. If the top valve is feeding the outdoor coil during heating as it would, then it would be inoperable during cooling. Refrigerant will not flow back through the valve, so a check valve would push open and allow liquid from the outdoor coil to go to the metering device at the indoor coil. Shown at the bottom of the illustration, in order to meter refrigerant to the indoor coil, the check valve would be pushed shut and refrigerant would feed through the valve into the indoor coil. When the reversing valve is energized and put into the heating mode the refrigerant flow is reversed. The top illustration would now be feeding the outdoor coil and bypassing the metering device on the indoor coil.

DRIERS

Remember that the refrigerant flow in a heat pump is reversible, so it will be necessary to use a drier that has non-directional flow. If a directional flow drier is used and the direction is reversed, particles from the drier could wash back into the system and cause restrictions (**Fig. 13-15**).

(**Fig. 13-15**)

TYPES OF REFRIGERANT USED - AND PRESSURES

Today's heat pumps use Refrigerant 22 and their pressures are directly related to varying temperatures, especially outdoor temperatures. When operating in the cooling mode, you could expect the pressure from the indoor coil at the compressor to be around 69 psi, give or take ± 5 psi. The pressure from the compressor to the outdoor coil should be from 250 psi to 275 psi. A drastic change in temperatures or conditions could change these pressures. For instance on a 90° day a 250 psi pressure reading from the discharge of the compressor could change and go to 300 psi, should the outdoor temperature reach say 105°. You could expect the pressure to go down to approximately 200 psi, if the outdoor temperature dropped to 70°. Either coil, indoor or outdoor, is directly affected by the amount of heat across them.

In the heating mode a more drastic change of outdoor temperatures can be expected. The pressure on the outdoor coil can range from 30 psi, approximately 20° air crossing the coil, up to 60 psi with 55° air across the coil. The indoor coil pressure will stay more constant, since the building temperature is more constant. A pressure of 225 psi to 250 psi would be close to normal.

Charging — You will find that getting the correct charge for the heat pump is not an easy task. The outdoor temperatures will vary so much with conditions changing. You can't use the old sniff and feel method that has been used in the past. If the charge is not correct then you can have problems. All manufacturers will have basically the same recommendations for charging. However, due to engineering aspects of the units, they will differ somewhat in pressures and temperatures. Later you will find in this chapter some manufacturers charging guides. They will help you get familiar with the way to do it according to the manufacturer. Don't worry too much about where you are going to get all the manuals and information about charging, because most manufacturers are putting a charging guide on the unit. If all fails and you can't seem to find the correct way of charging, use the dial-a-charge method.

Dial-A-Charge — With the dial-a-charge (**Fig. 13-16**) you can look on the nameplate of the unit and find the recommended charge in pounds or ounces. This charge is usually accurate up to 25 feet of liquid line. If you have longer lines try to estimate as close as possible the extra length of line and use the chart below.

LIQUID LINE DIAM (in.)	OUNCES OF R-22/FT. LENGTH OF LIQUID LINE
3/8	.58
5/16	.36
1/4	.21

Once you have determined how much refrigerant you need, then take liquid refrigerant from a refrigerant drum and let it fill the dial-a-charge to the correct amount. Now that you have filled the dial-a-charge, you will have to let all the refrigerant go out of the heat pump, making sure that none is left. It may be necessary to use a vacuum pump and pull all refrigerant out. After you have cleared the system, attach your gauge manifolds as they should be and let the liquid refriger-

(Fig. 13-16)

ant enter into the liquid line. You may not be able to get all the refrigerant from the dial-a-charge into the system. It will be important now to recommend a dial-a-charge with a built-in electric heater that you can plug into a convenient 120 volt outlet. After driving the refrigerant into the system, remove the charging apparatus. The system is charged. This is a fast, sure way of making the correct charge. If there is a disadvantage to this type of charging, it is that you had to let all the refrigerant that was left in the system go to the atmosphere. Whereas, if the system had been a little low, you could have added to it by using factory charging methods. There is also a good chance that you may have to let the charge go anyway, due to having to repair a leak. Remember, refrigerant does not just go away. If it is low, there has to be a leak. The only exception would be, if it had been left low for some reason.

Sight Glass — If a system employs a sight glass, your problems are greatly reduced. Simply add gas until the sight glass clears.

NOTE: Never put a sight glass on a system with a restrictor type metering device, such as a capillary tube. A bubbling sight glass, in this case, could mean a full charge. Charge by weight or manufacturers recommendations.

In **Figure 13-17** you see two sight glasses. The top sight glass shows bubbles indicating low refrigerant charge. The bottom sight glass shows clear indicating a correct charge. In each situation the unit must be in operation to make a diagnosis.

Look at **Table 13-1** closely. This allows you to charge for either the cooling mode or the heating

(Fig. 13-17)

mode. At the left of the table, find outdoor ambient, then find where it has "Cool Mode." These temperatures are the outside temperatures. Now find (suction pressures at the compressor-(psig) from 42 psi to 80 psi, then below that (suction line temperature at compressor - °F).

Let's see if your system is charged. As an example, let's say the outdoor temperature is 90°F and your suction pressure at the compressor is 68 psi and your suction line temperature is 60°. You would have a perfectly charged system. However, if your ambient was 90° the suction pressure was 60 psi and the suction line temperature was 65°, this would indicate a low charge. After correcting the leak problem, charge the system until the pressures and temperatures match correctly on the chart.

Now let's pretend you are in the heating mode. Find on the left side of the same table, "Heat Mode." Notice that this will only let you charge from outdoor temperatures from 62° down to 42°. For an example, say you had an outside ambient of 42°F, the suction pressure at the compressor is 50 psi and the suction line temperature is 32°. The heat pump would be accurately charged. Using the same ambient temperature of 42°, let's say the suction pressure at the compressor is 46 psi and the suction line temperature is 41°. You have a system that is low on refrigerant.

At temperatures below 40°, look to the right of the table and charge by this method. As an example, say the outdoor temperature is 27° and the indoor temperature is 70°, your head pressure should read approximately 193 psi.

Keep this in mind now. You can come close to charging systems by sniff, touch and feel, but you do not charge them accurately that way. The correct way may be more time consuming, but is more economical in the long run, to do it right. Equipment cannot be handled like in the old days.

To give you an indication of what the manufacturer expects in charging their units to the correct charge, we will exercise the courtesy of Rheem Manufacturing Co., using a five ton heat pump.

Suction Pressure At Compressor - PSIG / Suction Line Temperature At Compressor - °F

Mode	Outdoor Ambient °F	42	44	46	48	50	52	54	56	58	60	62	64	66	68	70	72	74	76	78	80
C	105+													42	44	45	46	48	49	50	52
O	105												44	45	47	48	49	51	52	53	55
O	100												48	49	51	52	53	55	56	57	59
L	95											50	52	53	55	56	57	59	60	61	
	90										54	55	57	58	60	61	62	64	65		
M	85									56	58	59	61	62	64	65	66	68			
O	80								60	62	63	65	66	68	69	70					
D	75							64	65	67	68	70	71	73	74						
E	70						66	68	69	71	72	74	75	77							
	65					69	70	72	73	75	76	78	79								
H M	62					51	52	54	55	57	58	60									
E O	57				44	46	47	49	50	52	53										
A D	52			38	40	42	43	45	46	48											
T E	47		31	33	34	36	38	39	41												
	42	25	27	29	30	32	34	35													

	Outdoor Ambient °F	Indoor Temp. °F — Head-PSIG* 60	70	80	Temp. °F Change Across Indoor Coil
H	37	207	227	243	24
E	32	190	208	228	22
A	27	176	193	215	20
T	22	164	181	204	18
	17	153	172	194	16
M	12	143	163	186	14
O	7	134	156	178	12
D	2	128	151	173	10
E	-3	123	147	169	8
	-8	120	143	166	6

Below 40°F in heating mode, use head pressure table. *PSIG at hot gas service port. Temperature change across indoor coil greater than shown indicates low indoor airflow.

(Table 13-1)

REFRIGERANT CHARGING PROCEDURE
Charging By Superheat — Capillary System

1. Attach thermocouple or thermometer to the suction line between compressor and reversing valve. Insulate to insure accurate measurement.

2. Attach suction gauge to suction port on compressor.

3. When suction line temperature and pressure have stabilized (15 to 30 minutes after start up or after defrost), record suction line pressure, outdoor temperature and suction line temperature. Enter **Table 13-1** at the intersection of suction line pressure and ambient temperature. The suction line temperature should coincide with the Table reading. Adding R-22 will raise the suction pressure and lower the suction line temperature. **CAUTION:** If adding R-22 raises both pressure and temperature, the unit is overcharged. If unit is in the heat mode and has frost on the outdoor coil, it should be run through a defrost cycle before adjusting charge. Should the intersection of the suction pressure and ambient temperature fall in the open area to left of the numbers in the Table, the following are likely causes:

IN COOLING MODE	IN HEATING MODE
A. Low indoor airflow.	A. Low outdoor airflow. Check for frost or dirt on coil.
B. Restricted refrigerant line.	B. Restricted refrigerant line.
C. Low charge.	C. Low charge.

Should the intersection of the suction pressure and ambient temperature fall in the open area to right of the number, the most likely causes are gross overcharge, low indoor airflow on the heating cycle or defective compressor.

Charging By Head Pressure — Heating Mode

Attach high pressure gauge to gauge port on large refrigerant line (hot gas) at unit, measure indoor air temperature and outdoor ambient. Run unit through defrost cycle and wait 15 minutes for unit to stabilize. (The head pressure should coincide with pressure shown in **Table 13-1**.) Adding charge raises head pressure. **CAUTION:** DO NOT add 30°F, or any other fixed number of degrees, to ambient temperature and charge to that condensing temperature. This will almost always result in a mischarged system. Quite often a system charged correctly in either the heat or cool mode will, when switched to the other mode, appear to be either a little over or a little undercharged. This is due to actual conditions at the installation being different from design conditions.

THE COOLING CYCLE

In **Figure 13-18**, we see a cooling cycle as it would be during normal operation. Follow along as the cycle is reviewed, starting with the discharge line entering the four way reversing valve. Note that the solenoid that controls the slide to position the valve for heating or cooling is energized. This would mean that control voltage is supplied to the coil. In this position the hot gas travels through the reversing valve to the condenser (outdoor coil). Outside air is forced across the outdoor coil and the refrigerant vapor or gas is cooled and condensed to a liquid. As liquid leaves the outdoor coil it enters the check valve and metering device arrangement. During the cooling mode the metering device will push closed and not let any liquid pass through. The check valve will be pushed open and liquid refrigerant will continue to flow through the liquid line without obstruction. However, where the liquid refrigerant enters the metering device at the

---

The information below indicates direction of charge adjustment required for a system due to conditions changing from design conditions.

Condition Change	R-22 Cooling Mode	Heating Mode	Condition Change	R-22 Cooling Mode	Heating Mode
Decrease Indoor Airflow	Decrease	Increase	Raise Indoor Temp.	Increase	Decrease
Increase Indoor Airflow	Increase	Decrease	75° Ind. Heat or Cool	Decrease	Decrease
Lower Indoor Temp.	Decrease	Increase	65° Ind. Heat or Cool	Decrease	Increase

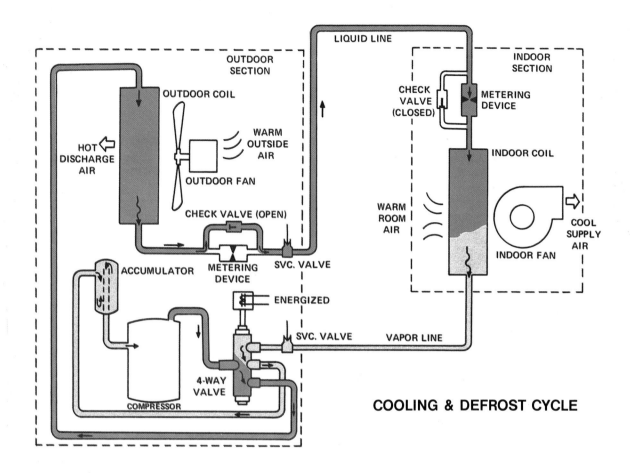

COOLING & DEFROST CYCLE

(Fig. 13-18)

indoor coil, the check valve now pushes closed, and refrigerant has to go through the metering device. Once the liquid refrigerant passes through the metering device, there is a drastic reduction in pressure and temperature, whereas the liquid immediately starts to boil and flash off. Since the refrigerant has a low boiling point, the evaporator (indoor coil) becomes cold and removes heat from the air as it is forced across by the indoor fan. The cool vapor leaves the indoor coil, travels down the vapor line (suction line) and back through the reversing valve, out again and on to the compressor.

The Service Valve — There are two service valves usually located on the outside of the outdoor unit. One service valve is located in the liquid line and the other in the vapor line. While in the cooling mode the service valve in the liquid line is to read condenser pressure or pressure on the high side, (high side meaning pressure on the high pressure side of the system). The other service valve is located in the vapor line (suction line) and is there to read the indoor coil pressure. The service ports should be installed with

cover caps. To read pressures, remove the caps and install your serviceman gauges.

The Accumulator — The accumulator is in the system for protection of the compressor. Should any liquid refrigerant happen to come through the reversing valve back to the compressor, the accumulator will catch it and allow it to boil off into a vapor. Liquid refrigerant can do damage to the compressor.

THE HEATING CYCLE

The solenoid coil is de-energized causing the refrigerant flow to be reversed (**Fig. 13-19**). The discharge gas is now going through the repositioned reversing valve and into the indoor coil. The fan blows the warm air from the coil into the area to be warmed and as heat is being removed from the coil, the refrigerant vapor is cooled and changed to a liquid. The check valve at this point is now pushed open and the refrigerant goes through the open check valve and not the metering device. The liquid goes to the outdoor coil where the check valve is pushed closed and liquid travels through the metering device. A reduction in pressure and

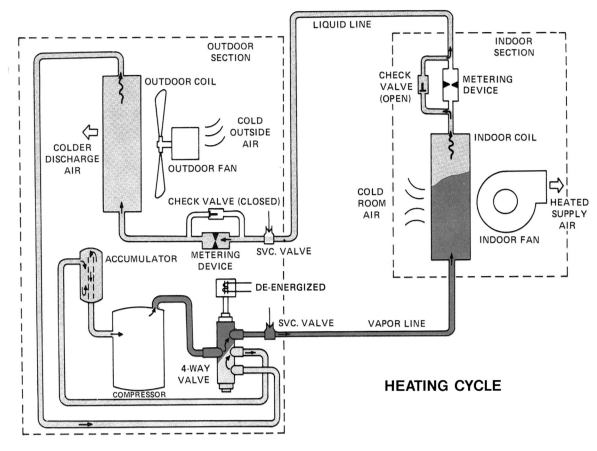

(Fig. 13-19)

temperature causes the refrigerant to boil and causes the outdoor coil to get cold and pick up heat from the outdoor air. The vapors from the outdoor coil come back through the reversing valve and to the compressor.

CHECKING PRESSURES

With the heat pump in the cooling mode, pressures will be checked as if it were an air conditioner. The service valve in the vapor line would be to check the pressure of the indoor coil (evaporator coil) and the service valve in the liquid line is to check the pressure of the outdoor coil (condenser coil).

When the valve reverses to the heating mode, you will now be reading high pressure on the indoor coil and low pressure on the outdoor coil. **Figures 13-20 and 13-21** highlight how the metering devices perform in both the heating and cooling cycle using the expansion valve for a metering device.

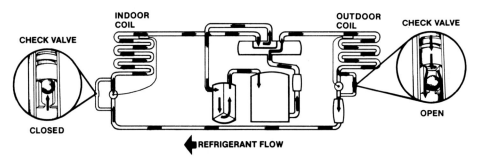

COOLING MODE

(Fig. 13-20)

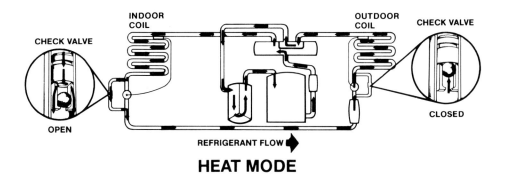

HEAT MODE

(Fig. 13-21)

ELECTRICAL

Electrical is usually as complicated as one makes it. If the controls and their specific operations are known, then the wiring diagram can be easier to understand. Most wiring diagrams are in a ladder type diagram. For most, this diagram is easier to follow as a guide and when trouble shooting. A ladder diagram is shown in **Figure 13-22**.

Figure 13-22 is about as basic as one can be, but let's go through this simple illustration to see if we know where we are. Notice that the diagram does have similarity to a ladder. Notice the L1 & L2 wire. Follow them down. The L2 wire goes directly to a motor (MTR). The L1 wire goes directly to the motor also, but a circuit is not complete to the motor until the R1 contact is made. This contact is controlled by a relay (R) which in this case is controlled by a push button. The relay "R" is supplied with a 24 volt circuit, stepped down by a 240 V/24 volt transformer. Notice the transformer is supplied by 240 volts.

Since there are 240 volts to the transformer, the voltage will be stepped down to 24 volts. If the push button is pushed and held, a 24 volt circuit is sent to the coil of the relay. It magnetizes the coil and pulls R1 contacts closed. This in turn completes a 240 volt circuit to the motor.

Most electrical circuits in the heat pump will be 208/240 volts and 24 volts. The compressor and fan motors will normally be 208/240 volts and the control circuit 24 volts.

Before moving into more complex wiring, refer to the wiring legend on the following page.

Most wiring diagrams are abbreviated simply because the wiring diagrams are too small to be written out in long hand. You will need the wiring legend. It will usually always be in the outdoor and indoor units, stuck on, along with the wiring diagrams.

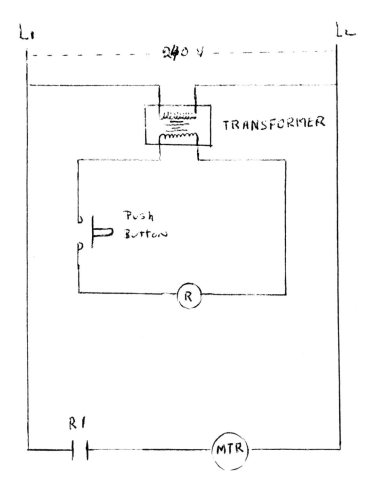

(Fig. 13-22)

AUTO	Automatic	IFR	Indoor Fan Relay
BLK	Black - Hi Fan Speed	L1, L2	Line Voltage
	IFM	LLPS	Liquid Line Pessure Switch
BLU	Blue - Medium Fan Speed IFM	LS	Limit Switch
C	Compressor Contactor	ODT	Outdoor Thermostat
CH	Crankcase Heater	OFT	Outdoor Fan Thermostat
COMP	Compressor	OFM	Outdoor Fan Motor
DFR	Defrost Relay	ON	System Selector Switch
DFT	Defrost Thermostat		(Auto - Heat/Cool)
DT	Defrost Timer	ON	Continuous Run - Indoor Fan
EH	Emergency Heat	RC	Run Capacitor
EHL	Emergency Heat Light	RVS	Reversing Valve Solenoid
FL	Fusible Link	SA	Start Assist (PTC or Start Capaci-
tor)			
FS	Fan Switch (Indoor)	SEQ	Sequencer
FT	Fan Thermostate (Outdoor)	TC0	Cooling Thermostat, First Stage
FU	Fuse	TC1	Cooling Thermostat, Second Stage
HTR	Heater	TC1A	Cooling Anticipator
IFM	Indoor Fan Motor	TH1	Heating Thermostate, First Stage
IFMC	Indoor Fan Motor Capacitor		

CONTROLS AND ACCESSORIES

Crankcase Heater — The crankcase heater is a device used to heat the crankcase of a compressor (**Fig. 13-25**).

During the off cycle of the compressor liquid refrigerant can migrate to the oil of the compressor. The crankcase heater will tend to boil away the liquid for a safe start of the compressor. If the liquid gets to the pistons of the compressor, damage can be done.

Compressor Contactor — The contactor (**Fig. 13-26a**), is a device which has a magnetic coil that magnetizes and pulls a set of contacts closed in order to start the compressor (**Fig. 13-26b**).

Start Assist — Some compressors have to have assistance to get started. One way is to use a start capacitor and a start relay. Another is to use a PSC

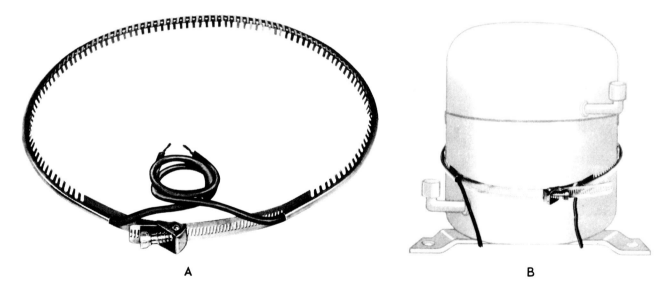

A   B

(Fig. 13-25)

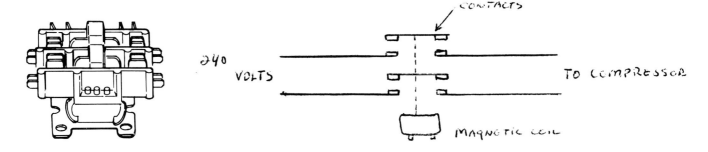

(Fig. 13-26a)   **(Fig. 13-26b)**

capacitor and a rapidly developing new way is to use what is called a Start Assist marked "A" in **Figure 13-27**.

Dual Capacitor — The dual capacitor has a two fold purpose. See **Figure 13-27** marked "B." Notice there are three terminals. The middle terminal is common "C," the right terminal is for fan and the left one for the compressor. This one capacitor can operate a fan motor and compressor. The fan capacitor side will usually have a micro-farad rating of around 5 and the compressor side 20 to 30 micro-farads. The terminals will be marked H-F-C. H stands for Hermetic compressor, F for Fan and C for Common.

Defrost Timer/Defrost Relay/Defrost Thermostat — Three controls are listed here in one, because they are very closely related to one another in their purpose **(Fig. 13-28)**. In **Figure 13-28** the relay to the left is the defrost relay.

Through the operation of the timer shown in the middle and the defrost thermostat, the defrost relay will either position the heat pump for normal operation or by means of contact, put the system into defrost. The timer times a cycle approximately every 1 1/2 hours to go into defrost.

The defrost thermostat is located in the outdoor unit attached to the liquid line. This control can actually terminate the defrost cycle and decide if a defrost cycle is needed.

Outdoor Thermostat — The outdoor thermostat is a control located in the outside unit, sensing outdoor temperature. Should the outdoor temperature be warm enough to not need auxiliary heat, the contact will open and kill the circuit to the auxiliary heat.

Reversing Valve — The reversing valve as we have already discovered is to change the direction of flow inside the refrigeration cycle.

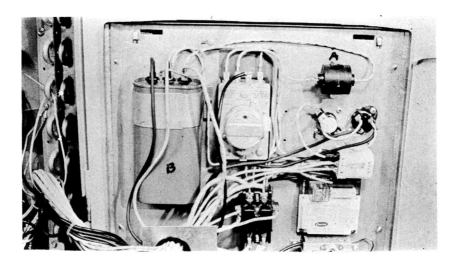

(Fig. 13-27)

(Fig. 13-28)

CONTROLS USED IN THE INDOOR UNIT

Emergency Heat — When the balance point of a heat pump is reached (when the heat from the outside air absorbed by the outdoor coil is equal to the heat load inside) the heat pump can no longer heat the house or building. Additional heat is needed from emergency heat **(Fig. 13-29)**.

The emergency heat is a bank of resistance heaters that are energized from the thermostat through a set a sequencers. Emergency heat is used only when the heat pump cannot put out enough heat in colder weather, where the system goes into the defrost cycle or if there is a compressor failure. A switch is in the thermostat to turn the system to emergency heat should it be necessary.

Sequencers — A sequencer in the case of emergency heat is used to close a circuit to the resistance heaters **(Fig. 13-30a and 13-30b)**.

Figure 13-30a shows the sequencer as it would appear in the indoor unit. **Figure 13-30b** shows the contact arrangement. The heater coil is a high resistance coil, usually 24 volts and when applied with voltage will create heat. The two contacts are bimetallic, sensitive to the heat and will close when heated. When the contacts are closed a circuit will be completed to the heaters usually being 240 volts.

The 24 volt circuit to the heater coil is controlled on demand from the room thermostat. There are two contacts, which can control two heaters. The contacts have a delay before they close, because it takes a short period of time for the heater coil to expel enough heat. See **Figure 13-31** for a simplified diagram of the Sequencers and Emergency Heat.

Limit Control — The limit controls are mounted near the heater strips as a safety device, should they over-heat due to lack of air or any reason that would cause over-heating. See **Figure 13-31** and you can see the limit switches (LS) would open and kill a circuit to the heaters. **Figure 13-32** shows a typical limit switch.

Fan Relay — The fan relay **(Fig. 13-33)** is in the circuit to provide either continuous or automatic fan (indoor fan) operation. When at the thermostat, the fan

(Fig. 13-29)

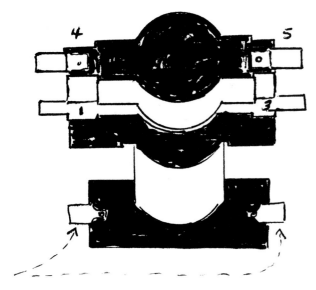

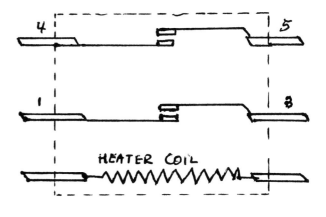

(Fig. 13-30b)

24 VOLT
CONTROL

Sequencer

(Fig. 13-30a)

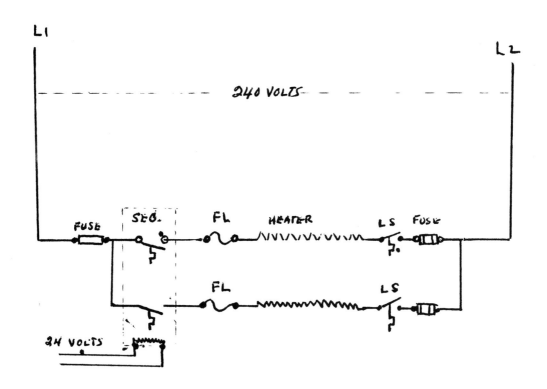

(Fig. 13-31)

211

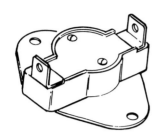

(Fig. 13-32)

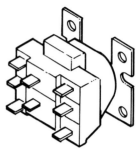

(Fig. 13-33)

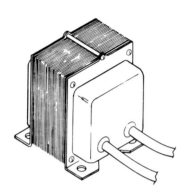

(Fig. 13-34)

Transformers — The transformer supplies 24 volts to the control circuit. It is a step down transformer usually stepping 240 volts to 24 volts (**Fig. 13-34**).

WIRING DIAGRAMS

Refer to **Wiring Diagram 13-1** and identify some of the controls and accessories we have discussed. The Wiring Diagram is not going to be completely mastered at this time. What is needed for you to do is identify the controls and their contact arrangement and just familiarize yourself with an actual Wiring Diagram.

First look at the thermostat to the left of the diagram. Notice especially the terminals R - G - E - L - C - W2 - 0 and Y. These terminals are ones you will field connect should you ever install a heat pump. The main thing to remember is that the terminals are fairly standard as far as their coloring and lettering go. As each terminal is identified, trace from that particular terminal to the right and see where the wire leads.

switch is turned to "auto" the relay will pull in and out starting and stopping the indoor fan with the system. In the "on" position at the thermostat, a continuous 24 volt circuit will be made to the relay coil and the fan will run continuously whether the system is calling or not.

TERMINAL R	The power wire to the Stat from the 24 volt transformer.
TERMINAL G	To the fan relay (FR)
TERMINAL E	To the emergency heat sequencers
TERMINAL L	The opposite power terminal from the TRANSFORMER to turn on the emergency heat light when in the emergency heat position.
TERMINAL W2	2nd stage heat - if 1st stage does not keep up, 2nd stage contacts will close and bring in emergency heat.
TERMINAL 0	To the reversing valve.
TERMINAL Y	Known as the cooling terminal and goes to the contactor coil, which closes and starts the compressor. Also runs compressor during heating cycle by way of H1 contact to Y. During heating the "O" terminal is not made through the H1 anticipator but back the other direction to the Y terminal.

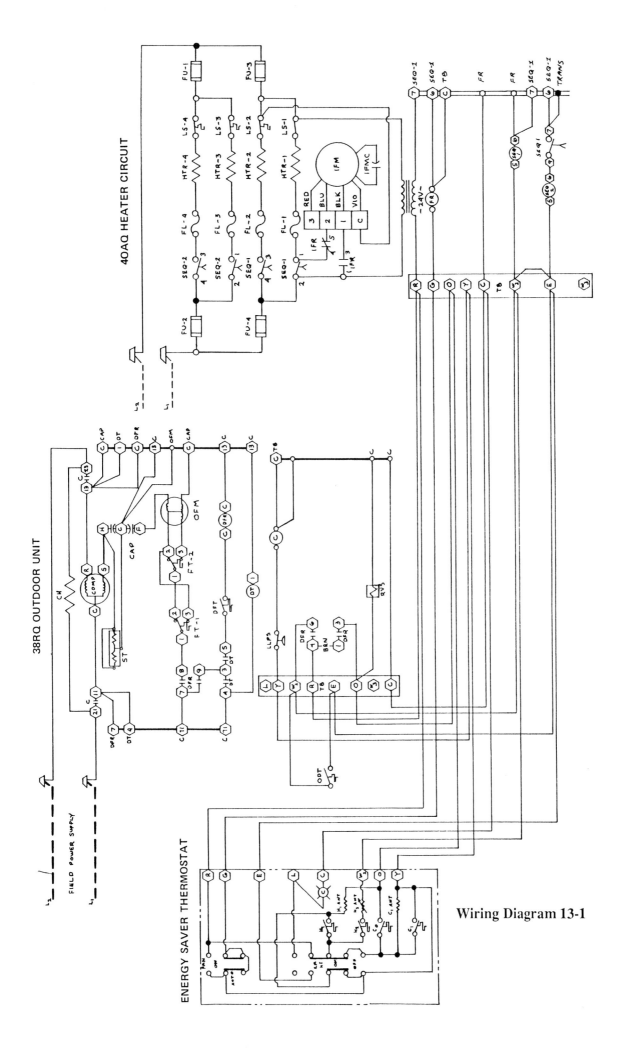

Wiring Diagram 13-1

213

Now notice the contact arrangement in the thermostat. There are the H1 contact (first stage heating), the H2 contact (for auxiliary heat), the (CO) contacts, which is the change over contacts, and the C1 contacts for cooling. The thermostat that is being shown here is an automatic one. It automatically changes from heat to cool should the demand be for either. Take a look at **Wiring Diagram 13-2** and let's see how the cooling cycle is set up through the thermostat. There is approximately 3 degrees between when the heat cuts off and the cool comes on. The first contact to close will be the (CO) (change over contact). Trace from the transformer and find where the red wire from the transformer ties to the R Terminal on TB-1 of the indoor unit. From there, it goes to R on the thermostat. From this point the circuit feeds through the thermostat by way of the selector switch and to the (CO) contact. If the room temperature is set at the cool setting, the (CO) contact will close and complete a circuit to the O Terminal on the stat. From there, trace that circuit to the "O" Terminal on the TB-1 board, then on to the TB-2 board and to one side of the reversing valve. This completes one half of the circuit. The other side of the reversing valve feeds to C (common) and completes a 24 volt circuit.

Now the system is positioned for the cooling cycle. When the temperature of the space rises another one degree, the C1 contact will close making the "Y" Terminal hot. Follow the blue wire. The "R" Terminal in the thermostat continues to feed internally through the now closed contact to "Y" on the stat. From there it goes to "Y" on TB-1 (Terminal board one) back out to "Y" on TB-2. Now notice at "Y" we go to a control lettered (LLPS). If you refer back to the wiring legend you see that this is the liquid line pressure switch. It senses pressure and should the pressure become exceedingly high, the control will open and stop the compressor. You can see the (LLPS), would open and kill one side of the contactor stopping the compressor. C in the wiring diagram is the magnetic coil for the contactor, but where are the contacts that stop the compressor. Find the Field Power Supply at the top left of the diagram. Start with L1 of the Field Power Supply. It goes to a "C" contact. This is now a closed contact since the magnetic coil of the contactor has been energized. At the same time from L2 the other side of the contactor is being closed. This completes a 240 volt circuit to the compressor. Notice also that (CH) (crankcase heater) is attached to terminal 21 and 23 of the contactor and as long as power is on, the crankcase heater is on. This is always true unless the CH has a CH thermostat.

A circuit has been completed to the compressor, but there has to be some type of help in starting the compressor. There is an (ST) (start assist) being used as shown on the diagram. Note the orange wire coming from the contactor to common, the common to the capacitor, then to the start assist, to H on the capacitor, then to S (start) on the compressor.

The start assist has the ability to store a large amount of energy and when it does, gives it up in the start winding of the compressor and assists the compressor in starting. The compressor is not capable of starting on its own.

Now that the cooling circuit has been completed to the reversing valve and the compressor, notice what else comes on. The outdoor fan motor must run to dissipate heat from the outdoor coil. See yellow wire circuit.

With the contactor contacts closed, one side of the circuit to the fan will be made from terminal 11 down to terminal 7 on the defrost relay. The relay contact will be closed between 7 and 8 on the relay, setting a circuit through FT-1 (fan thermostat one) and FT-2 (fan thermostat two) and on to the outdoor fan motor OFM. The circuit is completed to the fan motor back to the capacitor.

The fan thermostats control the speed of the outdoor fan. The fan has two speeds, low and high. During cooling the high speed is needed during hotter weather. If the outdoor temperature drops below 90°, the contacts in FT-2 will open across 1 and 2 and across 1 and 3. The fan is now on low speed. The FT-1 fan thermostat is used mainly for the heating. The control closes around 55°F outdoor temperature across 1 and 3 contacts and puts the outdoor fan in high speed so more air can be passed over the outdoor coil to pick up more heat.

What about the fan on the indoor coil (IFM)? What is it doing? It will have to be operating since it moves air across the indoor coil. Follow the circuit by way of the heavy black wire in the thermostat. Remember the fan is in the "auto" position at the thermostat, making the fan cycle with the outdoor compressor. When the C1 in the thermostat contact is closed the "Y" terminal is hot causing the large black wire to be hot. Trace it through the thermostat and out the "G" terminal of the thermostat to G on terminal board 1 (TB-1) and on to the fan relay coil. The circuit is completed to the other side of the coil to common (C).

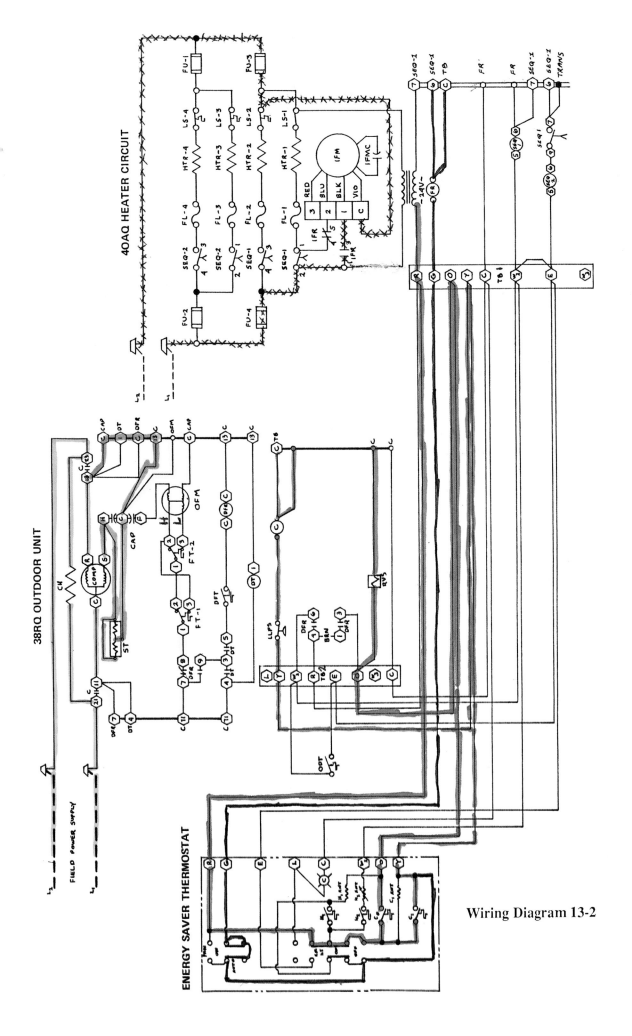

Wiring Diagram 13-2

215

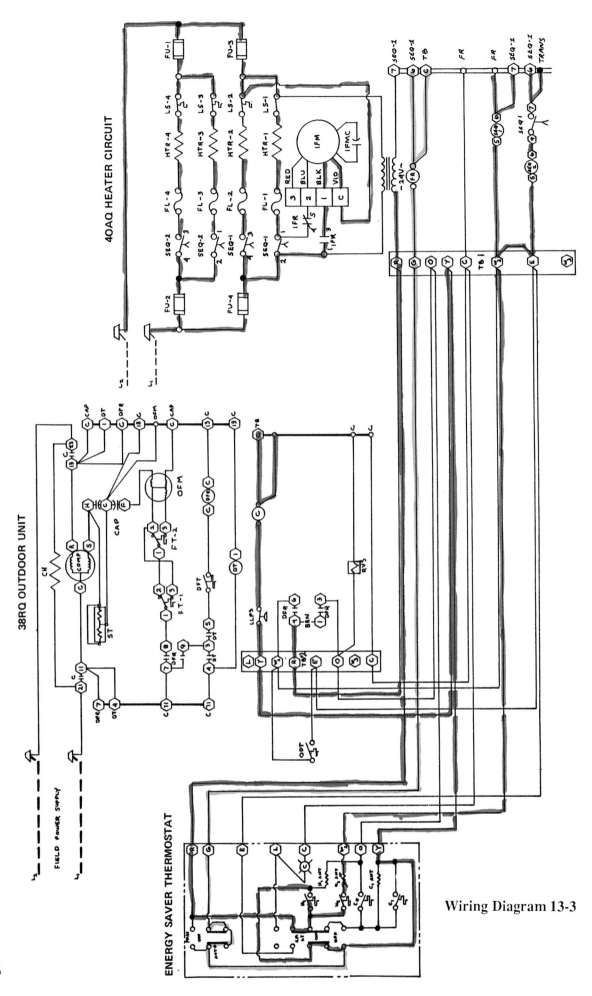

Wiring Diagram 13-3

216

The magnetic coil controls the IFR (Indoor Fan Relay) contacts. Since there is no call for heat, SEQ-1 (sequencer 1) will be open, therefore the closed contact on the IFR will not start the fan motor, but with the FR coil energized IFR contact between 4 and 5 opens and closes between IFR contacts 1 and 3 and starts the indoor fan motor, provided you have power on L1 and L2 of that circuit. Trace the X'd wires for the circuit. When the C1 contacts in the thermostat open, the circuit to the compressor is killed, stopping the compressor and at the same time stopping the indoor fan. The fan can be made to run continuously by switching the fan switch to "on."

The cooling circuit has been shown, now take a look at the heating cycle in **Wiring Diagram 13-3**. Heating through the thermostat is controlled through contacts H1 and H2. H1 contact will be for normal heating with the heat pump. The H2 contact is for auxiliary heat should there not be enough heat from the heat pump.

Start at terminal board at "R" Terminal where power is being fed from the transformer and trace the red wire to R on the stat. A circuit is made internally through the stat to the H1 contact. When H1 closes, a circuit is made to the "Y" Terminal. Keep tracing and you will find it goes to the contactor. Since the compressor is operating during heating you may have a difficult time understanding how we are heating. The compressor operates all the time, the reversing valve decides whether we are heating or cooling. The (CO) contacts in the thermostat are open and the RVS is dead. With the RVS dead the cycle has been reversed where the indoor coil is heating and the outdoor coil is cold taking heat out of the outdoor air and giving it up in the conditioned area.

In the event the heat pump cannot stay up H2 contact will close and complete a circuit through the blue wire to W2 on TB-1, which makes a circuit to the SEQ 1 (sequencer 1). This is a high resistance coil in the sequencer and produces heat when 24 volts are applied. This closes a bimetallic contact between 1 and 2 and 3 and 4 in the auxiliary heat circuit. Since H1 in the stat is already closed and the fan is operating, IFR contacts 1 and 3 are closed and IFR 4 and 5 are open. The fan would be in the high speed. So the sequencer will close after a delay of a few minutes and start HTR-1 and HTR-2 to heating. Note FL-1 and 2 are fuse links and LS-1 and LS-2 are limit switches against over heating.

We are not finished with the auxiliary heat yet.

Note that there is a jumper wire on TB-1 from terminals W2 and E. If W2 has power so will E and so will SEQ 2. But note that sequencer 1 has a contact between terminals 9 and 7. Sequencer 1 has to be closed before a circuit can be made to Sequencer 2 heater circuit.

Should you need emergency heat, simply switch the thermostat to Emergency Heat and terminal E and L will be made hot. Terminal "E" if you will follow over to TB-1 feeds E to sequencer 2 and by-passes from E to W2 on TB-1, setting up a circuit on sequencer 1. Remember sequencer 2 circuit cannot be energized until sequencer 1 has made. Terminal L feeds one side of the emergency heat light on the thermostat and completes a circuit by leaving C on the stat and going back to the common side of the transformer.

Now let's trace out the defrost circuit and see how it operates. See **Wiring Diagram No. 13-4**. First notice the ODT (outdoor thermostat). This thermostat is set so that at warmer outdoor temperatures it will open and not allow the auxiliary heat to come on during the defrost cycle. In milder weather the outdoor coil may not even accumulate frost.

The DT (defrost timer) operates when in cooling or heating and controls DT contacts between 4-3 and 5. The DFT (defrost thermostat) terminates the defrost cycle after the timer has initiated it. The DFT will open at approximately 80°F. Should the DFT not terminate the defrost cycle, the timer will.

During defrost the compressor continues to operate. The outdoor fan motor will cut off and the auxiliary heat will come on if the ODT (outdoor thermostat) is closed. The system will go into the cooling cycle, where the outdoor coil becomes warm and melts the accumulated frost away. Defrost will continue until the temperature at the DFT reaches approximately 80°F. It will open, drop out the DFR (Defrost Relay) and put the system back into a reverse cycle (the heating cycle).

Now take a look at **Wiring Diagram No. 13-5**. When the contactor contacts (c) are closed, there is continuous power (240 V) to the defrost timer (DT). Every 90 minutes the defrost timer (DT), momentarily closes contacts 4 & 3 at 3 for approximately 20 seconds. **Wiring Diagrams 13-5, 13-6 & 13-7** are mini-sections from the larger **Wiring Diagram No. 13-4**.

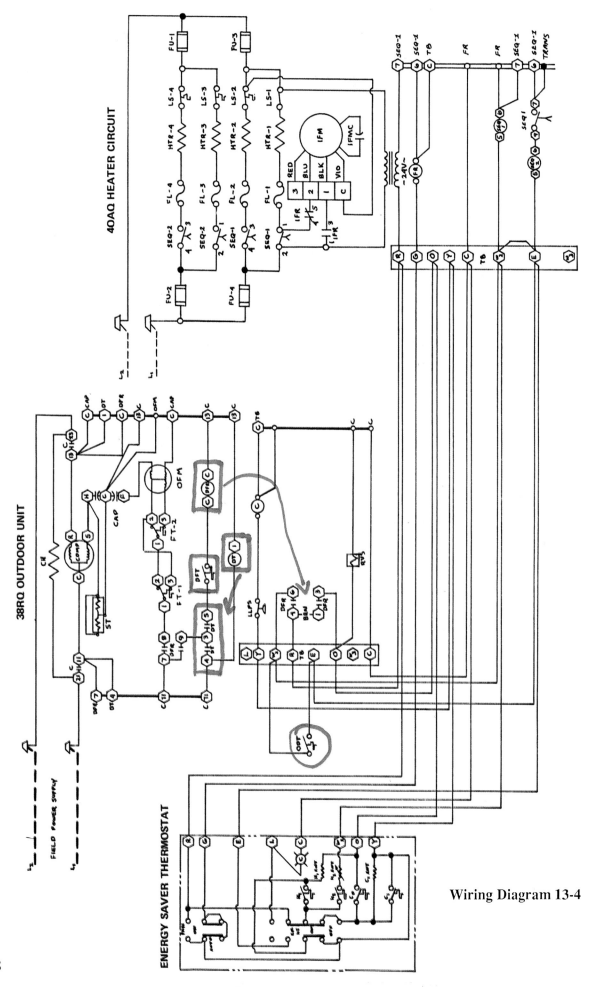

Wiring Diagram 13-4

DEFROST

When the defrost timer contacts between 4-3 close, contacts of the defrost timer between terminals 3 and 5 are already closed at the 90 minute interval (**Wiring Diagram No. 13-5**). The defrost thermostat at 4 is also closed due to a low refrigerant temperature in the outdoor coil and a circuit is provided to energize the defrost relay (DFR). This puts the unit into the defrost cycle. The normally open defrost relay contacts at 2, close providing a holding circuit for the defrost relay.

Upon energizing the defrost relay, a set of normally open contacts at 1 was opened, therefore stop-ping the outside fan motor. The fan motor is stopped so cold air cannot be pulled across the coil while in defrost.

Defrost Initiation — When the defrost relay (**Wiring Diagram 13-6**) was energized, contacts 4 & 6 and 1 & 3 of the defrost relay were energized. Contacts 1 & 3 completed a circuit to the RVS (reversing valve) which puts the system into the cooling mode. Hot gas from the compressor is now going to the outdoor coil, the outdoor fan is off and the outdoor coil is defrosting.

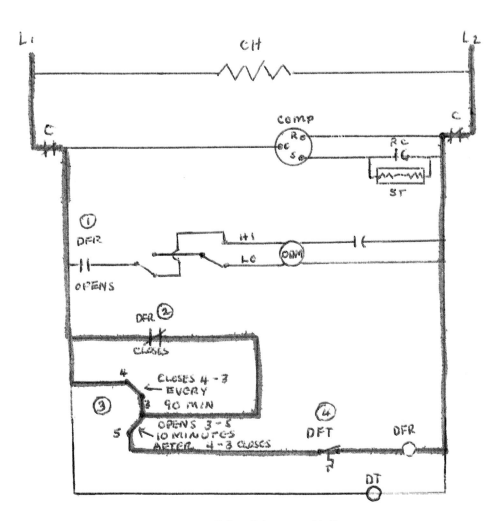

Wiring Diagram 13-5

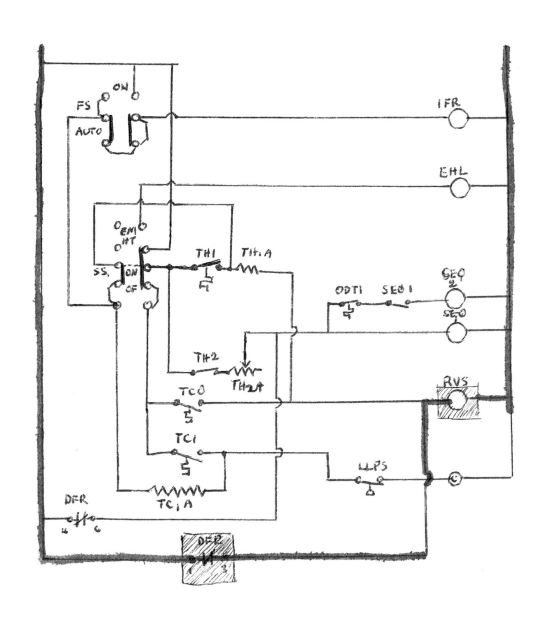

Wiring Diagram 13-6

LEGEND

C	Contactor Coil	ODT1	Outdoor Thermostat
COMP	Compressor	RC	Running Capacitor
CH	Crankcase Heater	RVS	Reversing Solenoid
DFR	Defrost Relay	SEQ	Sequencer
DT	Defrost Timer	ST	Start Assist
EHL	Emergency Heat Light	SS	System Switch on Thermostat
EMHT	Emergency Heat	TH1A	Anticipator 1st Stage (Heat)
FS	Fan Switch on Thermostat	TH2A	Anticipator 2nd Stage (Heat) (ADJ)
FT1	Fan Thermostat 1	TC1A	Anticipator 1st Stage (Cool)
FT2	Fan Thermostat 2	TH1	Thermostat, Heat, 1st Stage
IFR	Indoor Fan Relay	TH2	Thermostat, Heat, 2nd Stage
LLPS	Liquid Line Pressure Switch	TC0	Thermostat Change Over
OFM	Outdoor Fan Motor	TC1	Thermostat Cool, 1st Stage
OFMC	Outdoor Fan Motor Capacitor		

DEFROST TERMINATION BY TEMPERATURE

When the defrost thermostat (**Wiring Diagram 13-7**) senses a rise in liquid line temperature at 3 it will open at approximately 80°F. The de-energized defrost relay coil opens the DFR contacts at 2 and closes the DFR contacts at 1 which starts the outdoor fan.

Refer back to **Wiring Diagram 13-6** and take note that 4 & 6 of the DFR opens and kills sequencer 1 for auxiliary heat and opens DFR contacts 1 and 3 which kills the RVS (reversing valve). While on this diagram, note the ODT 1 (outdoor thermostat) and sequencer 2. If the outdoor temperature is colder, say under 40°F, the outdoor thermostat will close and upon going into defrost will also allow sequencer 2 to become energized and bring on all the auxiliary heat during defrost.

DEFROST TERMINATION BY TIME

Ten minutes after initiation of the defrost cycle, the defrost timer opens the 3 & 5 contacts for about 20 seconds. If the defrost thermostat hasn't opened at this point, the action of the defrost timer contacts de-energizes the defrost relay coil and the unit goes back to heating.

When the defrost cycle is over, the defrost relay contacts 1 & 3 will open, de-energizing the reversing valve. The unit will then continue its reverse cycle heating. **Wiring Diagram No. 13-4**.

If the outdoor coil was not completely defrosted due to the defrost thermostat staying closed, the system will go through another defrost at the next 90 minute interval of the defrost timer.

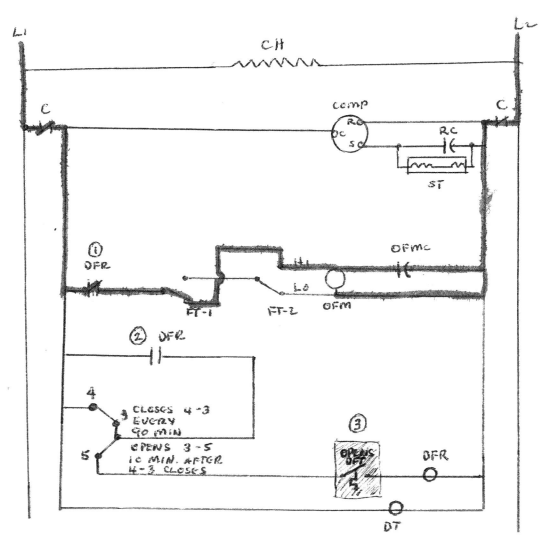

Wiring Diagram 13-7

TROUBLE SHOOTING THE HEAT PUMP

In previous chapters and studies, you have been taken through the many troubles of air conditioning. When the heat pump is in the cooling mode, you basically have just that, an air conditioner. But before we get into the system in the heating mode, let's recap a few things.

COOLING MODE

Operating Pressures: The operating pressures in the cooling mode using Refrigerant 22 during the summer months will be approximately 69 psi on the low side and 250 to 275 on the high side.

The indoor coil at this time is cool and removing heat from indoors. The outdoor coil is rejecting heat that was absorbed from indoors plus the heat of compression to the outside.

To refamiliarize yourself with operating pressures, in respect to their temperatures, refer to **Figure 13-35**. These pressures and temperatures will apply to the low side of the system or the indoor coil in the cooling mode.

In looking at **Figure 13-35**, you can see the temperatures varying from 30°F to 50°F and the pressures varying from 55 psi to 84 psi. The intent is to make you realize that the pressures and temperatures of the cooling coil are directly affected by different temperatures indoor and outdoor as well.

For example, an ideal situation for an air conditioner would be for the outdoor temperature to be around 95°F and the indoor temperature around 76°F. Under these conditions, you could expect the low side pressure to be around 69 psi and the high side pressure 250 psi.

However, let's say the outdoor temperature drops to 75°F and 70°F indoors. The lower temperatures outdoors and indoors would cause a lower pressure reading on both the low side and the high side.

Now refer to **Figure 13-36**. This shows the relationship of pressures and temperatures on the high pressure side of the system. The temperatures shown will be outdoor air temperatures and the pressures will be of the high side as we said.

By this time it is assumed that you understand about trouble shooting, but just for a review refer to **Figure 13-37** and recap on trouble shooting the heat pump in the cooling mode.

Evaporator Temp. °F	Refrigerant 22 PSIG
30	55
32	57
34	60
36	63
38	66
40	69
45	76
50	84

(Fig. 13-35)

Temperature	R-22
90°	245 psi
95°	255 psi
100°	290 psi
105°	310 psi
110°	350 psi

(Fig. 13-36)

TROUBLE SHOOTING GUIDE FOR THE HEAT PUMP

SYMPTOM	CAUSE	REMEDY
A. Suction Pressure Too Low	1. Low on gas 2. Dirty air filter at indoor coil 3. Evaporator fan belt slipping 4. Wrong rotation of evaporator fan 5. Evaporator fins dirty 6. Restriction in metering device 7. Restricted drier 8. Condenser fan thermostat stuck in high position	1. Leak check, repair leak and recharge 2. Clean or replace filter 3. Tighten fan belt 4. Make fan turn in correct direction 5. Clean evaporator 6. Clear restriction 7. Replace drier 8. Replace condenser fan stat
B. Suction Pressure Too High	1. Heat load too high 2. Outside temperature too high 3. Check valve at indoor coil leaking by 4. Valves in compressor bad 5. Too much air across evaporator 6. Reversing valve stuck in mid position	1. Reduce heat load or give system time to pull down. 2. Shade condensing unit during hot periods of the day, or relocate condensing unit. 3. Replace check valve 4. Replace compressor if hermetic. Replace valve plate if head is exposed. 5. Slow evaporator fan down 6. Slightly tap valve. Replace valve if it is sticking or stuck.
C. Head Pressure Too High	1. Condenser fins dirty 2. System over-charged 3. Air in system 4. Rotation of condenser fan wrong 5. Condenser fan belt slipping 6. Bad location of unit 7. Condenser fan in low speed	1. Clean condenser 2. Bleed gas off to proper charge. 3. Release charge, evacuate and recharge 4. Correct rotation 5. Tighten fan belt 6. Relocate unit 7. Replace condenser fan thermostat
D. Evaporator Ices Up	1. Dirty filter 2. Evaporator fan belt slipping 3. Rotation of evaporator fan wrong 4. Evaporator fins dirty 5. Fan running too slow 6. Condenser fan staying in high speed in mild weather	1. Clean or replace filter 2. Tighten fan belt 3. Correct rotation 4. Clean evaporator 5. Speed up fan 6. Replace condenser fan stat

BASIC TROUBLE SHOOTING
OF THE HEAT PUMP

Now let's get into some basic trouble shooting procedures. We can't cover the subject completely, but we will hit on some of the major points and hope the techniques taught to you will apply to other situations you may come across.

First let's mount our serviceman's gauges (**Fig. 13-38**).

With the heat pump in the cooling mode and the low side gauge on the suction line gauge service port you will be reading indoor coil pressure and the suction line is cool bringing cool vapor back to the compressor through the reversing valve. The high side gauge will be mounted on the liquid line reading pressure of the high side of the system. Under normal conditions with the system in the cooling mode you would read approximately 69 psi at the suction gauge port and 250 psi at the liquid line service port. Remember the indoor coil is cool and the outdoor coil would be hot.

Now let's go into the heating mode. When this happens the circuit reverses itself by way of the reversing valve. The direction of flow reverses itself as shown in **Figure 13-38**. What we are calling the suction line is now carrying hot gas from the compressor through the reversing valve and to the indoor coil. The hot gas will condense to a liquid in the indoor coil and travel back out to the condensing unit where it will enter a metering device. The flow of refrigerant will be metered into the outdoor coil where it will boil, make

the outdoor coil cold and remove heat from the air outdoors. It is best in the heating mode to place your gauges opposite, that is the low side gauge is now on the liquid line service port and the high side gauge is on the suction service port. So remember, when the system reverses itself, so do the pressure readings.

The pressures of the outdoor coil and indoor coil can vary greatly in the heating mode because of such a varying degree of temperatures. However, see **Figure 13-39** and these pressures will be close enough to where you can be close to what the pressures should be on your serviceman's gauges.

See **Figure 13-40** for outdoor coil pressures under different conditions.

When needing to know whether or not the pressure readings you find are accurate or not refer back to your charging guide **Table 13-1**. At a given outdoor temperature and knowing your pressure, you can then check the pressure versus the temperature to see if they are what they should be.

Let's now cover some of the most common complaints of heat pumps during the heating mode. In most cases a customer will call and complain that there is not enough heat. If during the heating season the complaint is that the air coming out of the register isn't hot and the customer is new to heat pumps, then chances are that there is no problem at all. It's probably just that no one explained to the customer that in the heating cycle a heat pump delivers a larger quantity of air at a lower temperature than does a gas or oil furnace.

In fact, if over a period of time the customer can

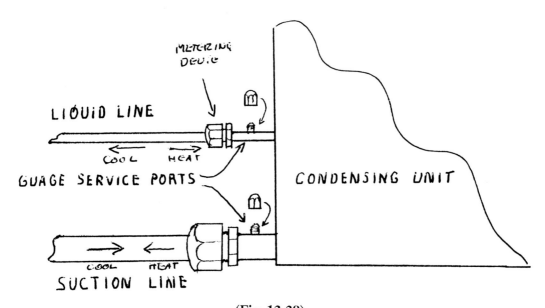

(Fig. 13-38)

INDOOR COIL PRESSURE

OUTDOOR AMBIENT	INDOOR TEMP OF		
	60°F	70°F	80°F
DEGREES FAHRENHEIT	INDOOR PRESSURE (PSIG)		
37	168	190	211
32	158	180	200
27	150	171	191
22	143	164	184
17	137	159	178
12	133	154	174
7	129	151	170
2	126	147	167
-3	123	145	164
-8	123	143	162

(Fig. 13-39)

OUTDOOR COIL PRESSURE

OUTDOOR AMBIENT	DEGREES FAHRENHEIT	OUTDOOR COIL PRESSURES (PSIG)
	62	57
	57	51
	52	46
	47	41
	42	38
	37	34
	32	30
	30	25
	25	20

(Fig. 13-40)

COMPLAINT "LACK OF HEAT!"

SITUATION 1. AIR FROM REGISTER ISN'T HOT

recognize a gradual rise in supply air temperature during heating, the unit is probably operating under restricted air conditions. This can cause increased compressor loading, cycling on compressor motor overload and compressor failure. Explain to the customer that if the discharge air feels hot, they should check the grilles, both inlet and outlet to see if they are blocked and the filter for cleanliness. The indoor coil could have enough dirt or lint on it to block air flow.

The customer may complain that the unit cuts off too soon before the space is warm or that it runs too long on heating.

One thing that should be checked is the heating anticipator setting. The heat anticipator is found when the cover of the thermostat is removed and you see a scale in 1/100th readings. You will see a pointer that can be moved to the proper setting.

The best method of knowing what valve to set the anticipator is to wrap 10 rounds of thermostat wire around one clamp as shown in **Figure 13-41**. Go to the

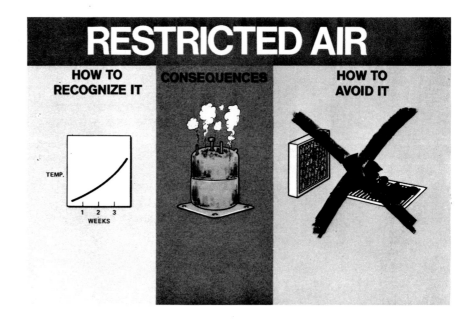

(Fig. 13-41)

heater terminal board and remove the wires from R and W2. Connect the wire on the clamp of the meter to R and W2. This puts the amprobe in series between the transformer and sequence 1. Start the heater and measure the current and divide by 10 (10 wraps of wire). For example: If the current reading was 1.8 then set the anticipator at .18.

Still in the heating mode, let's say you have a heavy coating of frost on the outdoor coil and the compressor is running, but the unit is not going into the defrost mode and the liquid line temperature is blowing 24°F at the defrost thermostat (the defrost thermostat at 24°F should be closed). Frost cuts down on the amount of air crossing the coil, which reduces heat transfer, so the defrost cycle is important.

Let's take a look at a method for trouble shooting this situation.

The unit may be overcharged or a component in the defrost system defective. The system charge cannot be checked with a heavy build up of ice, so let's

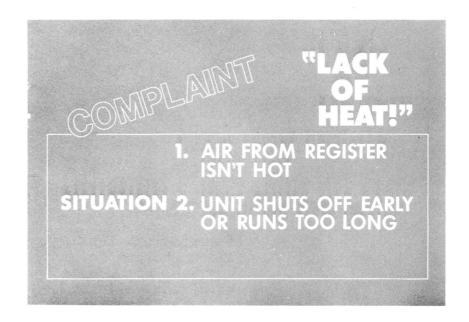

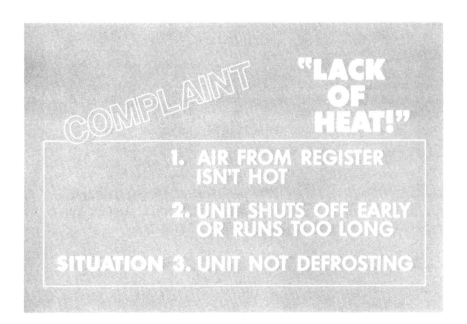

check the defrost electrical controls.

During the time the coil is iced and iced to the point that there is little heat rejection into the conditioned space, the area or space should be warm since the second stage heat (auxiliary heat) would be on.

Let's continue on and check the defrost circuit. With the outside unit in operation, check to see if the defrost timer at 1 has voltage. If it does and you do not see a wheel turning through a peep window, the timer is bad.

Now turn off the power to eliminate shorts and place a wire with clips between 4 & 5 on the defrost timer at point 2 (**Fig. 13-42**).

Turn power on to the unit and start. If all the defrost controls are functioning, the unit should go into defrost. If the unit does not go into defrost, then check as shown in **Figure 13-43**.

Next, we will check the reversing valve solenoid. In the defrost cycle the solenoid should be energized with voltage to the coil. If the reversing valve solenoid is de-energized (no voltage to coil), replace the defrost relay. Contacts 1-3 of the defrost relay should be closed. If there is voltage up to the reversing valve solenoid, but the valve isn't reversing the reversing valve coil could be bad.

Also the reversing valve could be stuck. If the coil is found to magnetize, then it would be the valve. Before condemning the valve, check and see that there has been no external damage to the valve that can be corrected.

If you do not find voltage on the coil of the defrost relay, then remove the leads from the defrost thermostat and OHM it. If the ohmmeter does not read, the defrost thermostat is open where it should be closed. To check the defrost relay, place a jumper wire with alligator clips on each end from terminal 11 to terminal 5 (**Fig. 13-44**).

If the defrost thermostat is closed, the defrost relay coil will be energized. If it does not pull in, replace the defrost relay.

Another complaint of lack of heat could be that the outdoor fan is not operating or turning backwards. If the motor is running backwards, you may need to

DEFROST TROUBLESHOOTING

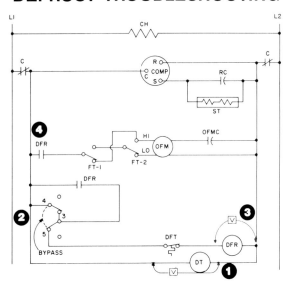

(Fig. 13-42)

DEFROST TROUBLESHOOTING

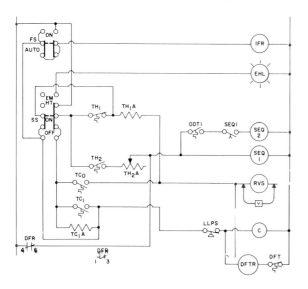

(Fig. 13-43)

DEFROST TERMINATION-TIME

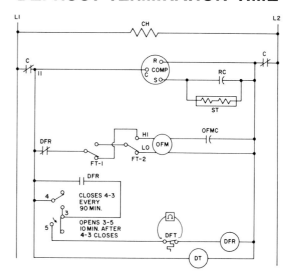

(Fig. 13-44)

reverse the capacitor leads or simply change the rotation as prescribed on the wiring diagram of the motor.

If the fan motor is not running up to speed or just buzzing and not starting, check the capacitor. Use a capacitor analyzer such as the one in **Figure 13-45**, or just simply substitute another capacitor of like voltage and micro-farads.

Another cause of lack of heating is that the refrigerant charge is low. Here is the way you can usually tell if a low charge of refrigerant is causing the outdoor coil to ice. The coil will partially ice and there will be a heavy build-up of ice, say on one half of the coil.

If the coil is iced solid, then the problem is usually in the defrost circuit. If the system is low on gas, leak check and recharge.

In the heating mode, it is best to charge with a dial-a-charge. The dial-a-charge (**Fig. 13-46**) is an accurate charging cylinder found at most refrigeration supply houses. It would be advisable to get a cylinder equipped with a 120 volt heater. When the heater is plugged in the heat will cause a pressure build up and push the refrigerant from the cylinder into the system. Before charging at all, pull a vacuum on the system. The dial-a-charge is not made to charge when there is pressure

(Fig. 13-45)

229

(Fig. 13-46)

on the system. **CAUTION:** do not over-charge the system. The heat pump system is a critically charged system.

If the heat pump is not heating, the compressor could be off or not running. If the compressor is idle and making no attempt to start, go back to Wiring Diagrams No. 13-1, 13-2 or 13-3 and find where the voltage comes to the contactor at 21 and 23. If you do not have voltage at that point you either have a tripped breaker or blown fuse. If there is voltage, then check at 11 and 23. If there is not any voltage there the contactor is open due to a burnt out contactor coil or no voltage to the coil. To check the coil, go directly to the coil terminals and see if you have 24 volts. If you do and the contacts are not pulling in, then there is a burnt out holding coil. If there is not any voltage to the coil you have a dead 24 volt circuit. Look at your **Wiring Diagram 13-2** and find the terminal board in the unit. You should have voltage to Y and C. If there is no voltage there, you have trouble elsewhere. However, if you do have voltage to Y and C and there is no voltage to the coil, notice that there is a LLPS switch in the line. If the unit has been off for some time this switch should be closed. It has been established that you have voltage at Y and C. Now check on the outside of the LLPS to the contactor coil and terminal "C." If there is no

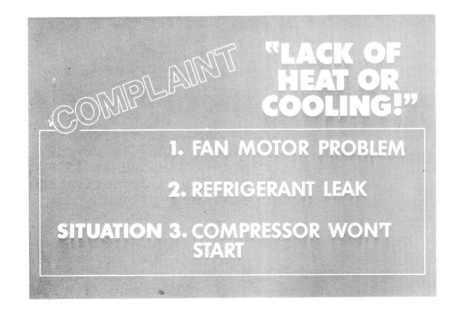

voltage the LLPS control is open. Before making up your mind to replace this switch, check and make sure it does not have a manual reset. If it does, reset and find the cause of its tripping. (Causes of this switch tripping on the LLPS reset in the heat mode is lack of air across the indoor coil or overcharged.) (Causes during the cooling mode is lack of air across the outdoor coil or overcharged.)

If the contactor is closed and there is voltage to 11 and 23 on the contactor and the compressor is making no attempt to start, there are two real possibilities that a wire is burnt off at the common terminal of the compressor, or the internal overload is open in the compressor. If there is voltage (240 volts) to C and S and C & R, the compressor should make some attempt to start. If the compressor is hot then the internal overload of the compressor is open. You should allow 30 to 40 minutes for this control to close. If the compressor is ambient temperature with voltage to the compressor, then the overload is open and probably never intends to close.

If the compressor does try to start, but only hums and kicks the overload, then check the start assist (ST). The quickest check for this is to get a start capacitor out of your truck and at the same time power is restored to the unit, touch the remote capacitor leads to S and R on the compressor. If the compressor starts and draws correct amperage, the start assist is bad. You may wish to install a new start assist, or install what is called a "hard start" kit which consists of a start capacitor and start relay. **Figure 13-47** shows the wiring for the "hard start" kit.

The red wires in **Figure 13-47** shows the wiring

necessary to replace the start assist. The rest of the diagram is as you find it in **Wiring Diagram No. 13-1**.

To understand electrical trouble shooting better, let's use **Wiring Diagram No. 13-4**, but first let's understand now that all wiring diagrams are not exactly the same, not all defrost means are the same, and neither are all trouble shooting the same. However throughout the chapter on heat pumps you have been familiarized enough to trouble shoot other units if you master this one. Now look at the Electrical Trouble Shooters Guide and do some electrical trouble shooting. You will be using the Trouble Shooters Guide and Wiring Diagram No. 13-4 in unison. See Page 239.

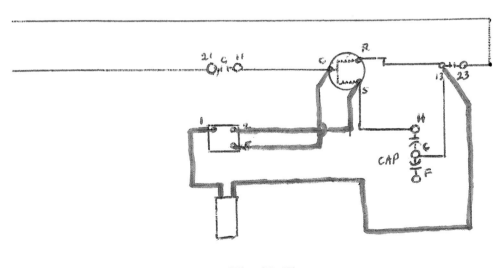

(Fig. 13-47)

TWO SPEED COMPRESSORS

Two speed compressors are becoming popular in heat pumps and even some conventional air conditioning systems. Lennox has for some time been using a two speed compressor in some of their air conditioners. The purpose of using the two speed compressor is to slow the compressor down when loads are not so heavy. For instance, if the outdoors is cool, but yet it is warm enough in the house or building, the compressor on low speed would cool the area. If the low speed didn't carry the load, the provisions are made in the control circuit to shift the system back to high speed.

There are several advantages to the two speed compressor. One is that in low speed the amperage draw of the compressor is less, therefore lowering the kilowatt consumption. Another is that the seasonal energy efficiency rating (SEER) is very high at that time. Finally, a variable speed compressor will vary in tonnage on a four ton unit for example from as low as 2 1/2 tons, to a full four tons. If there are any disadvantages at all with the two speed compressor, it will be that the humidity will rise in the home or business when the compressor is on low speed. This is because while the compressor is on low speed, the pressure in the cooling coil will rise, making the coil temperature warmer. The higher the temperature of the coil, the less humidity (moisture) will be removed. To offset this, just simply buy a control and install so that the fan motor can be slowed down across the cooling coil during high humidity situations.

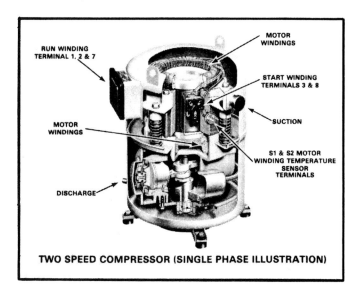

TWO SPEED COMPRESSOR (SINGLE PHASE ILLUSTRATION)

(Fig. 13-48)

Figure 13-48 shows the internal parts of a two speed compressor. The compressor has two terminal boxes, one for the run windings and the other for the start windings.

See **Figure 13-49** for the terminal hook-ups and arrangement of the windings. This is a single phase motor. The two pole winding is the high speed start circuit and the four pole windings are for the low speed starting circuit. A circuit to terminal one and three would put the high speed start winding in the circuit. A circuit to terminals one and eight would put the low speed start windings in the circuit. With a circuit to

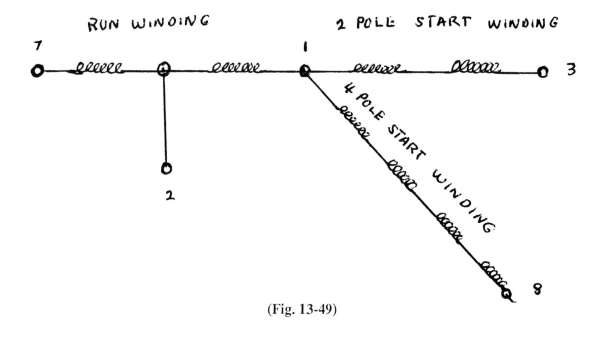

(Fig. 13-49)

terminals one and seven, you would have the two run windings in series and be on low speed. Jumper from terminal one to seven and run another line wire to terminal two and the compressor would be in high speed. To automatically start the compressor a start relay is used along with a start capacitor and a run capacitor.

THE START RELAY (POTENTIAL RELAY)

The starting relay is a device that is used to drop the start winding from the circuit after the compressor has reached approximately 75% of its speed. The relay contacts open causing the start winding and starting capacitor to be dropped from the circuit, leaving the compressor on the run winding. The starting devices are for both the high and low speed windings.

THE START CAPACITOR

THE START CAPACITOR — The starting capacitor is an energy storing device rated in microfarads (MFDs). On each start the capacitor collects enough force to increase the starting torque of the compressor. Without the start capacitor the compressor is not likely to start.

THE RUN CAPACITOR — The run capacitor on start is in parallel with the start capacitor, but when the start relay drops out, the run capacitor stays in the circuit making the compressor motor run as a two phase induction motor with improved power factor (decreases the operating amperage) and torque characteristics. See **Figure 13-50** for an illustration of the same starting devices for both high and low speeds.

THREE PHASE-TWO SPEED COMPRESSORS — The three phase compressor has the same function as the single phase compressor, but the electrical characteristics are different. **Figure 13-51** shows the winding arrangement.

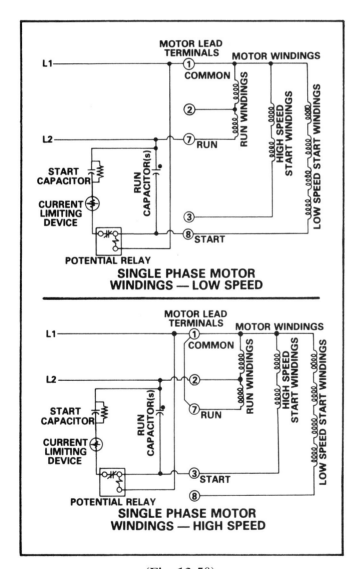

(Fig. 13-50)

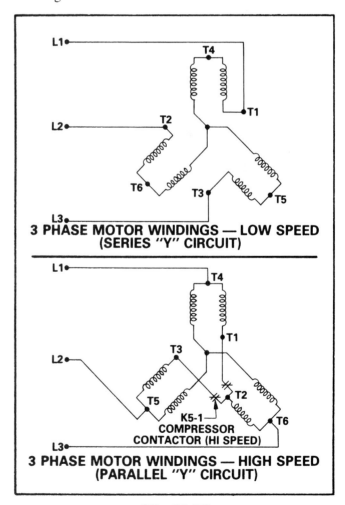

(Fig. 13-51)

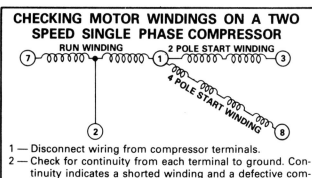

CHECKING MOTOR WINDINGS ON A TWO SPEED SINGLE PHASE COMPRESSOR

1 — Disconnect wiring from compressor terminals.
2 — Check for continuity from each terminal to ground. Continuity indicates a shorted winding and a defective compressor.
3 — Check for continuity from 7 to 3, 7 to 8 and 7 to 2. No continuity indicates an open winding and a defective compressor.

(Fig. 13-52)

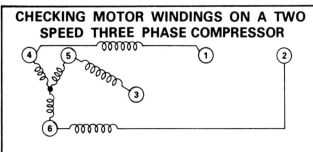

CHECKING MOTOR WINDINGS ON A TWO SPEED THREE PHASE COMPRESSOR

1 — Disconnect wiring from compressor terminals.
2 — Check for continuity from each terminal to ground. Continuity indictes a shorted winding and a defective compressor.
3 — Check for continuity from 1 to 3 and 1 to 2. No continuity indicates an open winding and a defective compressor.

(Fig. 13-53)

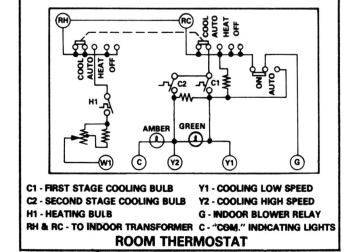

C1 - FIRST STAGE COOLING BULB	Y1 - COOLING LOW SPEED
C2 - SECOND STAGE COOLING BULB	Y2 - COOLING HIGH SPEED
H1 - HEATING BULB	G - INDOOR BLOWER RELAY
RH & RC - TO INDOOR TRANSFORMER	C - "COM." INDICATING LIGHTS

ROOM THERMOSTAT

(Fig. 13-54)

When the compressor is in low speed the contactor switches position themselves to put two windings on each power leg in series. This makes the compressor run on low speed. When the contactors reposition themselves for high speed the windings are then parallel with one another.

CHECKING TWO SPEED COMPRESSOR WINDINGS — Turn the power off to the unit and remove all wiring, being careful to mark the wires so you will know where they go back. Take your ohmmeter and set it on R X 10. You will check the windings for grounds and open windings as shown in **Figure 13-52** and **Figure 13-53**.

A SIMPLIFIED CONTROL CIRCUIT — Most thermostats controlling two stage compressors will be two stage (has two separate contacts) and single stage heat (only one contact for heat). These thermostats will usually have lights to indicate whether they are in low speed or high speed. A green light usually indicates low speed and an amber light, high speed. The thermostat is equipped with a temperature setting dial, system selector switch (off - heat - auto - cool) and fan control switch (auto - on), see **Figure 13-54**.

Figure 13-54 shows an internal schematic of a thermostat. Notice the terminals RH and RC are the power terminals. If only one transformer is used the RH and RC terminals will be jumpered, where one transformer will feed power to both heat and cool. Now see Y1 and Y2. A Y terminal is usually coded for cooling. Y1 is for one compressor speed and Y2 for the other speed. G is for the indoor fan, W1 for heat and C in this case is a common wire from the transformer for the high and low speed lights. C1 and C2 are the contacts that complete a circuit to Y1 and Y2. H1 is the heating contact that completes a circuit to W1.

Let's take a very simple diagram now using the thermostat in **Figure13-54**. Understand now the wiring diagram in **Figure 13-55** theoretically is wrong. There will be many more control devices needed to make the two speed compressor work properly as one will see later on. This diagram is just to give you a basic understanding how the thermostat controls the two speeds.

When looking at **Figure 13-55** you see the high and low speed which is in one compressor shell, not two compressors. The two speed only has one compressor, but two windings, one for low speed and the other for high speed. Note the transformer which

234

supplies 24 volts to the control circuit and the contactors which close to initiate the speeds. Power from the transformer is fed to RC on the thermostat through the mercury bulb C1 and C2, which feeds Y2 and Y1 out to the contactors. A circuit is completed back to the common side of the transformer. A 24 volt circuit has been completed to the contactors and both the high and low speeds have been energized. WRONG You can not put the high and low speeds together trying to operate at the same time. The only purpose of **Figure 13-55** is to give you a picture of how circuits are formed. Now let's look at some real circuits set up as they should be for two speed operation. Let's follow a diagram showing a high speed circuitry, but first take a look at **Figure 13-56** and locate the controls as they would look in the control panel of the unit.

We will now identify the controls one by one.

PROTECTION MODULE — The protection module connects to sensors that are embedded in the windings of the compressor and connect to S1 and S2 terminals. P1 and P2 terminals are the 24 volt terminals supplied by the transformer. The K1 and K2 terminals are the control circuit for the compressor. If the winding sensors detect overheating windings, the module breaks power to the compressor control circuit.

COMPRESSOR INTERLOCK — The compressor interlock is a timer that prevents compressor short cycling. When 24 volts is applied to R1 and R2 terminals a timing period of approximately 90 seconds begins and will close and start the compressor.

CONTROL RELAYS K1 AND K2 — Used for switching to high or low speed. Equipped with 24 volt control coils, controlled by the thermostat.

TRANSFORMER — 208/240 volts on primary/24 volt secondary - 70 VA for stepping down voltage for control circuit.

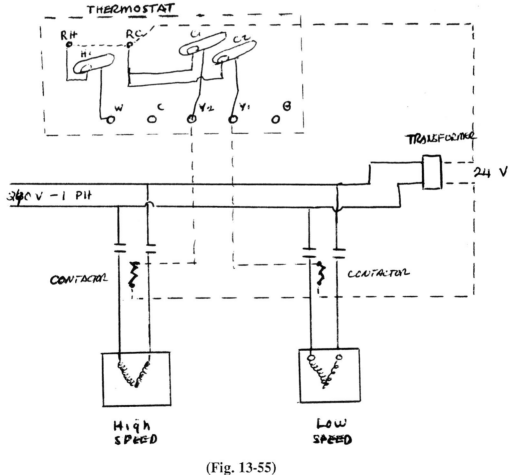

(Fig. 13-55)

CONTROL BOX

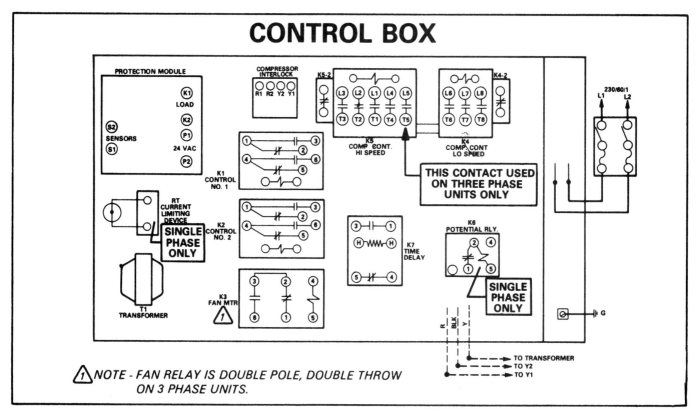

NOTE - FAN RELAY IS DOUBLE POLE, DOUBLE THROW ON 3 PHASE UNITS.

Courtesy of Lennox Industries, Inc.   (Fig. 13-56)

TIME DELAY K7 — Provides time delay between compressor speeds to allow complete stop, before changing to another speed. There is a 20 to 40 second time delay "on" time and 30 to 50 second "off" time.

COMPRESSOR CONTACTOR (K4 LOW SPEED AND K5 HIGH SPEED) — Two contactors are used here in combination with each having its own magnetic coil to pull the contacts closed. The K5 contactor has five poles and the K4 contactor three poles. There are two auxiliary contacts, one in the K4 and one on the K5. The auxiliary contacts are used to interlock the contact coils preventing simultaneous operation. K4 and K5 contactors should never be on at the same time. If they are put together it could result in a severe short, cause damage to the protective breakers or fuse circuit and possibly cause damage to the compressor windings.

THE CURRENT LIMITING DEVICE — This device is used to protect against current surges created when the potential relay contacts close and discharges the run capacitor. When the compressor stops running, the potential relay contacts close immediately. The RT

current limiting device absorbs the current surge and discharges the run capacitor(s). This prevents the relay contacts from welding.

K3 FAN MOTOR RELAY — This relay serves two purposes. One is to close the normally open contacts on demand of cooling to start the condenser fan motor and the other is to energize the crankcase heater when the relay drops out after the compressor stops.

THE POTENTIAL RELAY — This is the starting relay which energizes the start winding at start and drops it out when the compressor reaches approximately 75% of its speed. The start capacitor is dropped out of the circuit as well. The one starting relay is used for both high and low speeds. Now that you have become familiar with the control devices, let's go through the circuitry showing the high and low speeds. Start with **Figure 13-57** showing the system in low speed.

SINGLE PHASE LOW SPEED 24 VOLT CONTROL CIRCUIT (**FIGURE 13-57**) — The first stage of the thermostat Y1 energizes K1 relay coil No. 1

236

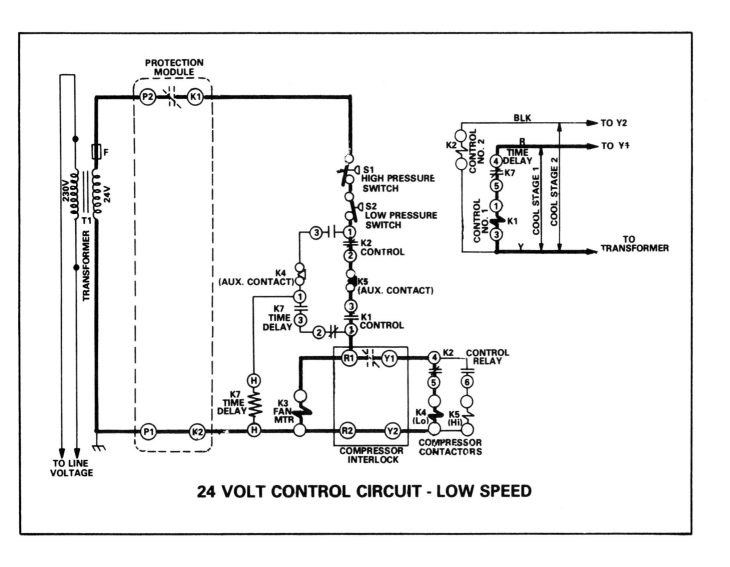

24 VOLT CONTROL CIRCUIT - LOW SPEED

Courtesy of Lennox Industries, Inc. (Fig. 13-57)

through the K7 time delay relay contacts. When K1 relay is energized, contacts are closed at K1 control. A circuit is now completed through the protection module P2 to K1, through the S1 switch, the S2 switch, through the normally closed K2 relay, through the K5 auxiliary contact, and the now closed K1 control relay. K3 fan motor relay coil is energized and the circuit continues through contact R1 to Y1. The K2 relay is de-energized allowing a circuit to travel to the K4 contactor coil. The low speed is energized. (The compressor interlock has a time delay to prevent compressor short cycling.)

SINGLE PHASE HIGH SPEED 24 VOLT CONTROL CIRCUIT (FIGURE 13-58)

The second stage of the thermostat Y2 closes energizing the K2 control coil No. 2, which opens contacts on the K2 control relay between terminals 1 and 2, and closes between 1 and 3. A circuit is completed from P2 to K1 on the protection module, through S1 and S2 pressure control, through the 1 and 3 terminals of the K2 control relay, through the K4 auxiliary contacts and to the K7 time delay relay. Now at this time notice the K7 time delay has two contacts, one between terminals 1 and 3 on the K7 time delay, and between

237

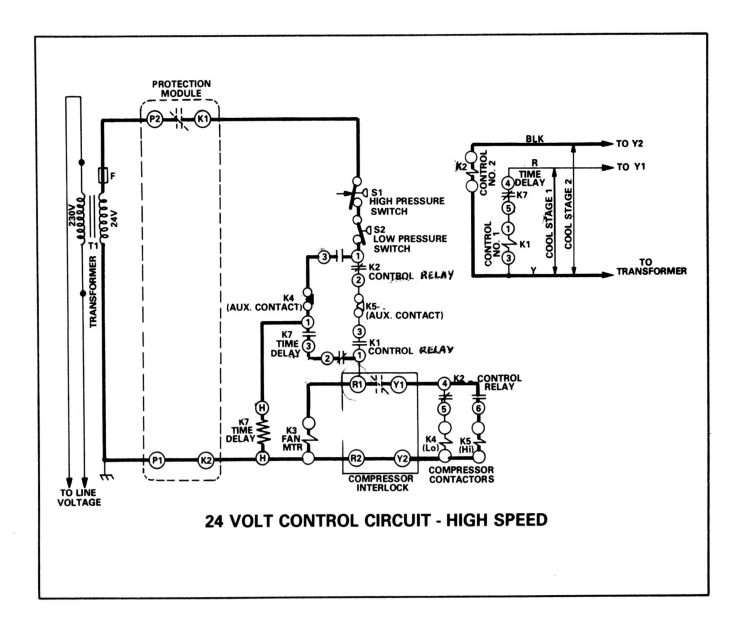

24 VOLT CONTROL CIRCUIT - HIGH SPEED

Courtesy of Lennox Industries, Inc.   (Fig. 13-58)

terminals 4 and 5. The K7 timer allows a time delay between the high and low stages. Before the high speed can be brought in, the low speed must be dropped out. The K7 timer performs that function. Now that the high speed is called for, the K7 timer delay coil is energized. After a period of time delay, the K7 time delay contacts to the K1 coil will open and drop out the low speed and the K7 delay contacts between 1 and 3 will close, allowing a circuit through the contacts in the K1 relay, through R1 to Y1, through the control relay contacts to the K5 contactor coil. The compressor is in high speed. There is always a delay of time between high and low speed on the K7 timer. On first stage demand there is no delay.

You have been given a limited review of how a two speed Lennox system operates. You have not been shown every aspect of all Lennox systems nor have we acquainted you with all other two speed systems. We can agree however, that you have been given a good general knowledge of the two speed. With your ability to understand what we have here will cause you to better understand other situations on other types of equipment.

■

238

ELECTRICAL TROUBLE SHOOTING GUIDE

SITUATION	CAUSE	CHECK	REMEDY
A. Compressor will not start (no hum)	1. No voltage	1. Check at terminal 21 and 23 at contactor. If no voltage check for blown fuses or tripped breaker 2. Voltage at terminal 21 & 23 but not at terminal 11 and 13 of contactor	1. Reset breaker or replace blown fuses 2. Bad contactor. Check voltage (24 volts) at terminal A & B. If voltage is there but contactor does not pull in, coil is bad in contactor, replace coil or contactor.
	2. Wire burnt off	1. Check for loose wire or burnt off wire at C terminal on compressor	1. Repair wiring
	3. Internal or external overload open	1. If overload is external check overload by ohming the overload with the unit power off. 2. If overload is internal check power at C & S and C & R.	1. If infinite ohms, overload is bad 2. If there is power and compressor is at ambient temperature, replace compressor. If compressor is hot, find reason for compressor overloading.
	4. Contactor not pulling in	1. Check for voltage at A & B on the contactor coil C. 2. If there is no voltage to contactor coil, trouble is somewhere else in circuit.	1. Replace contactor coil or contactor 2. Continue to check circuit
	5. Bad thermostat	1. With volt meter check across terminals Y & C at TB2. No voltage means open circuit in stat or loose wire between TB2 and stat.	1. Replace thermostat
B. Compressor will not start but hums	1. Low voltage	1. Check voltage. It must be within 10% ± of rated voltage.	1. Correct problem. If power is in distribution call power company.

ELECTRICAL TROUBLE SHOOTING GUIDE cont.

SITUATION	CAUSE	CHECK	REMEDY
B. cont.	2. Bad start assist	2. Use a remote capacitor with leads and jump across start and run terminals of the compressor. If compressor starts, start assist is bad.	2. Either replace start assist or install start relay and capacitor.
	3. Wire loose on start terminal	3. Check for loose wire	3. Make repairs
	4. Compressor stuck	4. Use a direct 240 volt circuit to common and run terminals on compressor. At the same time voltage is applied to common and run, touch start capacitor to start and run only temporarily. If compressor does not start it is stuck.	4. Replace compressor
C. Outdoor fan motor will not run	1. No voltage	1. Using diagram No. 13-4, check 11 and C.	1. If there is voltage, move on to next check.
		2. Check volts at 9 on the DFR & C. If there is voltage, move on to No. 3 check.	2. If no voltage, the relay is bad if the unit is out of defrost.
		3. Check FT-1 and FT-2. You should have voltage to 2 on FT-2 and C.	3. If the motor is not trying to start and cool to touch, replace motor. If hot, go to cause 2.
	2. Bad capacitor	4. Remove the wire from the (cap) on terminal C & F. Check with a capacitor checker or use an ohmmeter with the meter set at RX10. When the leads are touched to C&F there will be a rapid movement of the needle to the right, then a slow drop back. Capacitor is good.	4. Replace motor
		5. If capacitor shows bad	5. Replace capacitor if motor still does not start, replace motor.

ELECTRICAL TROUBLE SHOOTING GUIDE con't.

SITUATION	CAUSE	CHECK	REMEDY
D. Outdoor fan motor runs in high speed	1. FT-2 contacts stuck closed. If outdoor ambient is below 95°F contact should not be made across terminals 1 and 2 but across 1 and 3. 2. During defrost, outdoor fan runs all the time.	1. Check voltage across terminal 2 of FT-2 and C. If there is voltage the control is stuck in high position. 2. Check terminal 8 and C on the DFR with unit in operation. There should be no voltage.	1. Replace FT-2 control 2. If there is voltage, replace DFR
E. Outdoor fan motor runs in low speed constantly	1. FT-2 contacts stuck across 1 & 3. 2. Bad capacitor	1. Check voltage on terminal 3 of FT-2 and C. If there is voltage and the temperature is above 95°F the control is not going into high speed condition. 2. First establish that fan is in low speed. If you check at terminal 2 on FT-2 & C and get voltage the fan is in high speed but turning slow because of possibly a bad capacitor.	1. Replace FT-2 2. Check capacitor and replace if needed. Temporarily substitute a capacitor to verify your findings.
F. Indoor fan motor will not come on (outdoor unit operating)	1. No voltage	1. Check at point E & F in the heater unit. If no voltage check A & G. If voltage is there and not at E & F, see remedy 1. 2. If there is voltage at E & F and the unit is in heat pump mode, go to the IFR. Check voltage at terminal 3 of IFR and the common (C) hooked to LS-1. If there is voltage at 1 & C and not at 3 and C, then check further.	1. Replace fuses and determine why the fuses blew 2. The contacts being open does not mean bad relay. You must check what controls these contacts.

ELECTRICAL TROUBLE SHOOTING GUIDE cont.

SITUATION	CAUSE	CHECK	REMEDY
F. cont.	1. cont.	3. Check magnetic coil in relay FR in 24 volt circuit. Check at TB2 and C or 6. If there is voltage, see remedy 3. If not, go to cause 2.	3. Replace fan relay
	2. Bad thermostat	1. Place one lead of your volt meter to R on TB2 & C. You should get 24 volts directly off of the transformer. 2. Place one lead of the volt meter to "C" and G on TB2. If there is no voltage, see remedy 2. 3. Support your findings by jumpering R & G at the thermostat. If FR closes and fan comes on then without a doubt the thermostat is bad.	3. Move on to check 2 2. Replace thermostat 3. Replace thermostat
G. Indoor fan motor runs continuously	1. Fan relay stuck	1. Check voltage on terminal G & C on TB2. If no voltage the fan relay should be open. Further verify by checking the IFR contact at 3 & C. If there is no voltage then sequencer 1 could be stuck. Make sure that sequencer 1 is not calling by checking W2 & 7 on terminal board 2. If there is no voltage at this point, see remedy 1B.	1a. Replace relay (IFR) 1b. Replace sequencer 1
	2. Thermostat set in "ON" position	2. Make visual check and see	2. Turn fan switch to auto position
H. System will not go into reverse cycle	1. Stuck reversing valve	1. Remember that the reversing valve coil is de-energized when the system is in heating cycle. If there is no voltage to the reversing valve coil, then there is a problem in the valve itself.	1. Check system charge before replacing reversing valve. There may not be enough pressure differential across valve to make it operate. Also bad valves in the compressor can cause the same problem.

ELECTRICAL TROUBLE SHOOTING GUIDE cont.

SITUATION	CAUSE	CHECK	REMEDY
I. System will not defrost	1. Bad defrost timer	1. Check sight window on timer. If you see wheel not turning and there is voltage to 4 & 1 on the clock motor, the timer is bad.	1. Replace timer
	2. Bad defrost thermostat	2. Use volt meter at 5 on DFT and 13C. If there is voltage with the system in defrost this proves the DT contacts are made. If you do not get power on the outside of the DFT and 13C and the temperature is below 24°F at the DFT.	2. Replace the defrost thermostat
	3. Bad defrost relay	3. If there is voltage to the DFR and you do not have voltage to 9 & 13, see remedy 3.	3. Replace defrost relay
	4. Outdoor fan runs all the time	4. Check defrost relay (DR)	4. Replace if contact is stuck between terminals 7 & 8 of relay
J. Auxiliary heat will not come on during normal operation	1. Bad thermostat	1. Check W2 & C at TB-1. If the contact is made in the thermostat W2 & C should be hot (24 volt)	1. No power at W2 & C means replace thermostat
	2. Blown fuses	2. Check at point A & B. There should be power. If so and there is not voltage at point C & D or E & F, fuses are blown.	2. Replace fuses
	3. Bad sequencer	3. Bad sequencer or sequencers. First see if there is voltage to the sequencer coils. If there is but the contacts of the sequencers are not closing after a reasonable time, the sequencers are bad.	3. Replace sequencers
	4. Open limit switch	4. Cut off power and ohm across limit. No ohm reading means open limit.	4. Replace limit
	5. Blown heater fuse FL2-3-4	5. Ohm fuse. No continuity means fuse is blown.	5. Replace fuse

ELECTRICAL TROUBLE SHOOTING GUIDE cont.

SITUATION	CAUSE	CHECK	REMEDY
J. cont.	6. Burnt out element	6. Check for voltage at heater. If there is voltage up to the terminal of the heater and the heater is not drawing amperage, the element is bad.	6. Replace element
K. Nothing will come on when thermostat is switched to different positions.	1. No power to indoor unit 2. Bad transformer	1. Check voltage at L-1 & L-2 at indoor unit. 2. Check for line voltage at primary side of transformer. It should be 240 V. If there is no 24 volts out of secondary side of transformer, it is burnt out. *Note: Be sure there is not a built in fuse in transformer that can be replaced.*	1. Check for tripped breaker or blown fuse at distribution panel 2. Replace transformer

CHAPTER 14

ICE MACHINES

Ice machines are found in nearly every restaurant, motel and quick stop food store around the country. There are several manufacturers that make these machines. All have the same purpose and that is to make ice. However, the type of ice machine selected will depend upon what the ice is used for. There are many uses for ice. Some of these uses are to make ice for salad bars, drinks and packing down foods to name a few. So keeping this in mind the manufacturer has to satisfy the public demand in making machines that will produce ice in all configurations and quantities. Keep in mind that even though there are many configurations of ice, there is one common way to form the cubes and that is by refrigeration. Once the ice has formed into whatever configuration it is supposed to be in, it must be released, or harvested into a storage bin. You must know that it takes three things to make ice with a machine and that is (1) *water,* (2) *refrigeration,* and (3) *a chamber to freeze the ice in.* This chapter is to instruct you in how ice making is done with a machine automatically. You will first learn something about water, installing of the equipment and maintenance. From there we will move into how the refrigeration circuit works and something about electrical. Then a good session in trouble shooting the refrigeration cycle and the electrical circuit. By carefully studying this chapter you will be able to understand the ice machine and be capable of servicing it.

WATER

Water is a very important part of making ice. The condition of the water can affect the operation of an ice machine. A poor condition of water can cause nuisance break downs. Water carries with it impurities, the most common being calcium scale. As water freezes, mineral particles freeze out or remain in the sump water increasing in concentration. When the water can no longer hold them, they form a deposit on surfaces where they are not desired. Some troubles associated with scale is that it insulates. Another problem is that the scale can form on the water spray area, in the water sump and on the freezing area itself. In large deposits, the scale can flake off enough to restrict water passages. An insulating problem is caused when scale gets in the freezing plate between refrigeration and the water. Heat transfer is then cut off between the refrigeration and water resulting in lack of freezing. Periodic maintenance and cleaning can help control these problems, but it is not the cure. The best thing to eliminate the problem is to get with the owner and try to sell them on a good filtering system. Another problem with water is algae and slime. This problem is caused by airborne particles, mainly by smoke. In bars, night clubs and pubs, you will find an abundant growth of algae and slime. It can form so heavily on the machine that it actually gets thick and rubber-like. This problem cannot be removed with a filter. It did not come in with the water. Chlorine can reduce or control the growth of algae and slime, when added in proper quantities. However too much chlorine can be a problem by causing a bad taste in the water and damage to the metal of the freezing chamber. In any case, always consult with the manufacturer about the control of such things. They are usually always happy to assist you in keeping their product functioning at full capacity. It is very important to keep the ice cubes clear. This is accomplished by agitating the water. The water pump in moving the water causes an agitation, therefore usually always producing clear ice.

TYPES OF ICE

Cubes—Refer to **Figure 14-1** and you can see the many forms of ice after it is frozen and harvested. There is a *cube for every cause* you might say. The form of the evaporator plate is going to decide the configuration of these cubes. The cube will normally

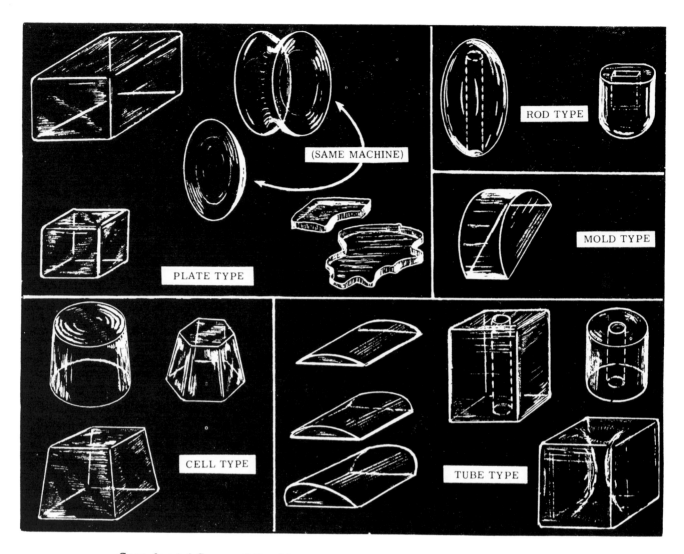

Samples of Some of the Many Types of Ice Cubes Produced Today.
Titles Indicate Type of Machine in Which Each is Produced.

(Fig. 14-1)

be formed in a plate type evaporator (freezing plate or chamber) and will usually be square or rectangular. (**Fig. 14-2**).

Flakes — To make flakes of ice there will generally be a tube type evaporator. Around the outside wall of the tube will be refrigeration tubes. Inside the tube there is an auger (**Fig. 14-3a**). The auger turns inside the tube and scrapes ice from the tube, pushes it upward and out the top of the tube in **Figure 14-3c**. From there it goes into the storage bin. Notice at the bottom of the tube, water is allowed to enter by gravity into the freezing tube. The auger is continually moving the water upward when removing the ice. By gravity the water level in the tube will equal the water level in the water reservoir. Note that in **Figure 14-3b** a gear

box, motor driven (motor not shown), drives the auger at a low speed.

HOW CUBES ARE FORMED

To form cubes you need three things: (1) *water,* (2) *refrigeration,* (3) and an *evaporator* formed to freeze cubes. Take the cube shaped evaporator, spray water up into the evaporator and start the refrigeration machine. In minutes you will have solid frozen cubes. **Figure 14-4** illustrates this action. Notice the pump pushing water up into a spray bar which turns by the velocity of the water, actually jet propelling it. Water sprays up into the freezing chamber where cubes are formed. Water drops down and drains back into the water sump. Notice the 1/4" make up line for water. As

246

(Fig. 14-2)

the water is frozen the level will go down. The float will open and maintain a water level in the sump. Refrigerant tubes are in contact with the cube chamber causing the plate to get cold enough to freeze and form the cubes.

At a determined time when the cubes have formed, the pump is automatically cut off and the cube chambers are heated. The cubes fall out and into a storage bin ready for use.

HOW FLAKES ARE FORMED

Flakes are usually made by a cylinder or tube type evaporator. **(Fig. 14-4a)**. There is a drive motor coupled to a gear box. The gear box has a gear ratio that turns the auger much slower than what the drive motor turns.

The evaporator tubes get cold (below freezing) and the water starts freezing on the inner wall of the cylinder. The thickness of the flake ice will be determined by the clearance between the auger blades and the cylinder wall. Before ice can be made however, there must be water and it comes through the water inlet tube into the freezing chamber. The water freezes to ice, the auger scrapes it off of the walls of the cylinder, pushes the ice upward because of the turn of the auger blades and pushes the ice out the ice drop.

THE REFRIGERATION CYCLE

In previous chapters you have been introduced to the refrigeration cycle many times. In studying the cycle for ice machines you will find the travel of refrigerant virtually the same as any other cycle. The components will even have the same functions, however the thing you need to see most about is that it is hooked a little different and the components are a little more unusual. You will be shown the single evaporator and two evaporator hook-up, but before we do that let's look at the types of refrigerants used in an ice machine.

REFRIGERANTS

All refrigerants used today are safe, harmless and odorless to the serviceman. The only problem that we have is that these refrigerants are highly carbonated and fluorinated. Many scientists today think that refrigerants such as R-11 and R-12 are attacking the earth's ozone layer. Dupont Company and others are in the process of developing alternate refrigerants. Until R-11 and R-12 are no longer used it will be your responsibility to be as conservative as possible before letting refrigerants into the atmosphere. The refrigerants you will find to be most common in ice machine

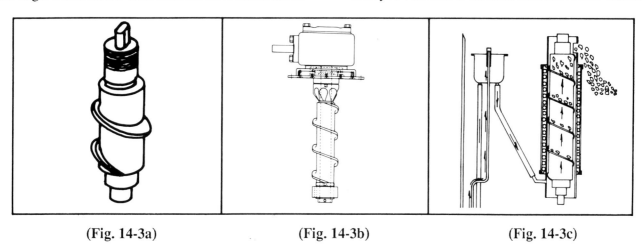

| (Fig. 14-3a) | (Fig. 14-3b) | (Fig. 14-3c) |

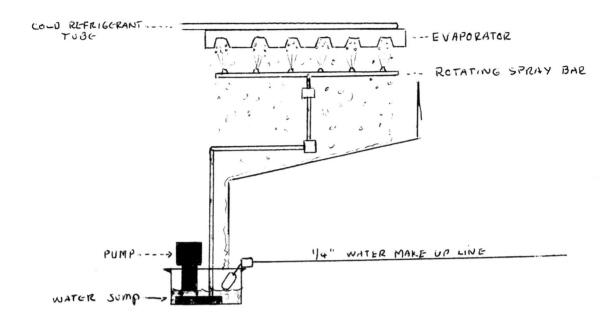

COLD REFRIGERANT TUBE
EVAPORATOR
ROTATING SPRAY BAR
PUMP
1/4" WATER MAKE UP LINE
WATER SUMP

(Fig. 14-4)

systems will be R-12, R-22 and R-502. The selection of refrigerants is made by the manufacturer. All of these have different operating pressures and evaporator temperatures. The thing for you to know is, what kind of refrigerant each manufacturer's machine uses and how much it holds. Do not interchange refrigerants. If a system uses R-12 then R-22 and R-502 cannot be substituted.

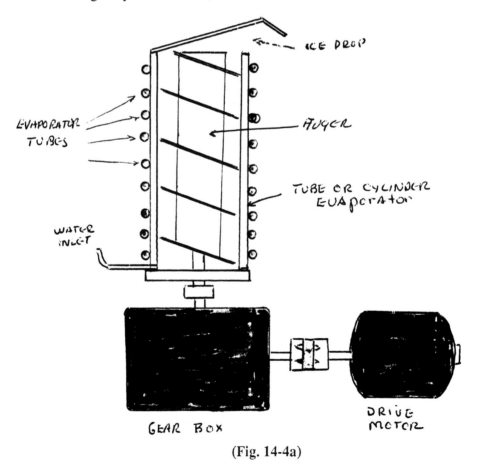

ICE DROP
EVAPORATOR TUBES
AUGER
TUBE OR CYLINDER EVAPORATOR
WATER INLET
GEAR BOX
DRIVE MOTOR

(Fig. 14-4a)

REFRIGERANT FLOW PATTERN
(USING CORE TYPE SERVICE VALVES)

(Fig. 14-5)

THE SINGLE EVAPORATOR

The single evaporator as shown in **Figure 14-5** has one freezing section. There will only be one evaporator making ice. When more ice is required, another evaporator is added to increase ice capacity as in **Figure 14-6**.

Let's review the refrigeration cycle a little closer as we follow in **Figure 14-5**. Find the compressor. The compressor is to compress low temperature, low pressure gases from the evaporator to a high pressure high temperature vapor so the refrigerant will condense in the condenser. As the vapor enters the condenser it is cooled by ambient air being forced across the condenser by means of a fan (not shown). By the time the refrigerant gets to the bottom of the condenser it is a high pressure liquid. From there the liquid drops to the bottom of the receiver. A tube toward the bottom of the receiver allows liquid to be pushed up through the drier

and over to the expansion valve. When the liquid passes through the expansion valve a tremendous drop in pressure and temperature takes place. With the refrigerant having a low boiling point, it will now absorb heat from the water circulating around the evaporator and will continue to do so until the water freezes. As the refrigerant changes state from a liquid to a vapor the evaporator continues to get cold as long as the compressor is in operation. The vapor then comes back to the compressor where it is compressed and the cycle starts over. Please note the hot gas solenoid valve. This valve opens during the harvest cycle and lets hot gas from the condenser enter into the evaporator. The water pump circulating water across the evaporator cuts off at the same time the solenoid opens. The hot gas then melts the cubes, slab or whatever configuration of ice you have loose and the ice deposits itself in the storage bin. The receiver you

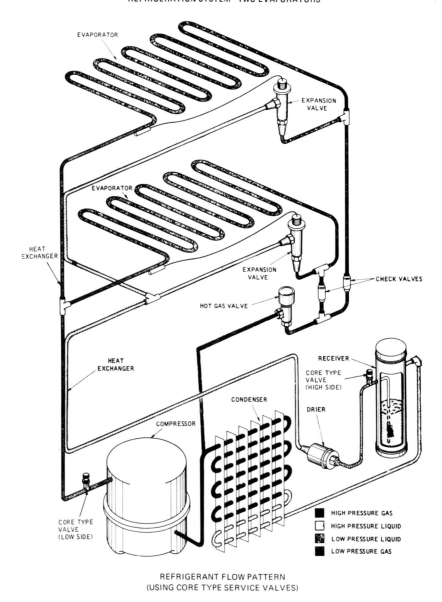

REFRIGERANT FLOW PATTERN
(USING CORE TYPE SERVICE VALVES)

(Fig. 14-6)

see is to store liquid refrigerant and a place to allow the varying levels of liquid with a system using an expansion valve. The receiver is also there when a system pump-down is needed. **Figure 14-6** shows a two evaporator hook-up. The only difference is that the added evaporator has its own expansion valve. The rest of the cycle is the same as **Figure 14-5**.

TYPES OF METERING DEVICES

To control the flow of refrigerant into the evaporator there has to be a metering device of some kind. There are several kinds, but the two most common are the Thermostatic Expansion Valve and the Capillary Tube with the expansion valve being the most common of the two. First let's cover the Thermostatic Expansion Valve (**Fig. 14-7**).

Notice the high pressure liquid coming into the expansion valve. There is a small orifice in the valve that causes a drastic reduction in pressure and temperature. Since the refrigerant is boiling at approximately 10°F, it will absorb heat from the water at the evaporator and continue to boil. As the refrigerant boils away, it changes into a cool low pressure vapor. The vapor then leaves the evaporator and goes to the

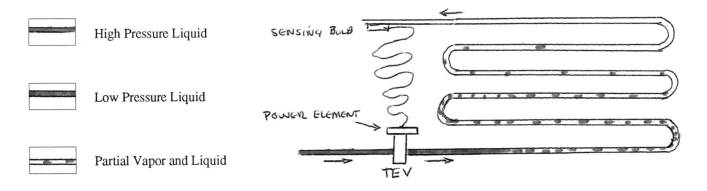

High Pressure Liquid

Low Pressure Liquid

Partial Vapor and Liquid

SENSING BULB

POWER ELEMENT

TEV

(Fig. 14-7)

compressor. Now let's see how the valve controls the flow of refrigerant into the evaporator. When any liquid has boiled and changed to a vapor, and you continue to add heat to the vapor, it is called *superheat*. Coming out of the evaporator there is vapor and any further heating of this vapor is called *superheat*. The valve is set to control at approximately 4°F superheat. The expansion valve will control the superheat by use of the power element, the sensing bulb and the connecting line between the power element and sensing bulb. These three parts are sealed together within themselves and hold their own charge of refrigerant. See **Figure 14-8**.

When there is a warming of the vapor at the point of the sensing bulb the bulb will sense it and cause a slight rise in pressure at the power element. The power element will expand and by means of some small rods will push the seat open, opposing the spring pressure exerted against the seat. When the sensing bulb senses a drop in temperature, the power element will retract and the spring will push the seat back toward the seated position, therefore cutting down on the flow of refrigerant into the evaporator. The valve will allow a 3 to 4°F superheat to take place in the evaporator. A wide superheat, say 25°F would be too high, therefore not cooling the evaporator enough. A low superheat, such as 1°F would overfeed and possibly cause damage to the compressor by letting some of the unboiled liquid back to the compressor. Keep this in mind now, the expansion valve does not snap open and shut to control the flow of refrigerant. It is a slow gradual process, very slow to respond. If you ever need to check your super heat, or adjust the superheat, here is what you do. See **Figure 14-9**.

With the unit in operation, put a gauge on the low pressure side of the compressor and strap a thermometer to the outlet of the evaporator. Take a pressure reading at the suction side of the compressor. Convert the pressure to temperature (use a pressure-temperature relationship chart). Next take a temperature reading at the outlet of the evaporator. In the case of **Figure 14-9**, the outlet of the evaporator is 34°F and at the compressor the converted temperature is 30°F. This makes a 4°F superheat. To raise the superheat, close the valve off. To lower the superheat, open the valve. The superheat adjustment is at the bottom of the valve with a cover nut.

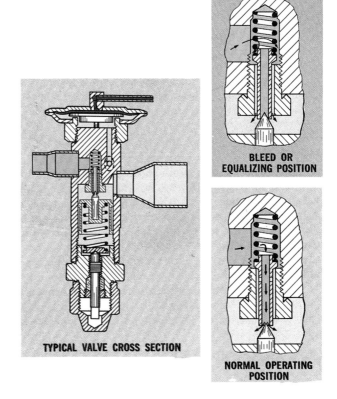

TYPICAL VALVE CROSS SECTION

BLEED OR EQUALIZING POSITION

NORMAL OPERATING POSITION

(Fig. 14-8)

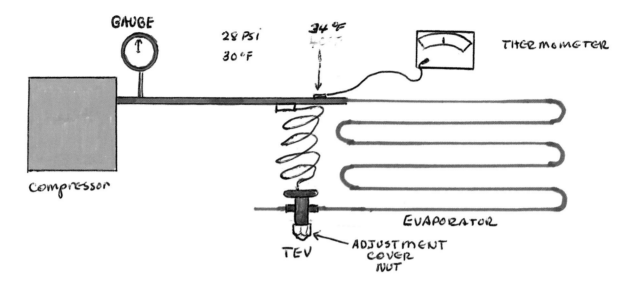

(Fig. 14-9)

LEAK CHECKING

If a system has lost its charge of refrigerant, there has to be a leak. There are two common ways of leak checking. One is to check with a soapy solution, the other is with a leak detector. The two types of leak detectors are the halide torch and the electronic leak detector. The halide torch is an old favorite (**Fig. 14-10**) and will get you by in most cases. But if you are wanting to find small leaks it would be best to use the electronic leak detector (**Fig. 14-11**).

In **Figure 14-10** the halide torch is lit. In the head of the leak detector there is an orifice and a reactor plate. A small flame burns across the reactor plate. When a trace of Freon is picked up through the sniffer hose it is sucked into the reactor plate and when burned across the reactor plate, the flame will turn green. Green is the color for a smaller leak and bright blue for a large leak. There are several makes of electronic leak detectors and they are all good. When these leak detectors find leaks they either cause a blinking light or an audible sound or both.

CHARGING

When does a system need charging, how do we know it needs charging and how much does it take to charge it. These three questions will be covered now. When does a system need charging? The answer is obvious, when the system is low on refrigerant. If the system is completely out of refrigerant there must be a very large leak. Pressurize the system with refrigerant. Leak check and find the leak, then repair. If the system

is only low on refrigerant, this can usually be determined by a lower than normal low side pressure and an evaporator that has one portion not icing. In this case you may add to the system by charging into the low side with vapor very slowly until the unfrozen portion starts icing, providing you have found the leak and repaired it without losing the charge. A lower than

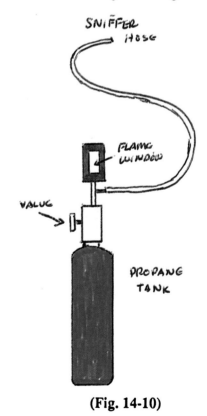

(Fig. 14-10)

252

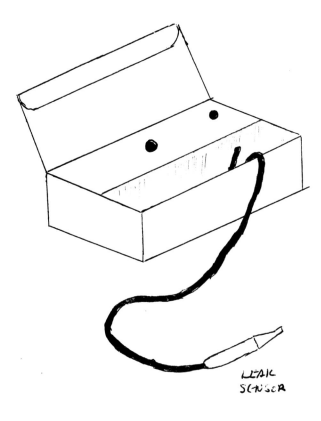

(Fig. 14-11)

normal low side pressure and a partially iced evaporator answers the question "How do I know it needs to be charged." "How much does it take to charge it?" It is recommended that you look on the nameplate of the unit. It will tell you what type of refrigerant to use and the amount of refrigerant to put in. Use a dial-a-charge and weigh in the amount of refrigerant you need, then dump it into the high side of the system. If the pressure rises in the unit to a point where you cannot get all the refrigerant out of the dial-a-charge, this would be a good time to have a built-in heater on the charging device. If you don't, then attach the charging hose to the low side of the system and start the compressor. Slowly let the liquid out of the dial-a-charge, say at a rate of 4 ounces per minute until it is emptied.

SYSTEM PRESSURES

With your serviceman's gauges mounted, using **Figure 14-12** as a guide and the system in operation, operating pressure will be reviewed. Find the manifold set (your serviceman's gauges). The gauge to the right is hooked to the high pressure side of the system and will read condenser pressure. The gauge to the left is hooked to the low side, reading evaporator pressure. You see also a vacuum pump and a dial-a-charge

charging cylinder. All of these could be hooked up at the same time and valved in the way they are to switch from one to the other. However, let's just talk about pressures now. Normally you would only have your gauges hooked up without the dial-a-charge and vacuum pump. Before you can know if a pressure reading is correct, you of course will have to know the type of refrigerant in the system. Look at the name plate to determine this. The most common types of refrigerants are R-502, R-22 and R-12. Let's take R-502 first, used in a self-contained ice maker (**Fig. 14-13**). Notice the ambient temperature. This will actually be the air being sucked across the condenser by the condenser fan. Then notice the freeze cycle and you have the head pressure, the suction pressure and approximate freezing time under "cycle time minutes." Then you have the harvest cycle giving you your pressure readings and cycle times in minutes. Cycle time minutes here mean the time it takes to melt the ice loose and for it to drop. **Figure 14-13** shows readings for an air cooled condenser. **Figure 14-14** shows readings for a remote air cooled condenser (remote meaning away from the unit) and **Figure 14-15** shows readings for a water cooled condenser.

Study these pressures for a minute. Notice that the higher the head pressure and ambient temperature goes, the longer the cycle time in minutes. Production of the ice machine is rated according to the degree of the ambient. Water cooled equipment has an advantage over air cooled units because you can normally get a cooler supply of water than air, more especially if you are in southern climates. The only problem with water over the condenser is that it usually has to be wasted and it causes scale problems in the condenser if the water is not treated. If the ice machine is to be located in an air conditioned store and the system is self-contained, selecting air cooled would probably be best. Air cooled is usually best all around probably because of the unavailability of water and the waste of the cooling water. When using refrigerant 22 your operating pressures will be lower than R-502. Rather than swamp you with charts and illustrations, we will say the system has formed ice and is about to harvest. Your suction pressure reading should read approximately 12 psi and the head pressure 225 psi with a 70°F ambient. When using refrigerant 12 your suction pressure would read approximately 8 psi and your head pressure 110 psi under the same conditions as R-22. Remember this, when the system goes into the harvest cycle, here is what you will see on your gauges. The hot

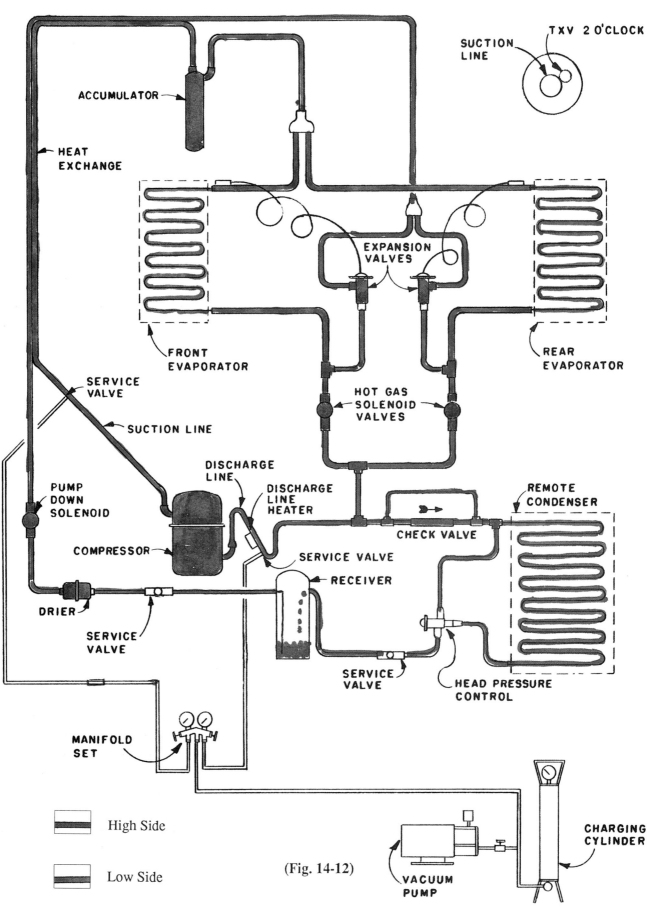

SUCTION
LINE
TXV 2 O'CLOCK

ACCUMULATOR

HEAT
EXCHANGE

EXPANSION
VALVES

FRONT
EVAPORATOR

REAR
EVAPORATOR

SERVICE
VALVE

SUCTION LINE

HOT GAS
SOLENOID
VALVES

PUMP
DOWN
SOLENOID

DISCHARGE
LINE

DISCHARGE
LINE
HEATER

REMOTE
CONDENSER

COMPRESSOR

CHECK VALVE

SERVICE VALVE

RECEIVER

DRIER

SERVICE
VALVE

SERVICE
VALVE

HEAD PRESSURE
CONTROL

MANIFOLD
SET

High Side

Low Side

(Fig. 14-12)

CHARGING
CYLINDER

VACUUM
PUMP

254

Ambient Temp. Degrees F.	Freeze Cycle			Harvest Cycle		
	Head Pressure PSI	Suction Pressure PSI	Cycle Time Minutes	Head Pressure PSI	Suction Pressure PSI	Cycle Time Minutes
50°	175-225	20-36	7-11	120-150	55-80	1.4-2.0
70°	180-220	22-40	8-12	140-170	65-85	1.2-1.8
80°	200-250	24-42	10-13	160-180	70-90	1.1-1.5
90°	240-280	26-44	12-15	170-200	80-100	1.0-1.5
100°	260-300	26-46	14-18	200-220	100-120	0.9-1.3
110°	300-350	28-48	18-24	225-250	120-130	0.8-1.2

*These are approximate characteristics that may vary depending on the operating conditions.

(Fig. 14-13 AIR COOLED)

Ambient Temp. Degrees F.	Freeze Cycle			Harvest Cycle		
	Head Pressure PSI	Suction Pressure PSI	Cycle Time Minutes	Head Pressure PSI	Suction Pressure PSI	Cycle Time Minutes
-20°	170-190	24-38	7-10	150-170	70-90	1.0-1.6
50°	170-190	24-38	8-11	150-170	70-90	1.0-1.6
70°	170-190	24-38	8-12	150-170	70-90	1.0-1.5
80°	180-220	24-38	8-13	160-180	80-100	1.0-1.5
90°	200-240	24-40	9-14	170-190	80-100	0.9-1.4
100°	230-280	26-42	10-15	190-220	90-110	0.8-1.3
110°	260-320	26-42	12-17	220-250	110-130	0.8-1.3
120°	280-340	28-44	14-19	240-270	120-140	0.8-1.3

*These are approximate characteristics that may vary depending on the operating conditions.

(Fig. 14-14 AIR COOLED (REMOTE))

Ambient Temp. Degrees F.	Freeze Cycle			Harvest Cycle		
	Head Pressure PSI	Suction Pressure PSI	Cycle Time Minutes	Head Pressure PSI	Suction Pressure PSI	Cycle Time Minutes
50°	215-225	24-38	8-12	130-160	65-85	1.4-2.0
70°	215-225	26-40	9-13	140-160	70-90	1.2-1.8
80°	215-225	26-42	9-14	150-170	75-95	1.0-1.6
90°	215-225	26-42	10-15	150-170	80-100	0.9-1.4
100°	215-225	26-44	10-16	155-175	80-100	0.8-1.3
110°	215-225	26-44	11-16	160-180	80-100	0.8-1.3

*These are approximate characteristics that may vary depending on the operating conditions.

(Fig. 14-15 WATER COOLED)

gas valve opens, and there is a rise in suction pressure and a drop in head pressure. The suction pressure will rise about 30 to 40 psi and the head pressure will fall about 40 to 50 psi.

HEAD PRESSURE CONTROL

Refer back to **Figure 14-12** and find the head pressure control. This device is to maintain a near constant head pressure during colder days. The control has a sensor on it and when it senses colder temperatures outside, the valve will start closing and by-pass some of the hot gas around the condenser and back to the receiver. The amount of gas that is by-passed depends on how cold it is outside.

In warmer weather the valve positions itself where most of the gas passes through the condenser. Another way to control head pressure is to use a pressure switch on the high pressure side of the system. Wire it so that it will cut off the condenser fan or fans. Set the cut-in and cut-out pressures depending on the kind of refrigerant gas you are using in the machine. The way this switch operates is, when the head pressure drops, the switch will open, stop the condenser fan, and allow the pressure to build back up. When the pressure reaches a set point the fan or fans will come back on and will continue to control like this. This keeps the head pressure from going too low. Low head pressure can cause the following problems:

1. Cause oil to leave the compressor
2. Lets liquid refrigerant cause damage to the compressor
3. Causes expansion valve to feed erratically
4. Goes into harvest cycle before cubes have fully developed

OPERATION OF CUBE EVAPORATORS

Most cube systems operate about the same, but their controls and settings will vary greatly from one to the other. This is one reason you can't go to the library and find a good technical book on the subject today because of such a wide range of conditions. Suppose the system just came out of the harvest cycle. The compressor will be operating, the hot gas valve is closed and the water pump for the evaporator plate is operating. After approximately 10-20 minutes of running time the freezing cycle will be terminated by a temperature sensing device or pressure device and the machine will go into harvest. The compressor still has to operate. The hot gas solenoid valve opens letting hot gas into the evaporator and the water pump cuts off.

The defrost only lasts for a couple of minutes. Then it goes back into the freezing cycle.

SYSTEM TROUBLE SHOOTING

Let's cover some of the most common troubles encountered with an ice machine outside of the water system being filled with impurities that cause refrigeration problems. **Figure 14-16** shows a system using a capillary tube. With the capillary tube, approximately two thirds of liquid refrigerant is in the evaporator and accumulator during the cooling cycle (under stabilized conditions). Only the last pass or two of the condenser contains high pressure liquid. The rest of the condenser is needed for condensation. The temperature at the strainer or drier outlet should be the same as the condenser outlet temperature. Under normal circumstances in room temperatures of 70°F and higher, evaporator temperature at inlet and outlet should be somewhere within 5°F of one another at cut-off.

During the off cycle, "high" and "low" side pressures will tend to equalize. How long it takes for the pressures to equalize depends on the length of cycle, usually about 3 to 5 minutes. During the start cycle for a few seconds there will be a noticeable temperature drop in the suction line near the compressor. There are three primary causes of system failures. They are as follows:

1. Restrictions (partial and complete)
2. Refrigerant charge (under and over)
3. Compressor failure (stuck, defective motor windings and inefficient compressor)

Diagnosing the Restricted System — Restrictions of refrigerant usually occur in the drier, see **Figure 14-17**. This limits the amount of refrigerant into the evaporator. The evaporator will therefore not be fully refrigerated.

Liquid refrigerant will back up in the condenser and start filling the upper passes. The tubes holding this back-up of refrigerant will feel cool compared with the condenser tubes containing high pressure vapor. In the case of a total restriction, the entire condenser will be cool since no work is being done by the compressor. You can determine liquid level in the condenser by adding heat to the passes. If liquid is in the tube it will not get hot when heated. The passes with vapor will. If there is a partial restriction at the strainer or inlet of the capillary tube, you will have a noticeable temperature drop from the inlet to outlet of the strainer. You sense this drop in temperature by touch.

There will be a wide temperature difference be-

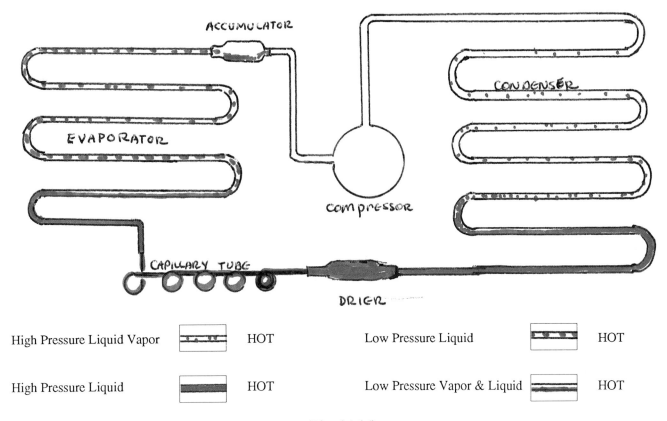

High Pressure Liquid Vapor	HOT	Low Pressure Liquid	HOT
High Pressure Liquid	HOT	Low Pressure Vapor & Liquid	HOT

(Fig. 14-16)

tween the inlet and outlet of the evaporator with a partial restriction. If there is a complete restriction the evaporator will be at ambient temperature. No hissing sound can be heard in the evaporator with a complete restriction. With a partial restriction the system will not equalize as rapidly as if it were not restricted and with a complete restriction the pressures would not equalize at all for several hours. With a partial restriction the compressor amperage will be higher than normal. With a complete restriction the amperage may exceed normal for several minutes and then pull to lower than normal draw and remain constant.

Diagnosing the Under Charged System — If a system is under charged it would act the same way as a partial restriction. A small amount of refrigerant would be in the evaporator, see **Figure 14-18**. You would notice also that a small portion of the evaporator ices just outside of the capillary tube and the rest doesn't. You would hear a low hissing at the inlet of the evaporator. With a severe undercharge, there would be hardly any refrigerant in the cycle. The condenser tubes would all be warm or hot all the way down since only hot vapor is there instead of liquid refrigerant. Again you could heat the tubes. If all of the tubes became warm easily, they would be empty of refriger-

ant. An under charged system would cause the amperage to be lower, the compressor would be hot (if suction cooled) and the suction line temperature would be warmer than normal.

Diagnosing the Over Charged System — With an over charged system there would be liquid refrigerant filling the evaporator, over-running the accumulator and going into the compressor. The suction line would probably be icing all the way back to the compressor, see **Figure 14-19**. The liquid level in the condenser would be higher than normal. If you use the heat test, most of the tubes would remain cool except for the uppermost of the condenser. Amperage draw would be considerably higher than normal and compressor temperature would be higher than normal.

Correcting the Restriction — If the system cannot be pumped down, discharge the system so there will not be any pressure. Heat the joints of the drier and melt the welds loose. Install a new drier, pull a vacuum and recharge. If the capillary tube is restricted, check with your local supply house. They have devices that can unclog the capillary. If you can't get to a supply house, then blow back through the capillary with refrigerant pressure. After opening a system for a long period of time, always replace the drier. Be careful and don't

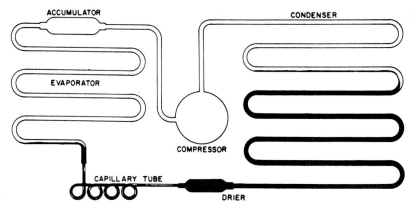

THE RESTRICTED SYSTEM

(Fig. 14-17)

overheat so much when welding, you could plug the capillary tube.

Correcting the Under Charge — If the system is low on refrigerant, then you have a leak. If the leak is found and can be repaired by tightening a nut or packing, you can slowly add to the existing charge. Be careful; if the pressure on the low side went into a vacuum and the leak was on the low side, air and moisture would be sucked into the system. Air is a non condensable gas in the system and will cause high head pressure. Moisture in the smallest amount can freeze in the capillary tube and cause a complete restriction. If you suspect the capillary tube to be restricted with moisture, apply heat at the outlet of the capillary and you will hear the liquid release into the evaporator when the moisture thaws. If the system has possibly run into a vacuum or is completely out of refrigerant, repair the leak, pull a vacuum, replace the drier and recharge.

THE ICE MACHINE WATER CYCLE

The major part of the water cycle is the pump (**Fig. 14-20**). These are low volume pumps, not designed to pump against high head. The more head pressure you put against the pump the smaller the quantity of water delivered. The pump will circulate water up to the freezing plate and return unfrozen water back to the sump where the pump is contained. As the water is frozen into ice, more water is needed in the sump. There will be a float valve in the sump to maintain a water level. **WARNING:** *Do not adjust the float where the water will overflow out of the sump. This will cause a continuation of warm water coming into the sump therefore cutting down on ice capacity.* After each harvest cycle it is necessary to get a changing of water. **Figure 14-21** shows one way this is accomplished.

When the system goes into a harvest cycle the pump cuts off. There is enough water in the tubing on

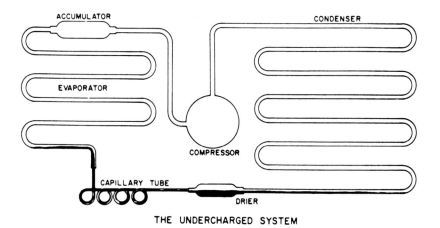

THE UNDERCHARGED SYSTEM

(Fig. 14-18)

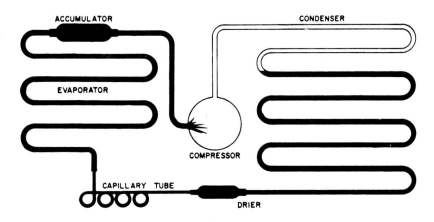

THE OVERCHARGED SYSTEM

(Fig. 14-19)

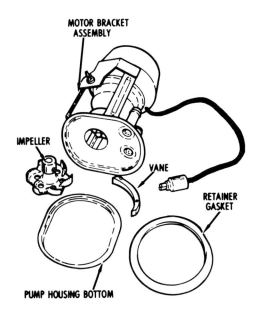

CAUTION: DO NOT OVER-OIL, since the excess oil may drip into the water reservoir and contaminate the water.

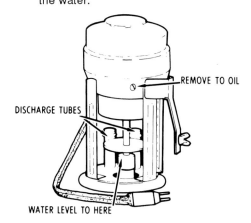

(Fig. 14-20)

the evaporator plates and in the area of distribution to drain down and make the water go over the overflow. Once the water goes over, the siphon causes the water to continue to flow until all the water is taken out of the sump. Once the siphon starts sucking air it will lose its prime and the float valve will fill the sump again with a fresh load of water.

THE HARVEST CYCLE

Hot Gas Defrost — As we have already established, the hot gas defrost is to open and let hot gas enter the evaporator. The hot gas will melt the ice away from its contact and drop it to the storage bin. **NOTE:** *When the storage bin is full of ice, a bin thermostat will sense the ice level and cut the machine off.* Remember this. If you ever change a hot gas solenoid, make an identical replacement. The reason is, if you buy one across the counter from anyone, they may give you one that does not have the proper orifice. This part should be considered an OEM part as will most other parts on the machine. Go back to the ice machine distributor to buy original OEM parts or substitute with their recommended replacements. **Grid Cutters** — There are a couple of machines that still use hot gas defrost but use grid wires heated with either 21 or 14 volts, see **Figure 14-22**. Notice the type of ice that is formed. It will be a slab of ice. As the water from the pump travels across the plate, ice builds up. When the slab thickens to about 1/2" to 5/8" inch thickness an ice thickness device will stop the pump and open the hot gas solenoid valve for harvest. The plate will warm up. The slab will release and slide down on the grid wires. The system will go back into the freezing cycle and while the next slab is being made, the grid wires will cut the slab.

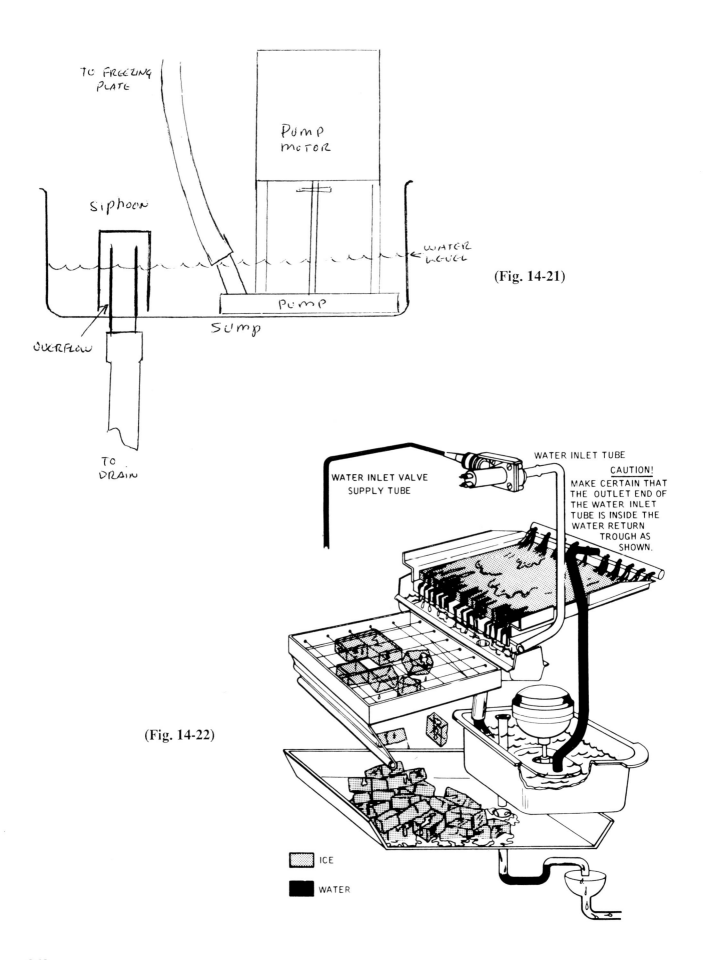

TO FREEZING PLATE

PUMP MOTOR

Siphoon

WATER LEVEL

(Fig. 14-21)

PUMP

SUMP

OVERFLOW

TO DRAIN

WATER INLET TUBE

WATER INLET VALVE SUPPLY TUBE

CAUTION!
MAKE CERTAIN THAT
THE OUTLET END OF
THE WATER INLET
TUBE IS INSIDE THE
WATER RETURN
TROUGH AS
SHOWN.

(Fig. 14-22)

ICE

WATER

REFRIGERATION TROUBLE SHOOTING GUIDE

SITUATION	CAUSE	CHECK	REMEDY
A. Compressor runs continuously; bin full of ice	1. Bin thermostat out of calibration	1. Put sensing element in glass of ice. Allow a few minutes for bulb to get cold. Now set thermostat to cut off compressor	1. Make proper setting
	2. Bin thermostat stuck	2. If above check fails, take one wire loose on control. If compressor stops, stat is bad.	2. Replace stat
	3. Incorrect wiring	3. Only if there is a reason to believe someone has changed the wiring or equipment is new from factory.	3. Use unit wiring diagram and check out
B. Low ice yield (be sure user is not expecting too much from the unit)	1. Located in cold area	1. Area could be too cold (below 32°F). System may not have head pressure control.	1. Move to warmer area above 55°F (for best results 70° to 90°F)
	2. Water falling on cubes	2. Make sure sump is not overflowing, drain stopped up, or distribution of water overflow into bin area.	2. Make corrections
	3. Thickness thermostat or pressure control set too thin or too thick	3. Depending on manufacturer, check for proper setting	3. Make proper adjustment. Run through several checks before leaving.
	4. Hot gas solenoid stuck partially open	4. Check voltage to solenoid coil. If no voltage, valve is stuck open.	4. Replace solenoid valve
	5. Insufficient refrigeration	5. Check sight glass for bubbles or system pressures	5. Repair leak. Recharge.
	6. Not enough water being circulated over evaporator plate causing unit to harvest early	6. Float valve stuck or stopped up	6. Clean float valve or strainer
	7. Drain water backs up in bin	7. Stopped up drain	7. Blow out or clean drain
	8. Head pressure too high	8. Mount gauges and read pressure	8. Blow out condenser or clean

REFRIGERATION TROUBLE SHOOTING GUIDE cont.

SITUATION	CAUSE	CHECK	REMEDY
C. Excessive water dripping on ice cubes	1. Water tank overflowing 2. Water trough out of position 3. Cube sheath not sliding out 4. Distribution pan at top of cuber stopped up 5. Water deflector over evaporator out of position	1. Check overflow tubes for stoppage 2. Check for broken brackets or alignment 3. Check for reason of hang-up 4. Check for algae and slime 5. Check for breakage or displacement	1. Repair tube 2. Repair 3. Repair 4. Clean 5. Repair where the deflector will not fall off
D. Ice cubes too thin	1. Thermostat or pressure control set for thin cubes 2. Not enough water circulated over evaporator 3. Evaporator thermostat or pressure control out of calibration 4. Low ambient temperature	1. Check for proper adjustment 2. Low water level in sump due to stopped up float valve, low float adjustment, or stopped up pump 3. Check actual pressure or temperature against actual control settings. 4. Check and see if system has head pressure control.	1. Readjust 2. After finding correct problem, repair 3. If badly off, replace control 4. If not, install head pressure control or relocate system
E. Ice cubes too thick	1. Evaporator thermostat or pressure set too thick 2. Evaporator thermostat or pressure control out of calibration 3. High head pressure	1. Check setting for actual setting. 2. Check actual pressure or temperature against control setting 3. Check for dirty condenser or overcharge	1. If setting can be made, reset. If not, replace control. 2. If badly off, replace control. 3. If condenser is dirty, clean. If overcharged, bleed system down to proper charge

REFRIGERATION TROUBLE SHOOTING GUIDE cont.

SITUATION	CAUSE	CHECK	REMEDY
F. Evaporator partially iced	1. Low on refrigerant 2. Expansion valve not feeding properly 3. Restricted drier in liquid line	1. Check for leaks 2. If charge is adequate, then check superheat 3. Checked by feel on the inlet of the drier, and then on the outlet. If you feel a temperature drop the drier is restricted	1. Repair leak and recharge 2. If superheat is too high and can be corrected by adjusting, then adjust. If the valve is restricted, replace valve. 3. Replace drier
G. Low suction pressure	1. Low on gas 2. Expansion valve partially restricted 3. Restricted drier 4. Low ambient temperature causing low head pressure	1. Check charge 2. If it is sure the gas charge is OK and the drier is not restricted, change expansion valve 3. Check drier inlet and outlet while unit is in operation. If outlet temp is lower than inlet, drier is restricted 4. If system has head pressure control, check why it is not controlling. If there is none, you may need to add control	1. Repair leak and recharge 2. Install new expansion valve 3. Change drier 4. Install some type of head pressure control
H. Suction pressure too high	1. Overcharged 2. Bad valves in compressor (inefficient compressor) 3. High head pressure	1. Check head pressure and suction. If both pressures are high there may be an overcharge. Feel of condenser tubes. If the tubes are cool almost to the top, there is a probability of overcharge. 2. Check head pressure, it would be low, suction pressure high, with low amperage draw on compressor. Pressures could tend to be equal if situation is bad enough. 3. See Section I	1. Slowly bleed gas down to proper charge. Let excess gas back into an empty drum if possible 2. Replace compressor if hermetic 3. See Section I

REFRIGERATION TROUBLE SHOOTING GUIDE cont.

SITUATION	CAUSE	CHECK	REMEDY
I. Head pressure too high	1. Dirty condenser	1. Visually check condenser	1. Blow out condenser
	2. Over charged	2. Check temp of condenser tubes. If tubes are cool, then there may be an overcharge	2. Bleed gas down to proper charge
	3. Non-condensable gas	3. Check and see if system has been opened and left open for a period of time, or the system may have a leak on the low side and sucked in air	3. Bleed down charge, evacuate, recharge
	4. Condenser fan motor off	4. Check voltage to fan motor. If there is voltage to motor, motor is bad	4. Replace motor

ELECTRICAL

The compressors you will be working on will either be single phase or three phase. If it is three phase then there will be no starting capacitors and no starting relay. It is self starting. With a single phase compressor you will need some type of starting device, but before we get on starting devices let's look inside the compressor and see how the windings are arranged and the terminals are identified, see **Figure 14-23**.

Notice there are three terminals and they are identified as (C) common (S) start and (R) run. Run is sometimes identified as (M) main, meaning main winding. The common terminal is common to the start and run windings. If the common terminal was made hot then one side of start and run would be made hot. The only thing we need to do now is run another wire to start and run and the circuit would be completed. The compressor would not start automatically however. Terminal arrangements may differ and will differ from compressor to compressor even if they are the same manufacturer. Generally Tecumseh products will have a terminal identification as in **Figure 14-24**.

The terminals are read as a book, always left to right, common, start and run. The Copeland compres-

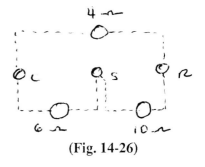

(Fig. 14-25)

sor will usually have common in the center if it is a "copelametic" (semi-hermetic) compressor, see **Figure 14-25**.

The "copelaweld" compressor (hermetic) usually has terminal identification similar to Tecumseh. Don't always have faith that these terminal arrangements will be true all the time. To be sure, use an ohm meter. If you check across the R and S terminals you would get the highest ohm reading. Across C and S would be the next highest and C to R would be the lowest of the three readings, see **Figure 14-26**.

(Fig. 14-26)

The readings in **Figure 14-26** do not mean that any compressor you may check will have these exact readings. These readings are an example only. For example the S and R terminals read 10 ohms. We know that is Start and Run because it gives us the highest. C and S the next highest and C and R the lowest. The

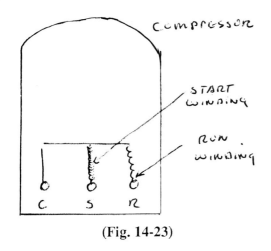

(Fig. 14-23)

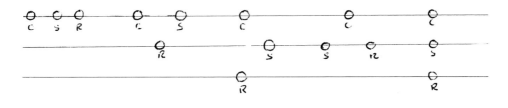

(Fig. 14-24)

value (ohm reading) you find across S and R should equal the sum of C + S and C + R.

The start winding in the compressor is a high torque winding used for starting the compressor and is only used temporarily to get the compressor started and up to speed. When the compressor has reached approximately 75% of its speed, this winding will be disengaged and left on the run winding. The run winding is a low torque winding incapable of starting on its own and will have to have the help of the start winding to get it up to speed. Once the run winding has been brought up to speed the start winding will drop out. The run winding is then able to carry itself. There are two devices used to drop out the start winding. These two common devices are the current relay and the potential relay. Before going further with starting devices, let's mention start capacitors and run capacitors. **Figure 14-27** shows a start capacitor and **Figure 14-28** a run capacitor.

The start capacitor stores energy rated in microfarads. On the initial start of the compressor the full energy of the capacitor is put into the starting winding to give it the necessary torque to start the compressor. This capacitor can only stay in the circuit for a short period of time. Longer periods will cause the capacitor to blow.

The run capacitor stays in the circuit all the time. It is there to actually lower the amperage draw of the compressor, make it run cooler and improve its power factor, which means money saved in operating cost. We will now return to starting relays.

The Potential Relay — This starting relay is an electrical device that serves the purpose of a starting switch. You will note that in **Figure 14-29** that the starting contacts are between terminals 1 and 2 and are normally closed when the compressor is not running. A potential coil between terminals 2 and 5 will open the start contacts between terminals 1 and 2 when the compressor reaches a predetermined speed. The induced voltage in the starting winding while the compressor is running will be applied across this potential coil and will keep the contacts open. When the contacts between terminals 1 and 2 open the start winding is dropped out of the circuit.

The Current Relay — The current relay shown in **Figure 14-30** also serves as a starting switch. The contacts between terminals L and S are normally open with the compressor off. When the compressor starts,

(Fig. 14-27)

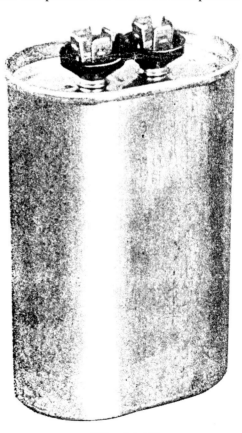

(Fig. 14-28)

266

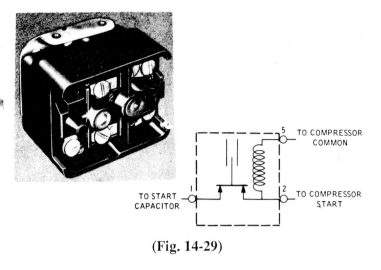

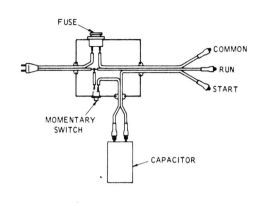

(Fig. 14-29)

(Fig. 14-31)

a circuit is made through the run windings of the compressor and the coil between L and M. The excess current draw on the initial start of the compressor will cause the contacts to close between L and S, therefore engaging the start winding and start capacitor in the circuit.

When the compressor reaches speed the current in the coil of the relay decreases, the contacts will open and disengage the start winding and start capacitor.

Checking the Compressor — When you want to know whether the compressor will run or not for sure, use the tester in **Figure 14-31**. You can easily make one of these or you can purchase one in a supply house. Simply hook the clips marked common, start and run to their proper terminals. The fuse should be a 30 amp fuse. Clip your capacitor to the clips provided. Plug in the starting device and at the same time push the momentary button. If the compressor starts and operates normally after releasing the momentary button the compressor should be good.

The Overload Protector — The overload protector will either be external or internal. When external you will usually see this overload near the terminals of the compressor. This device will open on either high

amperage draw or high compressor temperatures. The internal overload is embedded in the windings of the compressor and will open on high winding temperatures. When this overload opens it usually kills the common terminal internally and takes around 30 to 45 minutes to close. See **Figure 14-32a and 32b**.

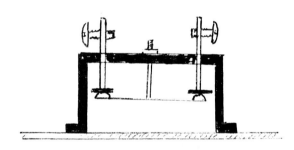

EXTERNAL OVERLOAD AGAINST BODY OF COMPRESSOR

(Fig. 14-32a)

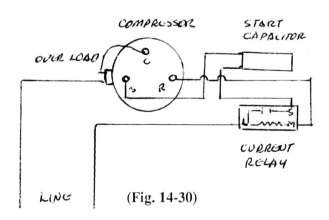

(Fig. 14-30)

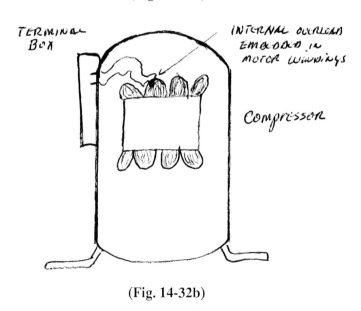

(Fig. 14-32b)

267

WIRING DIAGRAMS

Wiring diagrams are usually the most difficult for servicemen working on ice makers. It will be necessary for you to be able to understand them so that you will know how each one functions. We will go through a few diagrams so that you will see right away how each machine varies from one to another.

The first diagram you see (**Wiring Diagram 14-1**) is a Whirlpool ice maker. The wiring diagram at the top is the same as the one below it. The top diagram is a simplified ladder diagram. This ice maker makes ice in slabs. When the thickness switch changes position to the (NC) (normally closed) contact the (NO) (normally open) contact opens and stops the pump motor, the thick control motor and the condenser fan motor. The (NC) contact is closed which opens the hot gas solenoid valve. This lets hot refrigerant gas into the evaporator and melts the slab loose. The evaporator is

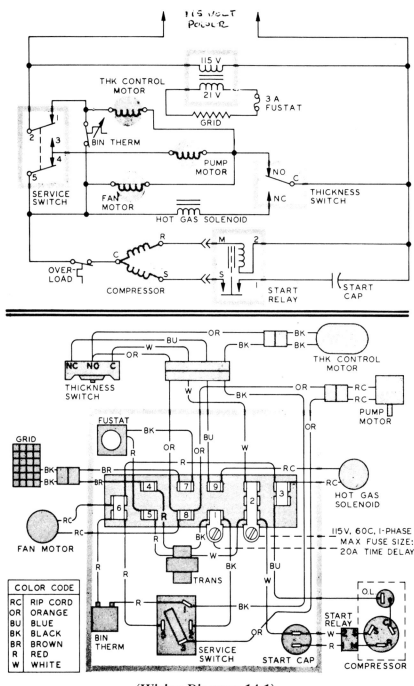

(Wiring Diagram 14-1)

268

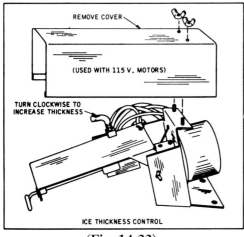

(Fig. 14-33)

at an incline, so the ice will slide out and into the cutting grid. Notice that the cutting grid transformer steps down voltage to 21 V to the grid. With this voltage on the grid a small amount of heat is created and the grid wires will slowly cut the slab into cubes and drop into the storage bin.

Notice the compressor. What kind of starting device is being used. It's a current relay and notice it is using a starting capacitor. Notice the overload which in this case is the external overload. Take a look at the service switch. In the position it is in, it will make ice. From terminal #1 you feed the thick control motor, the pump motor and the condenser fan motor. If you reposition the switch, contact 5 and 4 is opened, 2 and 1 is opened, but 2 and 3 are closed. Now the thick control motor has been cut off and the condenser fan motor and the compressor. The only thing operating is the water pump. Now find the bin thermostat. When the storage bin is full of ice the thermostat will open. When it does, a circuit is killed to the condenser fan motor, the pump motor, one side of the hot gas valve and the compressor. Power is still to the thick control and grid wire. **Figure 14-33** shows the ice thickness

control. It is a motorized arm operated motor and can be adjusted for proper thickness. There is a rotating offset arm that rotates and touches the ice. If the ice is thick enough, it pushes the assembly up and trips the ice thickness contacts which starts a harvest cycle.

The next ice machine is a Crystal Tips that will make either cubes or chips. This is done as shown in **Figure 14-34**.

The three views you see here are the different views of the evaporator. The capillary sheath has a thermostat sensing bulb inside and when touching the cubes will put the ice maker in the harvest cycle. If you want chips you adjust the sheath downward. The chips will come into contact with the capillary sheath and go into the harvest cycle and make only chips. If you want cubes, raise the sheath upward. This allows the chips on each side to freeze together and make cubes.

Now let's go through a wiring diagram that creates a little more of a challenge. Look **Wiring Diagram 14-2** over very closely noticing the motors, compressors, pump and relays. Find the wiring legend. This will help you identify the electrical components. Notice the incoming circuit and see L1, L2 and L3. It should tell you that it is a three phase circuit. There will be no starting components at all on the compressor.

The magnetic coil identified by K1 and K2 will start the compressor provided TC1-1 and TC1-2 thermostats are made. These controls are the bin thermostats (see the wiring legend). When K1 and K2 are energized they will magnetize and pull K1 and K2 contacts closed. This puts a three phase circuit to the compressor and the compressors will start. The control circuit will not function until the contacts to the compressor are closed. Find terminal 61. When the L1 circuit closes to T1 terminal 61 is made hot. Now

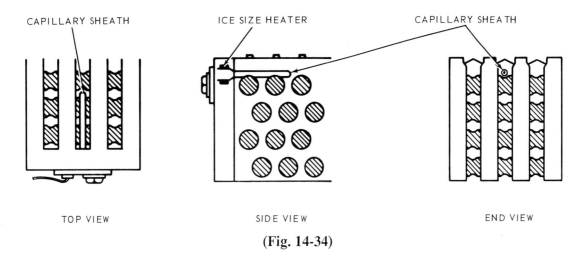

(Fig. 14-34)

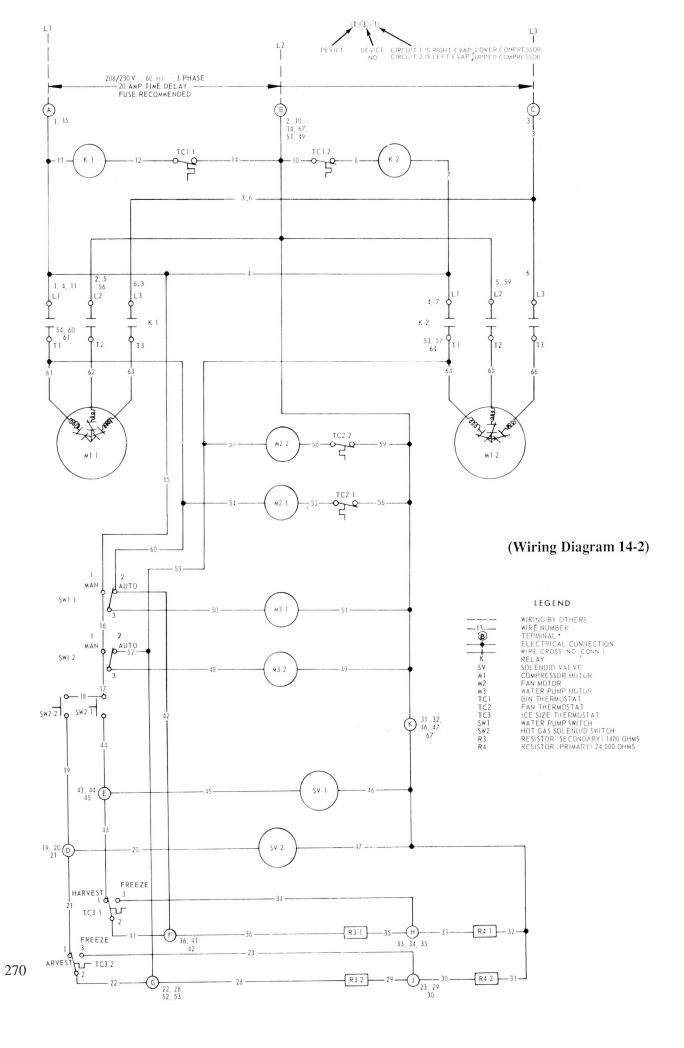

(Wiring Diagram 14-2)

LEGEND

— — —	WIRING BY OTHERS
—13—	WIRE NUMBER
Ⓑ	TERMINAL
•	ELECTRICAL CONNECTION
	WIRE CROSS (NO CONN)
K	RELAY
SV	SOLENOID VALVE
M1	COMPRESSOR MOTOR
M2	FAN MOTOR
M3	WATER PUMP MOTOR
TC1	BIN THERMOSTAT
TC2	FAN THERMOSTAT
TC3	ICE SIZE THERMOSTAT
SW1	WATER PUMP SWITCH
SW2	HOT GAS SOLENOID SWITCH
R3	RESISTOR (SECONDARY) 1400 OHMS
R4	RESISTOR (PRIMARY) 24,000 OHMS

follow this wire and you see the first tie point (wire 54) that goes to M2-1. M2-1 according to the legend is a fan motor and the M2-1 means it is for the No. 1 compressor circuit. So now we have the condenser fan motor operating on compressor No. 1. Continue following wire #60 and you come to SW1-1. SW1 we find according to the legend is a water pump switch and SW1-1 means it is for compressor No. 1. The SW1-1 switch can be set for manual or auto operation. You will want the auto position so the pump will cut off in the harvest cycle. So if the SW1-1 switch is in the auto position you will have a wire 50 leaving the No. 3 terminal going to M3-1. M3 is a pump motor and M3-1 means the pump motor for the No. 1 compressor. Now the water pump is operating. From terminal 2 on the SW1-1 switch you have a #42 wire going to terminal F, out wire #41 up to TC3-1 which is the ice size thermostat for the No. 1 compressor. As the switch is positioned in the diagram the system is in the harvest cycle, so follow wire #43 up to terminal E and across wire #45 to the solenoid valve which is the hot gas solenoid valve for the No. 1 compressor. Please take note that wire #60 comes from terminal 61 and is attached to L1. A circuit is completed, 240 V from L2 to the other side of the controls. SW2-1 hot gas solenoid switch is a manual switch, should you want the system to go into the harvest cycle on demand.

Before going any further notice R3-1 and R4-1. When the TC3-1 switch is in the freeze cycle a circuit is completed to these resistors. These resistors put out a small amount of heat and are mounted on the ice size thermostat sheath. Since the air temperature around the sensor is cold enough to give a false trip, the resistor will prevent it. When the thermostat sheath comes in contact with the ice the heat from the resistor will be overrode. Now let's go through the control circuit for compressor No. 2 starting at wire #64 after T1 of the compressor contactor and pick up wire 53. Trace this circuit exactly as you did on compressor No. 1. It is identical. On the wiring diagram we know now that according to the wiring legend M2-2 and M2-1 are the condenser fan motors and that TC2-2 and TC2-1 are going to control these motors. These are your head pressure control devices. The sensing bulbs of these controls strap to the condenser and cycles the motor off at 88°F condensing and cycles motor on at 98°F condensing. Referring again to the diagram you will see TC1-1 and TC1-2 controls, controlling K1 and K2. These are bin thermostats and if you will notice nothing operates when these controls open stopping the compressors.

Now we will review a wiring diagram from a Kold Draft brand machine. Follow **Wiring Diagram 14-3**. In following the wiring diagram let's first identify the controls and so forth.

Bin Thermostat — This is a thermostat located in the freezing area and has a long capillary line with a sensing bulb that extends 4 to 6 inches in the storage bin. When ice comes in contact with the bulb the contacts will open in the thermostat and cut the machine off.

Weight Control Switch At The Tank — The weight control switch is controlled by a tank (**Fig. 14-35**). After a harvest cycle, the water inlet valve will open and start filling the water reservoir. There is a connecting tube from the water reservoir connecting to the tank. As the reservoir fills you can see the level of water in the tank. When the water gets to the mark on the tank, the weight of the water will trip the weight control switch downward, dropping out the water fill valve.

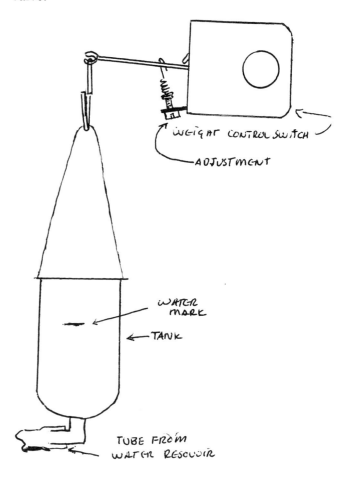

(Fig. 14-35)

Actuator Motor — The actuator motor drives the water plate down in time of harvest cycle and up after harvest cycle is complete. The actuator motor is a reversible motor and is geared to turn slow. When the panels covering the freezing section are removed, you can see the actuator turning.

Actuator Toggle Switch — The toggle switch will switch from one contact to the other by the use of a push rod attached to the water plate. One position of the switch is when the plate lowers and the other position is when it rises.

Actuator Thermostat — The actuator thermostat actually ends the harvest cycle after the slab has fallen from the evaporator.

Water Inlet Valve — The water inlet valve is to fill the water reservoir after the harvest cycle, getting ready for another freezing cycle. Just any solenoid valve is not used to replace this valve. The orifice in the valve is to allow so much water to flow through in a required amount of time (usually about 1/2 gallon per minute).

Cold Water Thermostat — The cold water thermostat is to prevent the water plate from dropping prematurely due to cold inlet water which could cause early resetting of the actuator thermostat before the weight control drops. If the incoming water is below 40°F the cold water thermostat will close a circuit to the defrost valve and keep the evaporator warm until the water fill cycle is complete.

Water Level — When the weight control tank goes down it breaks a circuit through the weight control switch and closes the fill valve. This maintains the water level in the weight control tank and the reservoir. The level can be seen in the plastic tank. As the ice machine operates and ice starts to form on the evaporator plate, the water will start going down and soon as enough water is gone from the tank, the weight control switch will reposition itself and go into the harvest cycle.

Control Stream — The control stream should be as shown in **Figure 14-36** and will vary some from machine to machine. Adjustments may have to be made at the adjusting screw to accommodate other model machines. However, under normal conditions you can see in **Figure 14-36** that water from the freezing chamber is being pushed out by pressure of the water pump and draining back into the reservoir. Remember also that the pump is pumping water into the water plate with small holes, directing water into the freezing chambers forming individual cubes. When the cubes have been formed it tends to close off the holes in the water plate. This creates more water pressure and causes the stream of water in **Figure 14-36** to increase and go over the dam.

The water goes out the drain tube and out the drain. This starts draining the reservoir and causes the water level to drop, which also takes the water out of the control tank. The weight of the tank lessens, the tank rises and trips the weight control switch. The system is now in the harvest cycle.

Wiring Diagrams — Now let's go through the wiring diagrams, starting with **Wiring Diagram 14-3**.

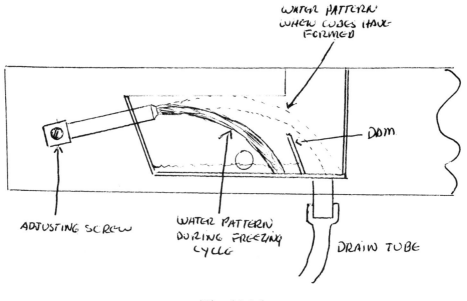

(Fig. 14-36)

The system has just completed a harvest cycle and the water plate is up. Follow the red lines, first through the bin switch (closed) to the compressor and fan (in operation). Now follow over to the weight control switch (in the up position-system is filling with water).

When the weight control tank fills with water the weight control switch will open and shut off the water valve. Continuing on the red line you see a circuit made to the pump. We are now in the freezing cycle.

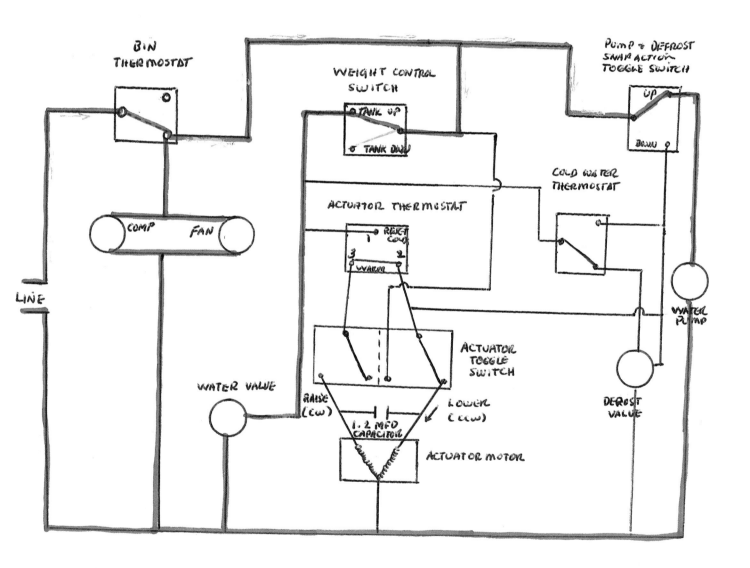

(CW) - Clockwise   (CCW) - Counter clockwise

(Wiring Diagram 14-3 IN WATER FILL POSITION)

In **Wiring Diagram 14-4** the system is in the freezing cycle and freezing cubes. The only parts of the machine that are in operation are the compressor, the fan and the water pump. After the cubes have formed, the water level will drop in the water tank. When enough water is gone, the lower weight will allow the tank to go up and trip the switch. Meanwhile during the freezing cycle the evaporator has reached approximately 20°F and the actuator switch has closed to another position (between 1 & 2). Go to **Wiring Diagram 14-5**.

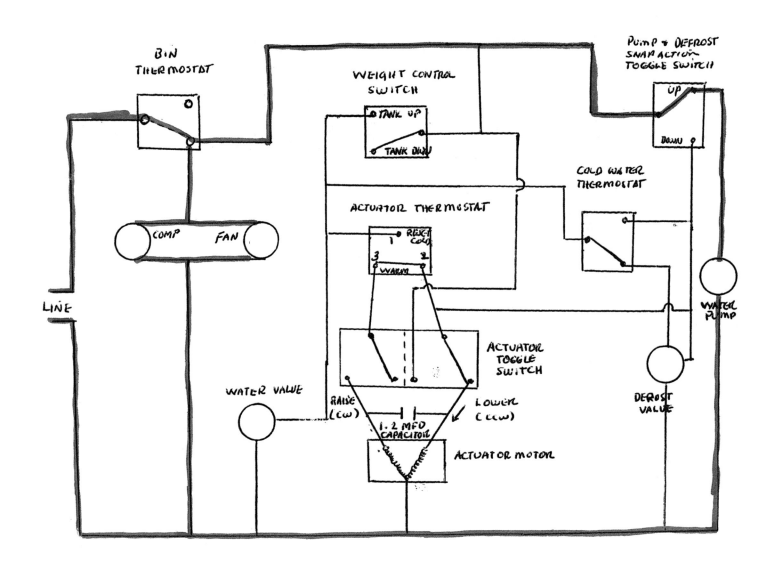

(Wiring Diagram 14-4)

In **Wiring Diagram 14-5** you can see that the control tank has emptied and made in upward position at the weight control switch. You now have a circuit to the water valve, through the now closed actuator thermostat, which now completes a circuit through the actuator toggle switch. The CCW actuator motor is now lowering the water plate. Also while the water plate is starting to lower, the defrost snap action toggle switch is made in the down position. This cuts off the pump. Notice there is also a circuit made through the cold water thermostat. This control only closes in case of the inlet water being too cold. If the water is warm enough the switch will remain open and the normal circuit would be through the pump and defrost toggle switch. During the harvest cycle, (and the water plate has made its travel down), the actuator will push the toggle switch over to the other side as shown by the blue-dotted line. This stops the actuator motor and there is no power to the actuator motor.

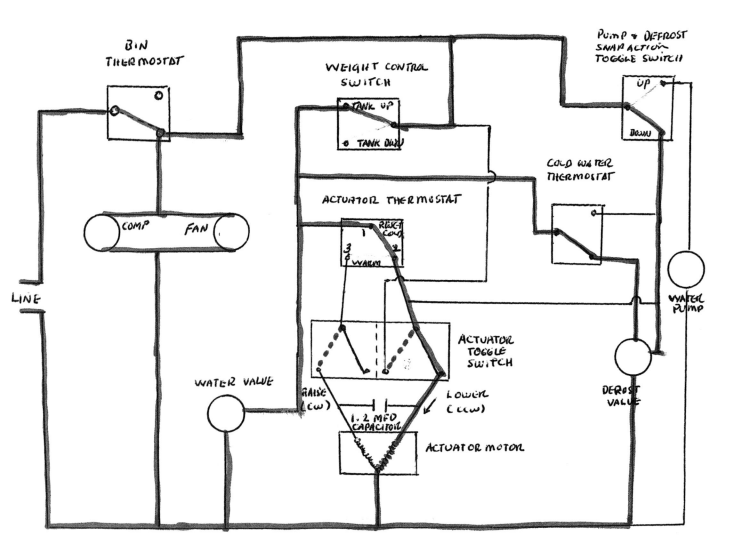

Water tank has emptied; ready to go into harvest

(Wiring Diagram 14-5)

In **Wiring Diagram 14-6** notice that the actuator thermostat has now closed contact between terminals 2 and 3 due to the freezing plate warming up because of the hot gas. A new circuitry has now been made through the actuator toggle switch on the right, through the contact of the actuator thermostat, through the left side of the actuator toggle switch and to the CW actuator motor. This brings the water plate up. When it does, the pump and defrost toggle switch is pushed up, killing the hot gas defrost valve (provided the cold water thermostat is not closed) and starting the water pump. Once the water tray has made its travel, the actuator toggle switch will move to the right and stop the actuator motor. The water valve will continue to fill until the weight control switch is pulled down and closes the water valve. You are not back in the freezing cycle and your circuitry would be as in **Wiring Diagram 14-4**.

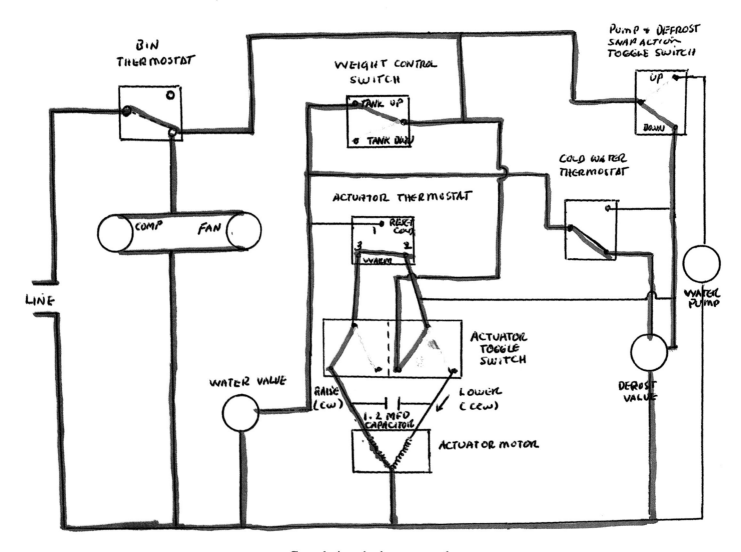

Completing the harvest cycle

(Wiring Diagram 14-6)

Kold Draft now has an electronic cuber. It has changed a great deal in its circuitry. It has a water tank which now has sensors in it and most of the controls are controlled through the circuit board. **Figure 14-38** shows a simulation of the water tank with the electronic sensors.

This operates by letting water during the fill cycle (water valve open) go up into the water tank and rise until it comes in contact with the fill sensor. The water valve is dropped out. During the freezing cycle the water level starts to drop and when it drops below the harvest sensor, the system is put into the harvest cycle. You shouldn't have any trouble mastering the electronic cuber. Just use a good wiring diagram from the machine, exercise a little patience and you should not have any problems.

MANITOWOC ICE MAKERS

Figure 14-39 shows a two evaporator type of machine. Ice cubes are formed in both plates. There are two water curtains, one in each evaporator. There is one compressor and two expansion valves, one feeding each evaporator. Water is pumped up by a water pump to the top of the evaporator out the distribution tubes. Water cascades down the evaporator plates while the refrigerant tubes are getting cold and freezing some of the water in the cube chambers. The water falls to the bottom, back to the sump where the pump picks it up and recirculates it. During defrost there will be two hot gas solenoid valves feeding the evaporators, one for each evaporator.

The Bridge Control — The bridge control (**Fig. 14-40**) is a low pressure control that is reverse acting and closes its contacts on a drop in pressure and opens

its contacts on a rise in pressure. The low pressure control initiates the timing circuit of the solid state timer. The timer is set at approximately 6 minutes. The bridge control controls the thickness of the slab. As the cubes freeze at the evaporator the plate gets colder and as the plates get colder the refrigerant pressure starts to drop. When the setting of the control is reached, it closes, and then feeds through a timer that times out for a few minutes then puts the system into the harvest cycle.

The operation of the machine during the harvest cycle and the refrigeration cycle, in combination with other controls, is controlled by the control relays marked A, B, C, and D (**Fig. 14-41**). You also see the delay timer in the same illustration.

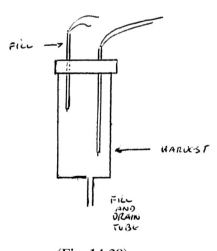

(Fig. 14-38)

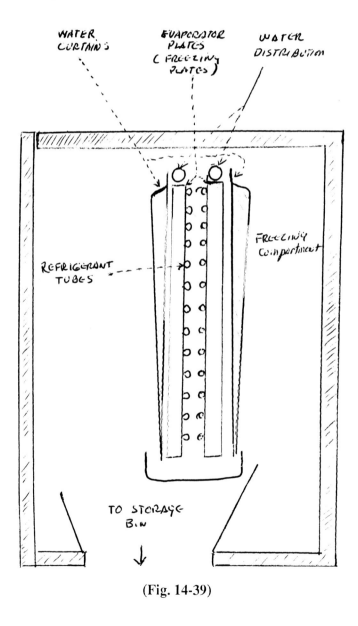

(Fig. 14-39)

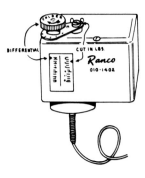

(Fig. 14-40)

The Bin Switch — There are two bin switches, one that is controlled by the front and back water curtains. When the ice is melted from the evaporator during the harvest cycle, the slab will fall out and push the water curtains out causing the bin switch to trip in an opposite position **(Fig. 14- 42)**. The bin switches serve their purpose in cutting off the machine when the ice slabs will no longer go down because of the bin being full. This causes the water curtain to stay open, therefore de-energizing contactor coil and stopping the compressor.

Safety Thermo Disc — If for some reason the bin switch fails, the machine would stay in the harvest cycle until the thermo disc (top right side of **Wiring Diagram 14-7**) opened due to the suction line becoming excessively warm. The system would then go back into the freezing cycle. The safety thermo-disc is strapped to the suction line.

Manual Harvest Switch — Upon pressing this switch (below bridge control, **Wiring Diagram 14-7**) you will manually put the system into the harvest cycle. The system will go through its normal harvest cycle and then automatically go back into the refrigeration cycle.

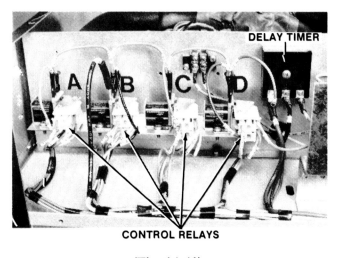

(Fig. 14-41)

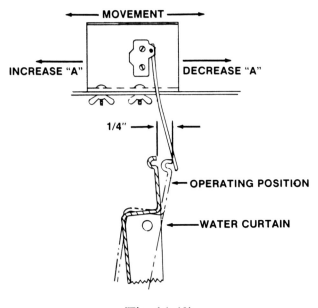

(Fig. 14-42)

278

Wiring Diagram 14-7 — Familiarize yourself with the diagram by locating the power coming in, the compressor, then locate the controls and relays we have mentioned in the preceding diagram. You will be guided through different situations such as when the system goes into the harvest cycle, when it comes out of the harvest cycle and when it is in the refrigeration cycle. By following the red colored lines we will trace out the circuit for the refrigeration cycle. The only things energized are the compressor, the water pump and the condenser fan motor. Please note the start wire is traced in red also. This circuit has been dropped out by the contacts in the start relay (terminals 1 and 2). There is some voltage traveling to the start winding by way of the run capacitor. This shift of power causes an amperage drop at the run winding and improves the power factor of the compressor.

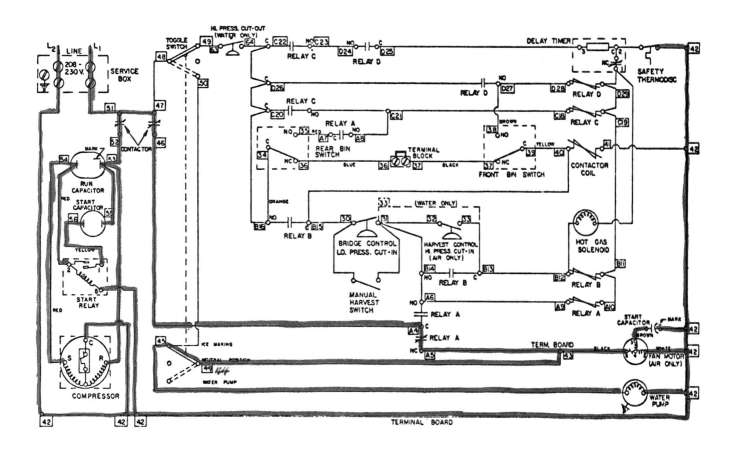

(Wiring Diagram 14-7)

Wiring Diagram 14-8 will show the compressor still operating during the harvest cycle because it will have to produce hot gas to defrost the evaporator. The compressor circuit will still be shown by red lines, but the controls that have now been energized will be shown by blue lines.

The defrost circuit will pick up one line of power from terminal 47 starting with the blue line, continue through the toggle switch (the switch for ice making position or just pump only) and through a high pressure cut-out (cuts the system off if the refrigerant pressure becomes too high at the condenser—used in water cooled condensers only) and continue on to terminal 30 on the bridge control low pressure control. Notice the circuit going from terminal B16, through the rear bin switch, through the front bin switch and then on to

the bridge control. Don't forget that the bin switches are controlled by the pushing out and closing of the water curtains when the ice is falling out. If you don't remember these small micro-switches called bin switches, refer back to **Figure 14-42**.

When due to making ice, the evaporator gets cold and the refrigerant pressure drops to the cut-in setting of the bridge control, power will go to relay "A" and relay "B." Relay "A" will open contact between terminal A2 and A5, stop the condenser fan motor and water pump. The contact will close between A2 and A6. This becomes a holding contact explained later. Relay "A" also closes a contact between terminals A7 and A8. This contact will be on stand-by at the time.

Relay "B" is energized when the high pressure cut-in (harvest control is closed) from terminal 32 to 33.

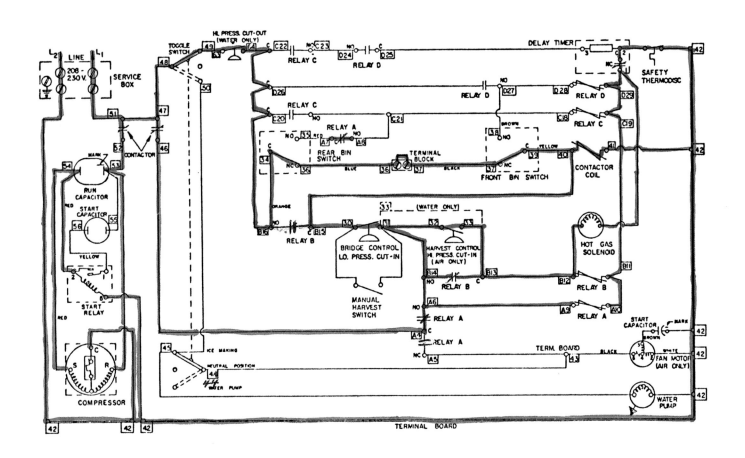

(Wiring Diagram 14-8)

The harvest control is there to assure you that the head pressure will be high enough to create enough heat to defrost. If the head pressure is low at the beginning of the harvest cycle it will build up because relay "A" has cut the condenser fan off . Once the pressure has built up, the harvest control high pressure switch will close and open the hot gas defrost. With relay "B" energized the contact between terminals B14 and B13 are closed and between B16 and B15.

During defrost the pressure at the Bridge control has risen and made the contacts open. This will not drop out relay A and B because now the contact of Relay A between A2 and A6 is closed feeding the relays. So you could say this was a holding circuit. See **Wiring Diagram 14-9**.

Now notice the position of the bin switches. They are closed. You must at this point put a picture in your mind, and this is it. The ice has just melted loose from the evaporator plates and in just a second or so the slab will fall down pushing the water curtains out and

tripping the bin switches in the position shown in **Wiring Diagram 14-9**. The bin switches will immediately go back to the (NC) position. Now look what has happened in that short period of time.

When the rear bin switch made to the normally open (NO) position relay "C" was made and when the front bin switch made to the normally open (NO) position relay "D" was made. All relays are made including the hot gas defrost. Relay "C" and "D" now have their contacts closed making the delay timer energized between terminals C22 to D25. When the timer has been in the circuit for a few seconds its contacts between terminals 1 and 2 will open. By looking at your diagram you can see that relay A, B, C, and D would be de-energized along with the hot gas solenoid valve. Another freezing cycle has begun.

The wiring diagrams you have just covered do not cover every wiring diagram you will be confronted with on Manitowoc ice makers, but it will give you a general sense of how the machines work. The dia-

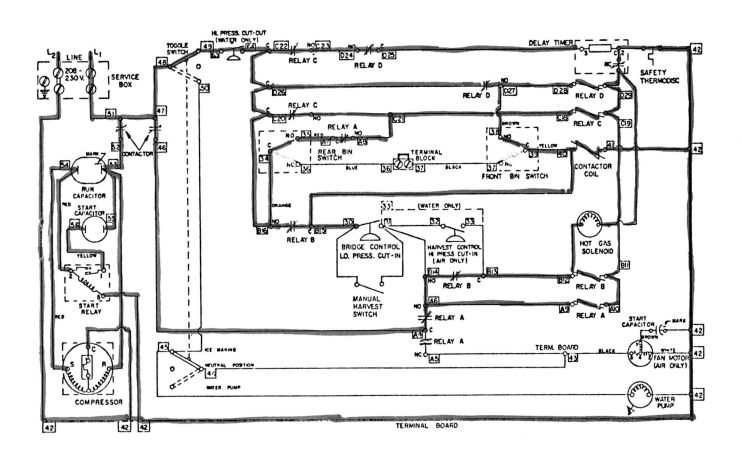

(Wiring Diagram 14-9)

grams will vary as to whether the unit is self-contained (meaning the condensing unit is in the same cabinet with the freezing section, but separated from one another) or if the unit is remote (the condenser is away from the ice maker) and whether it is air cooled or water cooled. Some of the units have two defrost solenoids and when this is the case the rear solenoid will be delayed to open a few seconds after the front solenoid to make sure the front slab slides off before the back slab. This is done because the only bin switch to bring the unit out of harvest is on the back section. You will also find in some units that the pump will be delayed for a few minutes until the evaporator becomes cold, before it starts pumping water. In any case, if you can master the diagram you just covered you shouldn't have too much problem with the other diagrams.

THE ELECTRONIC CONTROL

On some of the machines a different type of ice thickness control is used, such as in **Figure 14-43**. This sensor is not affected by system pressure. The control will call for defrost when the ice in the evaporator forces the water to contact it for 6 to 10 seconds. The control normally has 12 volts across it. The water contact will reduce the voltage to less than 7 volts. This voltage drop signals the module to place the cuber into hot gas defrost.

Transformer Board And Module — There are two transformer boards, one rated at 120 V and the other at 240 V. The boards contain a 5 amp quick blow fuse to protect the control circuit. Early production used a 3 amp fuse. Five (5) amp is now standard replacement fuse (**Fig. 14-44**).

Three Relay Board — Here in **Figure 14-45** you see a relay board with its relays and L.E.D. lights in conjunction with the delay timer to accomplish an

over-ride at the end of every completed harvest cycle. It also permits both front and rear ice slabs to fall at random. In addition the relays de-energize each hot gas solenoid as soon as the corresponding ice slabs have fallen. The relays are standard power midget relays which can be removed and replaced. The L.E.D. lights are to make an indication of the performance of the machine. When a relay is energized the L.E.D. light for that respective relay will light.

(Fig. 14-44)

Three Relay Board (E-1100 Series)

ICE THICKNESS CONTROL
Sensor (dual sensor standard on all models)

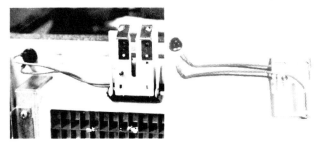

(Fig. 14-43)

(Fig. 14-45)

282

ELECTRICAL TROUBLE SHOOTING GUIDE

SITUATION	CAUSE	CHECK	REMEDY
A. Compressor hums but will not start	1. Low voltage	1a. Check voltage-must be within 10% of name plate rating. Check income power- if readings incorrect call power company.	1a. Call power company
		1b. Voltage rating on machine too high for income voltage	1b. Install proper machine with correct voltage rating
	2. Blown fuse (three phase models only)	2. Check all fuses. If one is blown compressor will single phase and hum.	2. Replace blown fuse
	3. Bad start capacitor	3. Check capacitor and see if it has indications of blowing, if not temporarily charge capacitor with direct line voltage (120 V). Cross terminals with jumper. If capacitor does not arc, capacitor is bad.	3. Replace capacitor with like MFD and voltage. (MFD rating can be varied ± 10%)
	4. Start relay contacts open	4. Jumper across terminals 1 & 2 on potential relay and L & S on current relay. If compressor starts, remove jumper.	4. Replace relay with proper rating
	5. Shorted run capacitor	5. Simply drop the run capacitor from the circuit. If the compressor then starts, run capacitor is bad	5. Replace run capacitor with exact MFD and voltage rating.
	6. Compressor stuck	6. Install straight line (**Fig. 14-31**). If it does not start, compressor is bad	6. Replace compressor
B. Compressor starts but will not get out of start winding	1. Low voltage	1. Same as 1A, Situation A	1. Same as 1A, Situation A
	2. Relay contacts stuck closed	2. On potential relay take wire from terminal 2 when unit starts. If compressor runs, relay was stuck. If the unit has a current relay remove wire on "S" terminal when compressor starts	2. Change relay, do not attempt to file or sandpaper contacts
	3. Bad run capacitor	3. Take capacitor out of circuit, turn unit on. If compressor operates ok, capacitor is bad.	3. Replace capacitor

SITUATION	CAUSE	CHECK	REMEDY
B. cont.	4. Bad compressor (open winding)	4. "Ohm out" compressor, check across C to S, C to R and S to R. Any combination that has lack of continuity indicates open windings.	4. Replace compressor
	5. Bad compressor (grounded compressor)	5. Make continuity check from each compressor terminal to shell of compressor or copper line on the unit. Any continuity reading means grounded compressor	5. Replace compressor
	6. Bad compressor (shorted winding)	6. Take ohm reading across C to S and C to R should equal ohm reading across S to R.	6. Replace compressor
	7. High head pressure-will not allow compressor to reach speed	7. Check head pressure. If head pressure is high, correct problem.	7. If air cooled, replace condenser fan motor if bad. Clean condenser if dirty. Correct charge if overcharged. If water cooled, correct water problem across condenser.
C. Unit blows fuses	1. Short	1. Check all wiring visually to see if there has been an arc to the housing of the unit.	1. If such is found, correct wiring
	2. Bad compressor	2. Check compressor for shorts and grounds.	2. Replace compressor
	3. Shorted pump motor	3. Isolate pump from circuit and see if fuse blows	3. Replace pump
	4. Condenser fan motor shorted	4. Isolate fan motor from circuit and turn on to see if fuse blows	4. Replace condenser fan motor
	5. Water splashing on controls due to a water curtain etc. being out of place	5. Check and see why curtain came loose	5. Repair problem with the water

ELECTRICAL TROUBLE SHOOTING GUIDE cont.

SITUATION	CAUSE	CHECK	REMEDY
D. System will not go into defrost (using Wiring Diagram 14-7)	1. Hot gas solenoid will not open	1. If system is not already in defrost, press the manual harvest switch and you should read voltage to the coil	1. Replace solenoid coil
	2. Bridge control will not close	2. If you see a full size slab and you know without a doubt that the machine should be in a harvest cycle, check across the terminals of the control with a volt meter with the unit in operation. A volt reading would indicate an open control	2. Make correct setting on control. If setting is ok, replace control
	3. Harvest control open with head pressure sufficiently high	3. Check head pressure. If it proves to be high enough the control is bad	3. Replace harvest control
	4. Relay "B" bad	4. Check voltage to coil of relay. If voltage is there but relay is not responding, relay is bad	4. Replace relay
	5. Open contacts in relay "B"	5. Install alligator clips and jumper both contacts. If system will go through a harvest cycle, relay is bad	5. Replace relay
E. Compressor, fan motor and water pump will not come on	1. Relay contact between terminals A2 and A5 open	1. Jumper A2 and A5 while in freezing cycle. If pump and fan motor come on, relay contact is not closing	1. Replace relay
	2. Contactor contact between terminal 47 & 46 not making contact	2. Jumper terminal 47 & 46. If pump and fan motor come on, contactor had bad contact	2. Replace contactor

ELECTRICAL TROUBLE SHOOTING GUIDE cont.

SITUATION	CAUSE	CHECK	REMEDY
F. System goes into defrost, but immediately goes out of de- frost	1. Relay contacts in relay "A" not closing due to dirty contact or coil in relay burnt out	1. Check voltage to the coil of relay "A." If there is voltage and the relay has pulled in, jumper terminals A6 and A2. If unit stays in defrost (NO)(normally open) con- tact in relay "A" was open.	1. Replace relay

You have been instructed to jumper controls to determine if they are open and this will make a determination. However, on a safer side it is suggested that you check across the terminals of these controls with a volt meter with the machine on. If the control is open you will get a voltge reading.

SITUATION	CAUSE	CHECK	REMEDY
G. System will not come out of harvest cycle	1. Delay timer contacts stuck closed	1. Disconnect wire on termi- nal 1 of the timer. If system goes back into harvest cycle, timer is bad.	1. Replace timer
	2. Safety thermo disc will not open. (This control should open and put the system back into the harvest cycle, should the timer not open.)	2. In a hypothetical situation we will pretend the timer and thermo disc are both stuck. Disconnect a wire on the thermo disc control. If the system goes back into the harvest cycle, the con- trol is bad.	2. Replace the thermo disc
H. Nothing will come on.	1. No power	1. Check at terminals 51 & 42. If power is there move on.	1. If not, replace blown fuses or check for tripped break- ers
	2. Open circuit in toggle switch	2. Check at terminals 48 & 42 and 45 & 42. Voltage should be there. Now check at ter- minals 44 & 49. If no power is there, toggle switch has open contact.	2. Replace toggle switch
	3. Bad contactor coil `	3. Check voltage to coil. If there is full voltage the coil is burnt out and not magnetizing.	3. Replace contactor or coil

ELECTRICAL TROUBLE SHOOTING GUIDE cont.

SITUATION	CAUSE	CHECK	REMEDY
I. Pump motor and condenser fan motor will come on, but compressor will not run	1. If unit is water cooled, high pressure control open (no condenser fan motor if water cooled)	1. Check and see if high pressure control has manual reset. If not, check across terminals with volt meter. If a reading shows, control is open.	1. Replace control
	2. One or both bin controls open	2. Turn off power. Install jumper between terminals B16 & 39. If compressor comes on when power is restored, one or both switches were open	2. Adjust or replace switches
	3. Contactor contact open	3. Check contact between terminals 51 & 52.	3. If contacts are bad, replace contactor
	4. Compressor overload open	4. If you check and get full voltage to common and run on the compressor, the overload is open	4. Determine why overload is open. If the compressor is at ambient temperature, the overload is bad and compressor will have to be replaced (internal overload). If it is hot then make checks on the compressor and its starting devices.

In completing electrical trouble shooting, you should remember relay C and D are to close and set up the delay timer circuit. If these two relays will never pull in, check relay A contact between A7 and A8. This contact has to be closed before relay C can pull in. With relay A contact closed, you will have to depend on the bin thermostat as the ice slab is dropped to bring in either relay C or D. The rear bin switch controls relay C and the front bin switch controls relay D. By now you should be able to trace that out on your own and see if that is so. The only other two most common things that could cause the relays not to close would be for the relay coils to be burnt out, or an open contact. If you will notice between terminals C-22 and D-25 that both relay C and D have to be closed before the timer circuit can be energized. The delay timer is to prevent the compressor from cycling off and on after defrost. ■

CHAPTER 15

REFRIGERATION RECOVERY AND NEW REFRIGERANTS

The air conditioning and refrigeration industry has been turned upside down with the EPA (Environmental Protection Agency) laws concerning recovery procedures of new types of refrigerants and lubricants. Most likely, many more changes are on the way until it is agreed that enough changes have been made to protect our ozone layer.

It has been found by accurate testing that the CFC refrigerants (chlorine-fluorine and carbon atoms) are the main culprit in destroying the ozone layer. These refrigerant types are being phased out and new refrigerants are at present being manufactured to replace them. When going to new refrigerants you will have to experience unusual changes within the system. This chapter will instruct you on performing these changes.

Most of us do not like the new laws imposed on us by the EPA or what other regulators involved, but like it or not it is here and we must comply. It is best to not look on this as an imposition, but an opportunity to make money. There are many units out in the field that will need to be retrofitted to make operable with the new ozone-free refrigerants. Many do not like changes, but this can be your chance to educate yourself on the issue so that you can instruct your customers as to the path they should take in complying also. Manufacturers are complying with the new law by installing the new refrigerants in their systems before they are shipped. They have been tested and balanced to assure the end-user of maximum performance and the lowest possible operating cost.

The longevity of the equipment will depend on your ability to properly maintain it as a service technician. It is my belief that the key to successful refrigerant management is education and the ability to remain current on all issues involving the new refrigerants. This chapter will give you a good understanding of what is going on now. There will be changes to come so stay on top and keep abreast of what is going on.

CATEGORIES OF REFRIGERANTS

We are familiar with the old refrigerants most commonly called Refrigerant 12-22-502. They were good refrigerants and we became very accustomed to them. The greatest problem we have with them today is that they are highly chlorinated and fluorinated and that is detrimental to the ozone layer. Refrigerants are classified in three main categories. There is the CFC (chlorofluorocarbons), the HCFC (hydrochlorofluorocarbons) and HFC (hydrofluorocarbons).

A. CHLOROFLUOROCARBONS (CFC)

Refrigerant - 11
Refrigerant - 12   *This category attacks the*
Refrigerant - 500   *ozone layer and are soon*
Refrigerant - 502   *to be prohibited.*

B. HYDROCHLOROFLUOROCARBONS (HCFC)

Refrigerant - 408A   *This category attacks*
Refrigerant - 22   *the ozone layer as well*
Refrigerant - 409A   *but not to the extent of*
Refrigerant - 123   *the CFCs.*

C. HYDROFLUOROCARBONS (HFC)

Refrigerant - 134A   *This category is*
Refrigerant - R-404A   *ozone safe.*

By no means are the refrigerants listed in these three categories showing you all the refrigerants on the mar-

ket. As you work with these refrigerants in the future you will have to go beyond the scope of this writing.

For alternative refrigerants please pay attention to the following charts.

ALTERNATIVE REFRIGERANTS

Low- and Medium-Temperature Commercial Refrigeration
Long-Term Replacements

ASHRAE #	Trade Name	Manufacturer	Replaces	Type	Lubricant [a]	Applications	Comments
R-507 (125/143a)	AZ-50	AlliedSignal Hoechst [b] Solvay [b]	R-502 & HCFC-22	Azeotrope	Polyol Ester	New Equipment & Retrofits	Close match to R-502
R-404A (125/143a/134a)	404A	AlliedSignal	R-502 & HCFC-22	Blend (small glide)	Polyol Ester	New Equipment & Retrofits	Close match to R-502
	HP62	DuPont					
	FX-70	Elf Atochem					
R-407A (32/125/134a)	60	ICI	R-502 & HCFC-22	Blend (high glide)	Polyol Ester	New Equipment & Retrofits	Higher discharge temperature than R-502
R-407B (32/125/134a)	61	ICI	R-502 & HCFC-22	Blend (high glide)	Polyol Ester	New Equipment & Retrofits	Lower efficiency than R-502

Low- and Medium-Temperature Commercial Refrigeration
Interim Replacements [c]

ASHRAE #	Trade Name	Manufacturer	Replaces	Type	Lubricant [a]	Applications	Comments
R-402A (22/125/290)	HP80	AlliedSignal DuPont	R-502 & HCFC-22	Blend (small glide)	Alkylbenzene or Polyol Ester	Retrofits	Higher discharge pressure than R-502
R-402B (22/125/290)	HP81	DuPont	R-502 & HCFC-22	Blend (small glide)	Alkylbenzene or Polyol Ester	Manitowoc Ice Machines	Higher discharge temperature than R-502
R-403A (22/218/290)	69S	Rhone Poulenc (NRI)	R-502 & HCFC-22	Blend (small glide)	Alkylbenzene or Polyol Ester	Retrofits	Higher discharge temperature than R-502
R-408A (125/143a/22)	FX-10	Elf Atochem	R-502 & HCFC-22	Blend (small glide)	Alkylbenzene or Polyol Ester	Retrofits	Higher Discharge temperature than R-502

(FIG. 15-1)

Medium-Temperature Commercial Refrigeration Long-Term Replacements

ASHRAE #	Trade Name	Manufacturer	Replaces	Type	Lubricant [a]	Applications	Comments
R-134a	HFC-134a	AlliedSignal DuPont Elf Atochem ICI	CFC-12	Pure Fluid	Polyol Ester	New Equipment & Retrofits	Close match to CFC-12

Medium-Temperature Commercial Refrigeration Interim Replacements [c]

ASHRAE #	Trade Name	Manufacturer	Replaces	Type	Lubricant [a]	Applications	Comments
R-401A (22/152a/124)	MP39	AlliedSignal DuPont	CFC-12	Blend (high glide)	Alkylbenzene or Polyol Ester	Retrofits	Close to CFC-12. Use where evap. temperature > -10° F
R-401B (22/152a/124)	MP66	AlliedSignal DuPont	CFC-12	Blend (high glide)	Alkylbenzene or Polyol Ester	For Transportation Refrigeration Retrofits	Close to CFC-12. Use where evap. temperature < -10° F
R-405A (22/152a/142b/318)	Green Cool 2015	GU	CFC-12	Blend (high glide)	Alkylbenzene or Polyol Ester	Retrofits	Higher capacity than CFC-12, similar to MP66
R-406A (22/142b/600a)	GHG	ICOR International	CFC-12	Blend (high glide)	Mineral Oil	Retrofits	Can segregate to flammable components
R-409A (22/124/142b)	R-409A	AlliedSignal	CFC-12	Blend (high glide)	Alkybenzene	Retrofits	Higher Capacity than CFC-12 similar to MP66
	FX-56	Elf Atochem					

Commercial and Residential Air-Conditioning Long-Term Replacements

ASHRAE #	Trade Name	Manufacturer	Replaces	Type	Lubricant [a]	Applications	Comments
R-123	HCFC-123	AlliedSignal DuPont Elf Atochem	CFC-11	Pure Fluid	Alkylbenzene or Mineral Oil	Centrifugal Chillers	Lower capacity than CFC-11
R-134a	HFC-134a	AlliedSignal DuPont Elf Atochem ICI	CFC-12	Pure Fluid	Polyol Ester	New Equipment & Retrofits	Close match to CFC-12
R-134a	HFC-134a	AlliedSignal DuPont Elf Atochem ICI	HCFC-22	Pure Fluid	Polyol Ester	New Equipment	Lower capacity – larger equipment needed
R-410A (32/125)	AZ-20	AlliedSignal	HCFC-22	Azeotropic Mixture	Polyol Ester	New Equipment	Higher efficiency than HCFC-22 and R-410B May require equipment redesign
R-410B (32/125)	9100	DuPont	HCFC-22	Azeotropic Mixture	Polyol Ester	New Equipment	Higher efficiency than HCFC-22 May require equipment redesign
R-407C (32/125/134a)	407C	AlliedSignal	HCFC-22	Blend (high glide)	Polyol Ester	New Equipment & Retrofits	Lower efficiency than HCFC-22, close capacity to HCFC-22
	9000	DuPont					
	66	ICI					

(a) Check with the compressor manufacturer for their recommended lubricant.
(b) Hoechst and Solvay both distribute AZ-50.
(c) Interim replacement, contains HCFC-22 which is scheduled for phaseout under the Montreal Protocol.

(FIG. 15-1)

THIS CHART PROVIDED BY COURTESY OF ALLIED SIGNAL.

Please note the headings of each section. First it shows the ASHRAE (American Society of Heating/ Refrig/Air Conditioning Engineers) number identification of the refrigerant, next the trade name, the manufacturer, next the type of old refrigerant it replaces, then the type covered in the next paragraph. Please note that R-507 is an Azeotrope refrigerant. This means the refrigerant does not change volumetric composition or saturation temperature as they evaporate or condense. For example, some refrigerants are mixed each with their own boiling point. In certain situations the lowest boiling point will be achieved therefore causing erratic temperature changes in the evaporator. Then you will notice that some refrigerants are the "blend type". By definition, an Azeotrope blend is one that cannot be separated into its components by distillation, thus having a single boiling point at a particular pressure and temperature - also note that the Blend can have a "glide". This means that a near Azeotrope blend does not have a single boiling point. When this type refrigerant enters the evaporator the lowest boiling point is achieved first, then the highest boiling point is next. There is not a constant temperature in the evaporator. There will be a variation in temperature from inlet to outlet known as the "glide" of the blend.

Next on the charts you will notice the type of lubricant (refrigeration oil) that must be used with each refrigerant. More will be covered on oils later in the chapter. Finally there will be comment on each of the refrigerants.

REFRIGERANT MANAGEMENT OPTIONS

You must initiate a refrigerant management plan for your equipment. This decision must be started by you. Not only will this information help you but you must also do some consulting on your own. When you are about to get involved in a project don't hesitate to ask questions from your supplier or wholesaler. They are as interested as you in making everything work and will be very useful in guiding you in the right direction.

When deciding on a course of action for a system, options become prevalent for handling the impending CFC phase-out. We will explain all four as follows:

1st Option - Ignore - When the issue is ignored you will find yourself beyond deadlines. You will have a loss of business and added expenses. There will be equipment downtime, lack of refrigerant supplies, loss of customers and the list goes on. This can come about simply by ignoring the change.

2nd Option - Containment - A good choice for equipment and systems targeted for replacement or phase out in the next three years or less for systems that are projected to remain in service longer than three years. Careful consideration must be given to the availability of economical refrigerant resources.

3rd Option - Retrofitting - A popular choice and what we will be covering. This extends the useful life of the equipment for a long period of time. Also we now have an environmentally safe refrigerant inside. Also the CFC recovered can be cleaned up and used in systems designed for containment.

4th Option - Replacement - This option is the most expensive and recommended for equipment that is so old it is not worthy of retrofitting or containment.

What you will be wanting to do is to retrofit the equipment by using Option 3. So you must prepare yourself to escort your customer into the new CFC - free environment. Retrofitting can be best defined as a process using existing equipment in conjunction with the new line of refrigerants and lubricants without major component replacement. When talking to your customer you should make them aware of several positive benefits.

1. The customer should be made "environmentally aware". Retrofitting is to employ their system with a low to zero ozone depleting refrigerant.

2. The customer should be made aware of "supply and availability". As the old refrigerant line is phased out, availability will dwindle and prices will escalate. By retrofitting, supply and availability is assured and cost is assumed to be stable.

3. Reduced Costs - It is cheaper to retrofit than to replace the equipment. Generally retrofitting requires only refrigerant, lubricant and labor costs.

4. Long term solution - Provides a continued operation of the equipment at a lower operating cost. Future tax increases on CFCs will mean increasing operating costs for non-retrofitted systems. It pays to retrofit - let's stay one step ahead.

Refrigerant	Manufacturer	ASHRAE#	Components	Application
FX 56	Elf Atochem	R-409A [2]	R-22/124/142b	R-12 Retrofit
FX 10	Elf Atochem	R-408A [2]	R-125/143a/22	R-502 Retrofit
FX 70	Elf Atochem	R-404A	R-125/143a/134a	R-502 New/Retrofit
FX 220	Elf Atochem		R-23/32/134a	R-22 alternative
AZ 50	Allied Signal	R-507 [1]	R-125/143a	R-502 New/Retrofit
AZ 20	Allied Signal	R-410A [2]	R-32/125	R-22 alternative
MP 39, 66	DuPont	R-401A,B	R-22/152a/124	R-12 Retrofit
HP 80, 81	DuPont	R-402A,B	R-125/290/22	R-502 Retrofit
HP 62	DuPont	R-404A	R-125/143a/134a	R-502 New/Retrofit
AC 9000	DuPont	R-407C [2]	R-32/125/134a	R-22 alternative
(Klea) 60, 61	ICI	R-407A,B [1]	R-32/125/134a	R-502 New/Retrofit
(Klea) 66	ICI	R-407C [2]	R-32/125/134a	R-22 alternative
(Isceon) 69L	Rhone Poulenc	R-403B	R-290/22/218	R-502 retrofit
GHG 12	Indianapolis Refrigeration Products	R-406A [1]	R-22/142b/600a	R-12 retrofit
Greencool 12	GU/Greencool	R-405A [1]	R-22/152a/142b/C318	R-12 retrofit
Greencool 2018 A,B	GU/Greencool	R-411A,B [2]	R-1270/22/152a	R-22 service
OZ 12	OZ Technologies		R-290/600	R-12 retrofit
HC 12A	OZ Technologies		Hydrocarbons	R-12 service

CFC Refrigerants	HCFC Refrigerants	HFC Refrigerants	Hydrocarbon Refrigerants	Perfluorocarbon Refrigerants
R-11	R-22	R-32 R-143a	R-290 (propane)	R-218
R-12	R-123	R-125 R-152a	R-600 (butane)	R-C318
R-115 (51% in 502 with 49% 22)	R-124	R-134a	R-600a (isobutane)	
	R-142b		R-1270 (propylene)	

1. These numbers have been approved for publication in an addendum to ASHRAE Standard 34-1992. Number Designation and Safety Classification of Refrigerants.

2. These numbers have been proposed for addition to ASHRAE Standard 34-1992. They will be accepted after completion of a public review process.

Refrigerant	ASHRAE#	Applications	Lubricant
R-11 ALTERNATIVES			
R-123	R-123	Retrofit or New Centrifugals	Mineral Oil
R-12 ALTERNATIVES			
R-134a	R-134a	R-12 New/Retrofit	Polyester (POE)
FX 56	R-409A	R-12 Retrofit	Mineral Oil/Alkylbenzene
MP 39, 66	R-401A,B	R-12 Retrofit	Alkylbenzene
GHG 12	R-406A	R-12 retrofit	Mineral Oil/Alkylbenzene
Greencool 12	R-405A	R-12 retrofit	Mineral Oil/Alkylbenzene
OZ 12		R-12 retrofit	Mineral Oil
HC 12A		R-12 applications	Mineral Oil
R-502 ALTERNATIVES			
FX 10	R-408A	R-502 Retrofit	Mineral/Alkylbenzene/POE
FX 70	R-404A	R-502 New/Retrofit	Polyester (POE)
AZ 50	R-507	R-502 New/Retrofit	Polyester (POE)
HP 80, 81	R-402A,B	R-502 Retrofit	Mineral Oil + Alkylbenzene
HP 62	R-404A	R-502 New/Retrofit	Polyester (POE)
(Klea) 60, 61	R-407A,B	R-502 New/Retrofit	Polyester (POE)
Isceon 69L	R-403B	R-502 retrofit	Mineral Oil + Alkylbenzene
R-22 ALTERNATIVES			
FX 220	R-410A	R-22 alternative	Polyester (POE)
AZ 20	R-407C	R-22 alternative	Polyester (POE)
AC 9000	R-407C	R-22 alternative	Polyester (POE)
(Klea) 66	R-407C	R-22 alternative	Polyester (POE)
Greencool 2018 A,B	R-411A,B	R-22 service	Mineral Oil/Alkylbenzene

ASHRAE NUMBERING SYSTEM

Pure Refrigerants (based on chemical makeup)

```
12      CCl₂F₂
134a    CHₓF - CF₃
         ├── given based on structure
         ├── number of fluorine atoms
         ├── number of hydrogen atoms, +1
         └── number of carbon atoms, -1
```

Blends (Azeotropes - 500 series, Zeotropes - 400 series)

Azeotropes are refrigerant blends which evaporate and condense exactly like a pure component at some temperature and pressure. Azeotropes may not behave this way at all temperatures and pressures, but they will be close.

Zeotropes will show some amount of temperature glide when evaporating or condensing. Some may act like azeotropes (glide is not noticeable in normal operation, less than 3°F). Zeotropes with glides greater than 3°F will have one end of the evaporator warmer than the other. This may impact system performance.

(FIG. 15-2)

SPORLAN PRESSURE-TEMPERATURE CHART

PSIG	Green	Purple	Yellow	Blue	White
	Temperature, °F				
	REFRIGERANT - (Sporlan Code)				
	22 (V)	502 (R)	12 (F)	134a (J)	717 (A)
5*	-48	-57	-29	-22	-34
4*	-47	-55	-28	-21	-33
3*	-45	-54	-26	-19	-32
2*	-44	-52	-25	-18	-30
1*	-43	-51	-23	-16	-29
0	-41	-50	-22	-15	-28
1	-39	-47	-19	-12	-26
2	-37	-45	-16	-10	-23
3	-34	-42	-14	-8	-21
4	-32	-40	-11	-5	-19
5	-30	-38	-9	-3	-17
6	-28	-36	-7	-1	-15
7	-26	-34	-4	1	-13
8	-24	-32	-2	3	-12
9	-22	-30	0	5	-10
10	-20	-29	2	7	-8
11	-19	-27	4	8	-7
12	-17	-25	5	10	-5
13	-15	-24	7	12	-4
14	-14	-22	9	13	-2
15	-12	-20	11	15	-1
16	-11	-19	12	16	1
17	-9	-18	14	18	2
18	-8	-16	15	19	3
19	-7	-15	17	21	4
20	-5	-13	18	22	6
21	-4	-12	20	24	7
22	-3	-11	21	25	8
23	-1	-9	23	26	9
24	0	-8	24	27	11
25	1	-7	25	29	12
26	2	-6	27	30	13
27	4	-5	28	31	14
28	5	-3	29	32	15
29	6	-2	31	33	16
30	7	-1	32	35	17
31	8	0	33	36	18
32	9	1	34	37	19
33	10	2	35	38	19
34	11	3	37	39	20
35	12	4	38	40	21
36	13	5	39	41	22
37	14	6	40	42	23
38	15	7	41	43	24
39	16	8	42	44	25
40	17	9	43	45	26
42	19	11	45	47	28
44	21	13	47	49	29
46	23	15	49	51	31
48	24	16	51	52	32
50	26	18	53	54	34
52	28	20	55	56	35
54	29	21	57	57	37
56	31	23	58	59	38
58	32	24	60	60	40
60	34	26	62	62	41
62	35	27	64	64	42
64	37	29	65	65	44
66	38	30	67	66	45
68	40	32	68	68	46
70	41	33	70	69	47
72	42	34	71	71	49
74	44	36	73	72	50
76	45	37	74	73	51
78	46	38	76	75	52
80	48	40	77	76	53
85	51	43	81	79	56
90	54	46	84	82	58
95	56	49	87	85	61
100	59	51	90	88	63
105	62	54	93	90	66
110	64	57	96	93	68
115	67	59	99	96	70
120	69	62	102	98	73
125	72	64	104	100	75
130	74	67	107	103	77
135	76	69	109	105	79
140	78	71	112	107	81
145	81	73	114	109	82
150	83	75	117	112	84
155	85	77	119	114	86
160	87	80	121	116	88
165	89	82	123	118	90
170	91	83	126	120	91
175	92	85	128	122	93
180	94	87	130	123	95
185	96	89	132	125	96
190	98	91	134	127	98
195	100	93	136	129	99
200	101	95	138	131	101
205	103	96	140	132	102
210	105	98	142	134	104
220	108	101	145	137	107
230	111	105	149	140	109
240	114	108	152	143	112
250	117	111	156	146	115
260	120	114	159	149	117
275	124	118	163	153	121
290	128	122	168	157	124
305	132	126	172	161	128
320	136	130	177	165	131
335	139	133	181	169	134
350	143	137	185	172	137
365	146	140	188	176	140

PRESSURE-TEMPERATURE CHART

PSIG	Pink	Orange	Sand	Teal	Green	Tan
	Temperature, °F					
	REFRIGERANT - (Sporlan Code)					
	MP39 (X) or 401A (X)	HP62 (S) or 404A (S)	HP80 (L) or 402A (L)	AZ-50 (P) or 507(P)	124 (Q)	125
5*	-23	-57	-59	-59	3	-63
4*	-22	-56	-58	-57	4	-61
3*	-20	-54	-56	-56	6	-60
2*	-19	-53	-55	-55	7	-58
1*	-17	-52	-54	-53	9	-57
0	-16	-51	-53	-52	10	-56
1	-13	-48	-50	-50	13	-53
2	-11	-46	-48	-47	16	-51
3	-9	-43	-45	-45	18	-49
4	-6	-41	-43	-43	21	-46
5	-4	-39	-41	-41	23	-44
6	-2	-37	-39	-39	26	-42
7	0	-35	-37	-37	28	-40
8	2	-33	-36	-35	30	-39
9	4	-32	-34	-34	32	-37
10	6	-30	-32	-32	34	-35
11	8	-28	-30	-30	36	-33
12	9	-27	-29	-29	38	-32
13	11	-25	-27	-27	40	-30
14	13	-23	-26	-25	41	-29
15	14	-22	-24	-24	43	-27
16	16	-20	-23	-23	45	-26
17	17	-19	-21	-21	46	-24
18	19	-18	-20	-20	48	-23
19	20	-16	-19	-18	49	-22
20	21	-15	-17	-17	51	-20
21	23	-14	-16	-16	52	-19
22	24	-12	-15	-15	54	-18
23	25	-11	-14	-13	55	-16
24	27	-10	-12	-12	57	-15
25	28	-9	-11	-11	58	-14
26	29	-8	-10	-10	59	-13
27	31	-6	-9	-9	61	-12
28	32	-5	-8	-8	62	-11
29	33	-4	-7	-6	63	-10
30	34	-3	-6	-5	65	-8
31	35	-2	-5	-4	66	-7
32	36	-1	-4	-3	67	-6
33	37	0	-2	-2	68	-5
34	38	1	-1	-1	69	-4
35	39	2	0	0	71	-3
36	40	3	1	1	72	-2
37	41	4	1	2	73	-1
38	43	5	2	3	74	0
39	44	6	3	4	75	0
40	45	8	4	5	76	1
42	47	10	6	6	78	3
44	48	10	8	8	80	5
46	50	12	10	10	82	7
48	42	14	11	12	84	8
50	44	16	13	13	86	10
52	45	17	14	15	88	11
54	47	19	16	16	90	13
56	49	20	18	18	91	15
58	50	22	19	19	93	16
60	52	23	20	21	95	17
62	53	26	23	22	97	19
64	55	26	23	24	98	20
66	56	27	25	25	100	22
68	58	29	26	27	101	23
70	59	30	29	28	103	24
72	61	32	31	29	104	26
74	62	33	32	30	106	27
76	64	34	33	32	107	28
78	65	35	34	33	109	29
80	66	37	36	34	110	31
85	69	40	39	37	114	33
90	73	42	42	40	117	36
95	76	44	42	46	120	39
100	78	48	47	46	123	42
105	81	50	48	51	126	44
110	84	52	50	53	129	47
115	87	55	53	55	132	49
120	89	57	53	56	135	51
125	92	59	55	58	138	54
130	94	62	57	60	140	56
135	96	64	60	62	143	58
140	99	66	62	64	145	60
145	101	68	64	67	148	62
150	103	70	66	69	150	64
155	105	72	68	71	152	66
160	108	74	70	73	155	68
165	110	76	72	74	157	70
170	112	78	74	76	159	72
175	114	80	75	78	161	73
180	116	82	77	80	163	75
185	117	83	79	82	165	77
190	119	85	81	83	167	79
195	121	87	82	85	169	80
200	123	88	84	87	171	82
205	125	90	86	88	173	83
210	127	92	87	90	175	85
220	130	95	91	93	178	88
230	133	98	94	96	182	91
240	136	101	97	99	185	94
250	140	104	99	102	188	97
260	143	107	102	105	192	99
275	147	111	106	109	196	103
290	151	115	110	112	201	107
305	155	118	114	116	205	111
320	159	122	118	120	209	114
335	163	126	121	123	213	118
350	167	129	125	126	217	121
365	170	132	128	129	221	124

(In the right-hand chart the MP39, HP62, HP80 and AZ-50 columns carry vertical labels "BUBBLE POINT" and "DEW POINT" with arrows, indicating the bubble-point region in the upper portion and the dew-point region in the lower portion.)

(FIG. 15-3)

As you delete the old refrigerants, we can remember being well acquainted with their operating pressures. Now we have a whole new league of refrigerants having different pressures and temperatures. Use the pressure-temperature charts shown in Figures 2 and 3. Familiarize yourself with these new refrigerants. Please note in the left hand column you will see (PSIG). This is pounds per square inch gauge pressure. Now notice at the top of the chart you will see (Temperature °F), then five different refrigerants listed. For ease of using this chart go to the 22(V) column and follow down to 40°F. This is the average temperature of an air conditioning evaporator. Using the 40°F temperature you can see that the pressure of this boiling refrigerant is 68 PSI. If the air conditioner is working properly and you have installed your service gauges, this is the pressure you would expect. However, move to the 502 (R) and you would see a corresponding pressure of 80 PSI with an evaporator temperature of 40°F. If you continue across you can see dramatic pressure and temperature changes. So it should be understood that all refrigerants have different pressures at the same evaporator temperatures. When making a retrofit it will be necessary to use the proper retrofit refrigerant to do the same job as the old refrigerant did with the least amount of changes.

LUBRICANTS

A lot of confusion comes with this issue, but we must understand that the new refrigerants require a new type of lubricant in the compressor. Turn back to Figure 1 and you see the recommended lubricants for different refrigerants and please note the comments. Most new refrigerants can be used in the same equipment the old refrigerants came out of without any component change. Some minor adjustments will have to be made such as reset the super-heat of an expansion valve. Most fixed orifice metering devices can stay with little effect to the operation. Mineral oils associated with old refrigerants are not accepted with the new refrigerants and must be removed or flushed from the system before the new refrigerant is added.

When adding the new oil, here are some tips to use:
1. Leave the old refrigerant in the system since it will mix with the new oil, but the new refrigerants cannot be mixed with the old oil.
2. Remove the present mineral oil from the system.
3. Add the new equivalent charge of new recommended lubricant.
4. Operate the system from 24 to 48 hours to allow the new oil and residue of old oil to mix.
5. Repeat recommendation 4 until the minimum desired percentage is reached.
6. Remove the old refrigerant and start retrofitting your system with the new refrigerant.

Number of Flushes versus Mineral oil Residual Levels for a typical System. Please note that different systems may experience different residual oil levels. These numbers are for reference only.

Flush #	50%*	40%*	30%*	20%*	10%*
1	25%	16%	9%	4%	1%
2	12.5%	6.4%	2.7%	.8%	.1%
3	6.25%	2.56%	.8%	.16%	.01%
4	3.1%	1.0%	.25%	.03%	0%

* Indicates the starting level of mineral oil remaining in system after initial draining of the original mineral oil charge.

(FIG. 15-4)

Please note that Figure 4 is to illustrate the percentages of mineral oil left in a typical system. To be sure of the residual of mineral oil left in a system there will be test kits and instruments to determine the residual. Your wholesale supplier will be a great help to you in you retrofitting jobs.

SYSTEM COMPONENTS

Most all system components will be suitable with the new refrigerants. There may be a slight decrease in capacity but usually not noticeable enough to become alarmed. Some of the things that will change is the resetting of pressure control settings since the new refrigerant may have more or less pressure. This would include high pressure controls, low pressure controls, and oil pressure regulators. On larger systems capacity controls and unloaders may have to be reset. Some larger compressors may have seals and "O" rings that may have to be replaced (contact the manufacturer about this). The thermostatic expansion valve will usually always work on the new compatible refrigerants. A change in the super-heat setting of the valve is possible. The capillary tube is a fixed orifice feeding device and should work alright with new refrigerants. The liquid line drier should be changed on retrofit and should also be selected to be compatible with the new refrigerants. It is recommended that solid core or compacted bead type sieves be used to keep the desiccant from breaking up and cause restrictions in the system.

PROCEDURES IN MAKING THE RETROFIT

Studying the following eight steps in retrofitting will be helpful. Use them when retrofitting.

STEP 1 - Gather all information, such as the current refrigerant charge, lubricant charge, operating temperatures and pressures and overall system performance. This is mostly necessary since we don't want to retrofit a system that has a problem existing.

STEP 2 - Recover the existing refrigerant charge using an EPA certified recovery machine.

STEP 3 - Remove the mineral oil and replace with new lubricant (note that even with new refrigerants mineral oils are compatible but at least try to change to new mineral oil).

STEP 4 - Replace the liquid line filter drier.

STEP 5 - Evacuate the system to 500 microns (using a micron indicator) to assure that any traces of moisture are out of the system. The new refrigerants have a high capability of retaining moisture.

STEP 6 - Charge the system with the new refrigerant. The new charge will be approximately 90% of what the old charge was. Due to the characteristics and chemical make-up of the refrigerants and their "blend" it may be necessary to charge as a liquid.

STEP 7 - Install proper indentifying markings on the system to show the system has been retrofitted and type of new refrigerant used.

STEP 8 - Start the system and make final adjustments of expansion valve, pressure controls and any other devices needed to be adjusted. The operating pressures of the new refrigerants can be from 5% to 25% higher than most typical refrigerant applications.

Throughout this chapter you have been shown a great deal about refrigerants, lubricants and retrofitting. Hopefully it was not too much for you to asorb. The question you may be asking now is this "Ok, but when I get on the job, how easy will it be to make the retrofit." In the next topic you are going to be shown how. The following questions will be answered for you.

1. How do I remove the refrigerant from the system?
2. How do I get the oil out of the compressor?
3. Why should I pull a vacuum on the system?
4. What are the changes in refrigerant charging procedures?

HOW TO RECOVER THE REFRIGERANT: Recovering refrigerant and storing it without emission into the atmosphere is not new. Where there was large amounts of refrigerant in larger systems it was not economically practical to blow the charge, so the freon was recovered into empty drums and re-used. However under new EPA guidelines the old way is not permissible. You will need a certified recovery machine and be certified yourself to go any further. If you do not have these two things, stop now. Not only can you damage our ozone layer by releasing gases, but you can bring heavy fines on yourself. Complying with the rules is good advice.

The recovery machine is actually a refrigeration machine itself. It has a compressor and a condensor. The compressor of the recovery machine sucks the refrigerant from the system, compresses it and pushes it into the condensor where the vapor is cooled and condensed to a liquid, then pushed into your certified recovery device. Take a look at Figure 5.

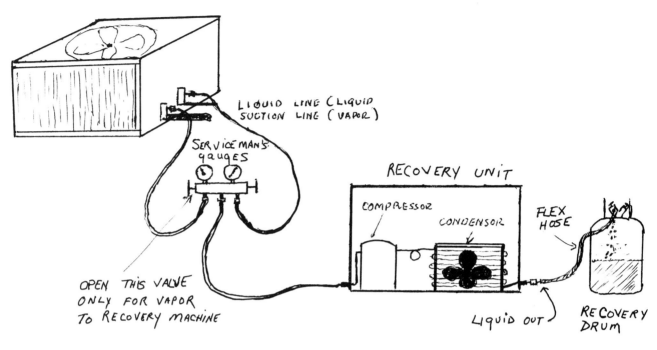

(FIG. 15-5)

Figure 5 is a crude drawing and does not show accumulators and oil separators and such integrated in the recovery machine but will help you get an understanding of its operation.

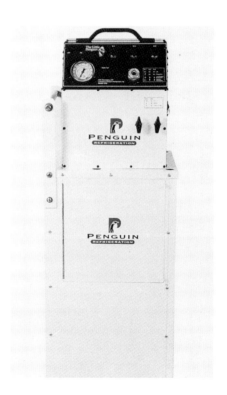

(FIG. 15-6)

Figure 15-6 shows a typical recovery machine.

There are many recovery units on the market. Your selection will be determined by the weight of the unit, its recovery rate and simplicity. The recovery unit will pull all of the refrigerant out of the unit in vapor form and put it into the recovery drum in a liquid form.

SAVE THE REFRIGERANT OR DISPOSE - You may reuse the refrigerant if you are taking it from the unit for repairs and put the same refrigerant back into the air conditioning system. If the compressor (hermetic or semi-hermetic) has burnt windings it will cause an acid formation and is not recommended for reuse. In this case you would recover the refrigerant into a recovery drum and when full take it to your nearest reclaim center (usually a local supply house). They will pay you a small amount of money per pound, clean it and put it back out for purchase. When a refrigerant is recovered it cannot be used in a different unit. Also, when recovering refrigerants, pull vapor only from the equipment into the recovery machine. Most recovery units are designed for vapor only and liquid will do damage to the recovery unit. There are however machines that will pump liquid.

RECOVERY DRUMS - All recovery drums are certified and used only for recovery. Figures 7a & 7b gives you a look at how the drum is constructed.

(FIG. 15-7B)

Usually you will have two valves, one for liquid and the other for vapor. They will usually be colored blue to indicate vapor and red to indicate liquid. When the red valve is opened the long tube connected to the valve will let liquid only out until empty. The pressure of the refrigerant will force it out. When the vapor valve is opened only vapor will leave the drum since the liquid is not to the top of the drum.

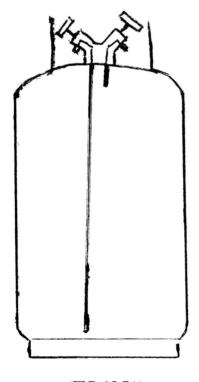

(FIG. 15-7A)

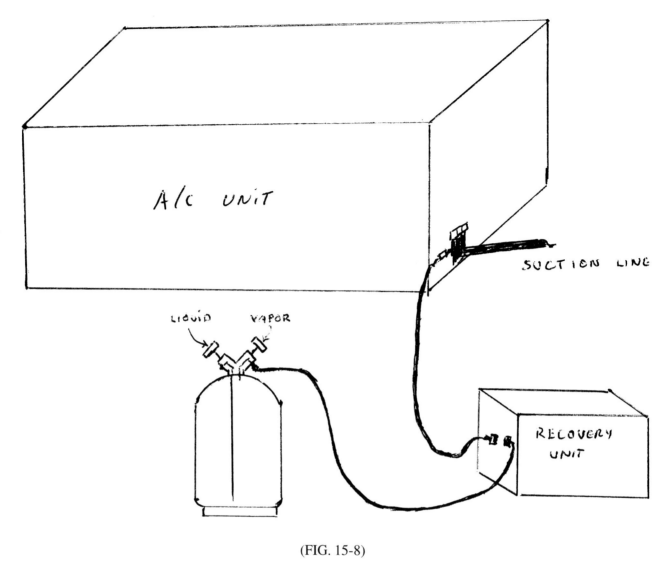

(FIG. 15-8)

Figure 15-8 shows a typical hook-up from the air conditioning or refrigeration unit into the recovery machine to the drum. The recovery connection on the recovery system goes to the suction line of the appliance (vapor only). After the unit has condensed the vapor to liquid it will push into the recovery drum. The recovery drums can be purchased in 30 lb. capacity, 50 lb. capacity and 145 lb. capacity. While doing service work the 30 lb. and 50 lb. drums will be found to be best.

WARNING: Use scales to weigh the refrigerant. Never over-fill the drums and never expose them to excessive heat.

REMOVING REFRIGERANT IN LIQUID FORM - There will be times when the appliance is large and has lots of refrigerant. You may then want to speed up the recovery process by taking the refrigerant out in liquid form. See the hook-up in Figure 15-9. If your recovery machine is designed to pump liquid - disregard this procedure.

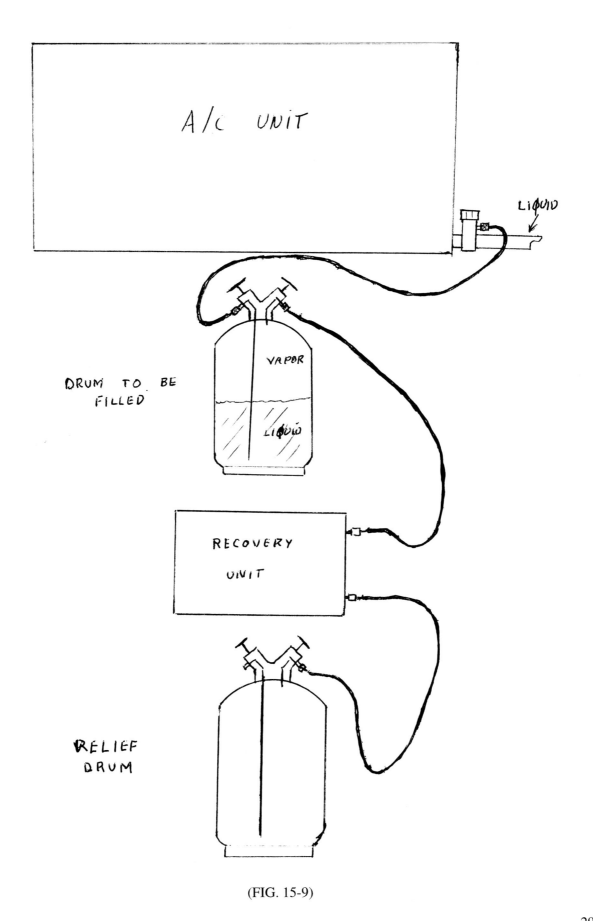

(FIG. 15-9)

In Figure 15-9 you see two drums. The recovery unit is pulling pressure down on DRUM A and allowing liquid to flow to that drum. From DRUM A you see vapor is being taken from the drum into the recovery machine, condensing the vapor to liquid and pushing it into DRUM B.

PRESSURES OF THE RECOVERY MACHINE - Recovery units have low pressure and high pressure controls. They are safety devices used to automatically cut the recovery unit off. You find gauges on the units to read the low pressure and high pressure. The low pressure gauge reads the pressure of the appliance being recovered in combination with the suction pressure of the compressor in the recovery unit. The head pressure gauge reads the discharge pressure of the compressor. Upon starting the unit with all valves being positioned correctly, you will notice the head pressure will rise. The suction pressure will be high because the appliance to be recovered is idle. The recovery system will have to operate for awhile until you see this pres-

sure start dropping. Just be patient.

SPEEDING UP VAPOR RECOVERY - If the ambient temperature is high around the recovery unit, the head pressure will run high and slow down the recovery. To speed up the process get something like a garbage can and put ice and water inside. Set the recovery drum in the ice and water. The pressure will drop and let the refrigerant move faster. In some cases just simply take a water hose and trickle water over the drum if you are in an area where you can.

LUBRICANTS - While retrofitting units you are going to find changing compressor oils to be the most nuisance. Most older units are charged with mineral oils and are not compatible with the new refrigerants. If the new refrigerant is compatible with the mineral oil, great. But if it has to be changed you need to know how. Here's how. If the compressor is a hermetic or semi-hermetic compressor the oil removal is difficult. The simplest way is to remove the compressor from the appliance. See Figure 10.

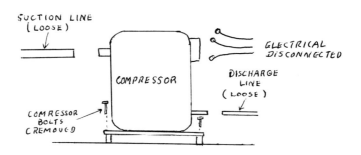

FIGURE 15 - 10

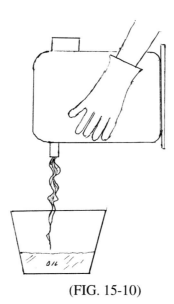

(FIG. 15-10)

Turn the compressor on its side and drain the oil into a container that will measure. Now take the new oil in the same quantity and pour it into the compressor suction port, then reinstall the compressor. Be sure to evacuate thoroughly because you have exposed the new oil to the atmosphere. This procedure is OK if one oil change was all it would take. However, some of the old oil is still in the component parts. This means there may have to be two, three or four oil flushings. Can you imagine having to remove the compressor this many times? NOTE: Test devices are available to indicate when all of the mineral oil is at a safe level within the new oil. Now that you understand the difficulty of changing the oil, look at the tip in Figure 11.

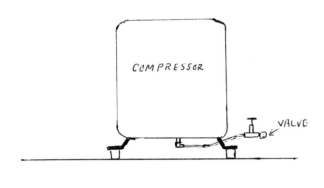

(FIG. 15-11)

After removing the oil from the compressor, drill a 1/4" hole in the bottom of the compressor or where convenient as close to the bottom of the compressor as possible. Tap threads into the hole. Screw a 1/4" MPT X 1/4" sweat adapter into the hole. Clean the paint away or sand off. Silver solder with silver solder flux the fitting into the compressor. Now you have a line with a valve that will drain the oil without having to pull the compressor from the system. I believe the manufacturers will see the difficulty the trade has here and in the very near future make provisions for you to drain the oil. Don't forget that if this process is used you must recover the refrigerant so no pressure is on the compressor. To speed up the oil drainage with the proper drain, put a little nitrogen pressure on the compressor. The oil will come out faster. Do not use refrigerant for this process and remember to measure the oil removed so you can put the exact amount back in.

If the compressor is a larger semi-hermetic the oil

change is not as much of a problem. It will have drain provisions. See Figure 12.

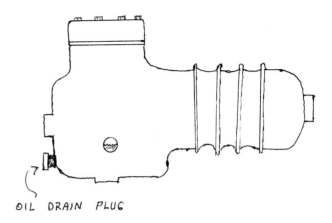

OIL DRAIN PLUG

(FIG. 15-12)

Not only does this compressor have a drain plug but an oil sight glass to tell you when there is enough oil in the compressor. It would not be necessary to measure the oil. The oil level should be half way up on the sight glass. To put the oil back in this compressor it will be best to purchase an oil hand pump stocked at your nearest supply house. The oil pump would be handy also for the set up in Figure 11 by pumping the oil back through the valve into the compressor.

DRIERS - If a retrofit is made, or if a system has the new refrigerant already in it the selection of the drier is important. Some driers are not suitable with new gases. Consult your supplier on the recommended driers.

AFTER THE RETROFIT - When a retrofit is made you must post on the unit the type of refrigerant and if the oil has been changed and the type. Record all pressures and temperatures. Make sure the unit is operating properly before departing the job.

MAKE FINAL ADJUSTMENTS - If the appliance employs a thermostatic expansion valve it may have to be reset by its super-heat. See pages 250-251 in Chapter 14 for setting super-heat or to see if it is working as it should. Let the system operate for at least 15 minutes before making this test or adjustment.

SETTING THE SUPER-HEAT - If the unit has pressure controls it will be necessary to adjust the pressures in accordance to the new refrigerant pressures.

Now you have a good knowledge of recovery and

retrofitting. It is not necessary to panic too much, but always stay educated on the matter. The deadlines are changed from time to time and I am sure more will be added. It is a while before some of our common refrigerants will be deleted completely. The best thing to do is use these refrigerants until you can't buy any more. However, if retrofitting a unit comes about you will know how to do it. Just remember we have a new law to comply with involving refrigerants. Let's do our part in keeping our ozone layer safe. Actually in this new adventure, money can be made with it. Always charge for the use of your recovery unit, usually $50.00 to $65.00 every time you take the unit from the truck and use it. I would advise you to have printed on all your invoices or work tickets a $5.00 EPA compliance fee. You paid money to buy recovery machines, recovery drums and for certification. The end user has to share in these expenses.

Figure 15-3 shows the varying sizes of refrigerant cylinders and their color codes.

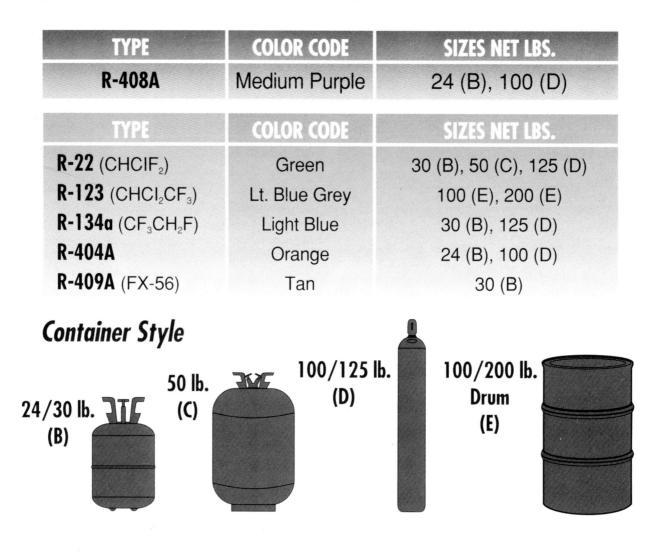

TYPE	COLOR CODE	SIZES NET LBS.
R-408A	Medium Purple	24 (B), 100 (D)

TYPE	COLOR CODE	SIZES NET LBS.
R-22 ($CHClF_2$)	Green	30 (B), 50 (C), 125 (D)
R-123 ($CHCl_2CF_3$)	Lt. Blue Grey	100 (E), 200 (E)
R-134a (CF_3CH_2F)	Light Blue	30 (B), 125 (D)
R-404A	Orange	24 (B), 100 (D)
R-409A (FX-56)	Tan	30 (B)

Container Style

24/30 lb. (B) 50 lb. (C) 100/125 lb. (D) 100/200 lb. Drum (E)

(FIG. 15-13)